Chemistry: Principles, Patterns, and Applications

Chemistry: Principles, Patterns, and Applications

Contributors

Miguel Valcárcel et al.

AURIS
Reference

www.aurisreference.com

Chemistry: Principles, Patterns, and Applications

Contributors: Miguel Valcárcel et al.

Published by Auris Reference Limited

www.aurisreference.com

United Kingdom

Chemistry: Principles, Patterns, and Applications

ISBN: 978-1-78154-858-5

British Library Cataloguing in Publication Data
A CIP record for this book is available from the British Library

Printed in the United Kingdom

Exclusively distributed by CBS Publishers & Distributors Pvt. Ltd.

Sales & Distribution Rights only for India, Pakistan, Bangladesh, Sri Lanka, Nepal and Bhutan. This book is not to be sold outside these territories.

Contents

List of Abbreviations

AFM	Atomic Force Microscopy
CE-MS	Capillary Electrophoresis – Mass Spectrometry coupling
CRM	Certified Reference Material
DOC	Dissolved organic carbon
DOM	dissolved organic matter
EEM	Excitation-emission matrix
FTIR	Fourier Transform Infrared Spectroscopy
GC-FTIR/MS	Gas Chromatography – Fourier Transform Infrared Spectroscopy / Mass Spectrometry coupling
GC-MS	Gas Chromatography – Mass Spectrometry coupling
GC-MS/MS	Gas Chromatography – Mass Spectrometry / Mass Spectrometry coupling
ISO	International Organization for Standardization
LC-ICP-MS	Liquid Chromatography – Inductively Coupled Plasma Spectrometry – Mass Spectrometry coupling
LC-MS	Liquid Chromatography – Mass Spectrometry coupling
MDPs	Method Defined Parameters
NIST	National Institute of Standards and Technology
PAHs	Polycyclic Aromatic Hydrocarbons
PCBs	Polychlorinated Biphenyls
POCTs	Point-of-Care-Testing
R&D&I	Research, Development and Innovation
SR	Social Responsibility
SRAC	Social Responsibility of Analytical Chemistry
SSS	Sample Screening Systems
aa	Amino acid
HDP	Host defense peptides
RCM	Rigid core micellar
THF	Tetrahydrofuran

List of Contributors

Miguel Valcárcel
Faculty of Sciences of the University of Córdoba, Spain

Aline Thaís Bruni
Departamento de Química, Faculdade de Filosofia, Ciências e Letras de Ribeirão Preto, Universidade de São Paulo

Vitor Barbanti Pereira Leite
Departamento de Física, Instituto de Biociências, Letras e Ciências Exatas, Universidade Estadual Paulista, São José do Rio Preto Brazil

Jacob D. Durrant
Department of Chemistry & Biochemistry, University of California San Diego, La Jolla, California, United States of America

J. Andrew McCammon
Department of Chemistry & Biochemistry, NSF Center for Theoretical Biological Physics, National Biomedical Computation Resource, University of California San Diego, La Jolla, California, United States of America
Department of Pharmacology, University of California San Diego, La Jolla, California, United States of America
Howard Hughes Medical Institute, University of California San Diego, La Jolla, California, United States of America

BalaKrishna Kolluru
National Centre for Text Mining, Manchester Interdisciplinary Biocentre, University of Manchester, Manchester, United Kingdom

Lezan Hawizy
Unilever Centre for Molecular Informatics, University of Cambridge, Cambridge, United Kingdom

Peter Murray-Rust
Unilever Centre for Molecular Informatics, University of Cambridge, Cambridge, United Kingdom

Junichi Tsujii
National Centre for Text Mining, Manchester Interdisciplinary Biocentre, University of Manchester, Manchester, United Kingdom

Sophia Ananiadou
National Centre for Text Mining, Manchester Interdisciplinary Biocentre, University of Manchester, Manchester, United Kingdom

Santosh K. Misra
Department of Bioengineering, University of Illinois at Urbana-Champaign, Urbana, IL, 61801, United States of America, Beckman Institute of Advanced Science and Technology, University of Illinois at UrbanaChampaign, Urbana, IL, 61801, United States of America, Department of Materials Science and Engineering University of Illinois at Urbana-Champaign, Urbana, IL, 61801, United States of America, Carle Foundation Hospital, Urbana, IL, 61801, United States of America

Mao Ye
Department of Bioengineering, University of Illinois at Urbana-Champaign, Urbana, IL, 61801, United States of America, Beckman Institute of Advanced Science and Technology, University of Illinois at UrbanaChampaign, Urbana, IL, 61801, United States of America, Department of Materials Science and Engineering University of Illinois at Urbana-Champaign, Urbana, IL, 61801, United States of America, Carle Foundation Hospital, Urbana, IL, 61801, United States of America

Sumin Kim
Department of Bioengineering, University of Illinois at Urbana-Champaign, Urbana, IL, 61801, United States of America, Beckman Institute of Advanced Science and Technology, University of Illinois at UrbanaChampaign, Urbana, IL, 61801, United States of America, Department of Materials Science and Engineering University of Illinois at Urbana-Champaign, Urbana, IL, 61801, United States of America, Carle Foundation Hospital, Urbana, IL, 61801, United States of America

Dipanjan Pan
Department of Bioengineering, University of Illinois at Urbana-Champaign, Urbana, IL, 61801, United States of America, Beckman Institute of Advanced Science and Technology, University of Illinois at UrbanaChampaign, Urbana, IL, 61801, United States of America, Department of Materials Science and Engineering University of Illinois at Urbana-Champaign, Urbana, IL, 61801, United States of America, Carle Foundation Hospital, Urbana, IL, 61801, United States of America

Anila Sarwar
Fuel Research Centre, Pakistan Council of Scientific & Industrial Research, Karachi, Pakistan

M. Nasiruddin Khan
Department of Chemistry, University of Karachi, Karachi, Pakistan

Kaniz Fizza Azhar
Scientific Information Centre, Pakistan Council of Scientific & Industrial Research,

Karachi, Pakistan

Kalliat T. Valsaraj
Department of Chemical Engineering, Louisiana State University, Baton Rouge, USA

Francesco G. Mutti
Dipartimento di Chimica Inorganica, Metallorganica e Analitica "Lamberto Malatesta", Università di Milano, Istituto ISTM-CNR, Via Venezian 21, 20133 Milano, Italy

Roberta Pievo
Dipartimento di Chimica Inorganica, Metallorganica e Analitica "Lamberto Malatesta", Università di Milano, Istituto ISTM-CNR, Via Venezian 21, 20133 Milano, Italy

Maila Sgobba
Dipartimento di Chimica Inorganica, Metallorganica e Analitica "Lamberto Malatesta", Università di Milano, Istituto ISTM-CNR, Via Venezian 21, 20133 Milano, Italy

Michele Gullotti
Dipartimento di Chimica Inorganica, Metallorganica e Analitica "Lamberto Malatesta", Università di Milano, Istituto ISTM-CNR, Via Venezian 21, 20133 Milano, Italy

Laura Santagostini
Dipartimento di Chimica Inorganica, Metallorganica e Analitica "Lamberto Malatesta", Università di Milano, Istituto ISTM-CNR, Via Venezian 21, 20133 Milano, Italy

Nicola A. McEnroe
Department of Biology, Trent University, Peterborough, Ontario, Canada

Clayton J. Williams
Department of Biology, Trent University, Peterborough, Ontario, Canada

Marguerite A. Xenopoulos
Department of Biology, Trent University, Peterborough, Ontario, Canada

Petr Porcal
Biology Centre of the Academy of Science of the Czech Republic, v.v.i., Institute of Hydrobiology, České Budějovice, Czech Republic

Paul C. Frost
Department of Biology, Trent University, Peterborough, Ontario, Canada

Rebecca M. Weisinger
Department of Chemistry, Stanford University, Stanford, California, United States of America, Department of Biochemistry, Stanford University, Stanford, California, United States of America

Robert J. Marinelli
Department of Biochemistry, Stanford University, Stanford, California,United States of America

S. Jarrett Wrenn
Department of Biochemistry, Stanford University, Stanford, California,United States of America

Pehr B. Harbury
Department of Biochemistry, Stanford University, Stanford, California,United States of America

Mario Ficker
Department of Chemistry, University of Copenhagen, Thorvaldsensvej 40, DK- 1871 Frederiksberg C, Denmark

Johannes F. Petersen
Department of Chemistry, University of Copenhagen, Thorvaldsensvej 40, DK- 1871 Frederiksberg C, Denmark

Jon S. Hansen
Department of Chemistry, University of Copenhagen, Thorvaldsensvej 40, DK- 1871 Frederiksberg C, Denmark

Jørn B. Christensen
Department of Chemistry, University of Copenhagen, Thorvaldsensvej 40, DK- 1871 Frederiksberg C, Denmark

Minmin Chu
School of Chemistry and State Key Laboratory of Fine Chemicals, Dalian University of Technology, Dalian 116024, China

Xin Liu, Yanhui Sui
School of Chemistry and State Key Laboratory of Fine Chemicals, Dalian University of Technology, Dalian 116024, China

Jie Luo
School of Chemistry and State Key Laboratory of Fine Chemicals, Dalian University of Technology, Dalian 116024, China

Changgong Meng
School of Chemistry and State Key Laboratory of Fine Chemicals, Dalian University of Technology, Dalian 116024, China

Ashriti Govender
Sasol Technology R&D, PO Box 1, Sasolburg 1947, South Africa

Daniel Curulla-Ferré
Gaz & Energies Nouvelles, Total S.A., Paris La Defense 6, France

Manuel Pérez-Jigato
Physical Chemistry of Surfaces, Eindhoven University of Technology, PO Box 513, 5600 MB, Eindhoven, The Netherlands

Hans Niemantsverdriet
Physical Chemistry of Surfaces, Eindhoven University of Technology, PO Box 513, 5600 MB, Eindhoven, The Netherlands

Yuanyuan Li
School of Chemistry and Chemical Engineering, Shanghai University of Engineering Science, Shanghai, China

Gaili Sun
School of Chemistry and Chemical Engineering, Shanghai University of Engineering Science, Shanghai, China

Yiming Mi
School of Chemistry and Chemical Engineering, Shanghai University of Engineering Science, Shanghai, China

Roozbeh Javad Kalbasi
Department of Chemistry, Shahreza Branch,Islamic Azad University, 311-86145 Shahreza, Isfahan
Razi Chemistry Research Center, Shahreza Branch,Islamic Azad University, Shahreza, Isfahan
Iran

Neda Mosaddegh
Department of Chemistry, Shahreza Branch,Islamic Azad University, 311-86145 Shahreza, Isfahan
Razi Chemistry Research Center, Shahreza Branch,Islamic Azad University, Shahreza, Isfahan
Iran

Haidong Zhang
Engineering Research Centre for Wasted Oil Recovery Technology of Ministry of Education, Key Laboratory of Catalysis Science and Technology of Chongqing Education Commission, Chongqing Technology and Business University, Chongqing 400067, China

Xiaohong Li
Shanghai Key Lab of Green Chemistry and Chemical Processes, School of Chemistry and Molecular Engineering, East China Normal University, Shanghai 200062, China

Preface

Chemistry is a branch of physical science that studies the composition, structure, properties and change of matter. The text *Chemistry: Principles, Patterns, and Applications* features modern applications, early integration of examples from chemistry, and a strong approach to problem solving that moves away from rote memorization to a thorough understanding of key concepts and recognition of important patterns. First chapter focuses on fundamentals of analytical chemistry. Quantum chemistry and chemometrics applied to conformational analysis have been discussed in second chapter. In third chapter, we present a novel computer algorithm, called AutoClickChem, capable of performing many click-chemistry reactions *in silico*. The aim of fourth chapter is to abstract the functionality of chemistry text mining tools from the underlying implementation via reconfigurable workflows for automatically identifying chemical names. Fifth chapter discusses how nanoscale chemistry influences delivery of peptido-toxins for cancer therapy. Sixth chapter focuses on coal chemistry and morphology of Thar reserves in Pakistan. In seventh chapter, the surface chemistry and properties of aqueous atmospheric aerosols have been explored. Biomimetic modeling of copper complexes has been discussed in eighth chapter. In ninth chapter, we examine the quantity and quality of dissolved organic matter (DOM) in urban ponds as it compares to DOM sampled from more natural aquatic ecosystems. Tenth chapter focuses on mesofluidic devices for DNA-programmed combinatorial chemistry. Eleventh chapter deals with guest-host chemistry with dendrimer. A first-principles investigation on unique reactivity of transition metal atoms embedded in graphene to CO, NO, O_2 and O adsorption has been presented in twelfth chapter. Thirteenth chapter deals with first-principles elucidation of the surface chemistry of the C_2H_x (x=0–6) adsorbate series on Fe(100). The structures and properties of y-substituted Mg_2Ni alloys and their hydrides have been discussed in last chapter.

Chapter 1

ANALYTICAL CHEMISTRY TODAY AND TOMORROW

Miguel Valcárcel

Faculty of Sciences of the University of Córdoba, Spain

INTRODUCTION

Dealing with Analytical Chemistry in isolation is a gross error [1]. In fact, real advances in Science and Technology —rather than redundancies with a low added value on similar topics— occur at interfaces, which are boundaries, crossroads —rather than barriers— between scientific and technical disciplines mutually profiting from their particular approaches and synergistic effects. Figure 1 depicts various types of interfaces involving Analytical Chemistry.

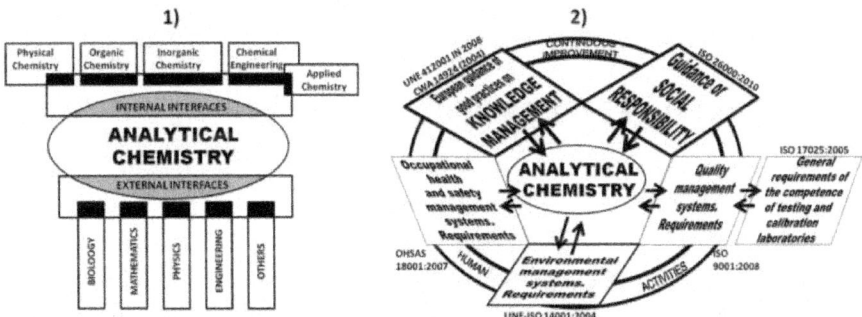

Figure 1: Analytical Chemistry at various interfaces. (1) Internal and interdisciplinary interfaces in the realm of Chemistry. (2) Interfaces with norms and guides. For details, see text.

Analytical Chemistry should in fact be present at a variety of interfaces such as those of Figure 1.1. Two belong to the realm of Chemistry (the framework of reference), namely:

Internal interfaces with other chemical areas (e.g. organic, inorganic, physical and applied chemistry, chemical engineering). Classifying Chemistry into these disciplines or subdisciplines, which are related via "fading" interfaces (1), has become obsolescent.

External interfaces with other scientific and technical disciplines such as biology, biochemistry, mathematics, physics or engineering, where Analytical Chemistry can play an active role (e.g. in the determination of enzyme activities or that of drugs of abuse in biological fluids) or a passive one (e.g. in chemometric developments for data processing or the use of immobilized enzymes in analytical processes).

Also, if Analytical Chemistry is to be coherent with its foundations, aims and objectives (see Section 2.2. of this chapter), it should establish two-way relationships with a variety of international written standards (norms and guides) in order to contribute to the continuous improvement of human activities (see Figure 1.2). The classical relationship between Analytical Chemistry and quality has materialized in ISO 17025:2005, which is the reference for laboratory accreditation. This norm contains technical requirements and other, management-related specifications that are shared with those in ISO 9001:2008, which is concerned with quality in general. Also, written standards dealing with knowledge management and social responsibility are highly relevant to the foundations and applications of Analytical Chemistry, even though they have rarely been considered jointly to date. In addition, Analytical Chemistry is very important for effective environmental protection, and occupational health and safety, since the (bio)chemical information it provides is crucial with a view to making correct decisions in these two complementary fields.

CORNERSTONES OF MODERN ANALYTICAL CHEMISTRY

Analytical Chemistry has evolved dramatically over the past few decades, from the traditional notion held for centuries to that of a modern, active discipline of Chemistry. Changes have revolved mainly around new ways of describing the discipline, and its aims and objectives, a broader notion of real basic references, the definition of the results of research and development activities and a holistic approach to analytical properties.

Definition

Analytical Chemistry can be defined in four simple ways as: (1) the discipline in charge of "Analysis" (the fourth component of Chemistry in addition to Theory, Synthesis and Applications, all of which are mutually related via the vertices of the tetrahedron in Figure 2); (2) the discipline in charge of the production of so named "(bio)chemical information" or "analytical information"; (3) the discipline of (bio)chemical measurements; and (4) the chemical metrological discipline, which is related to the previous definition.

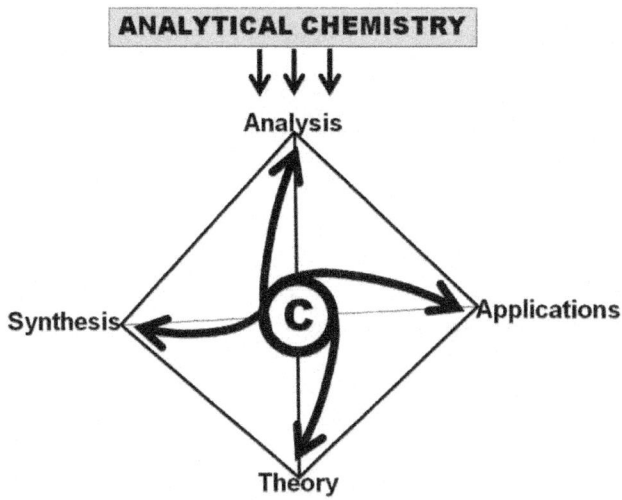

Figure 2: Analytical Chemistry is a discipline of Chemistry (C) inasmuch as it is responsible for "Analysis", an essential component of Chemistry in addition to theory, synthesis and applications in different fields (e.g. environmental science, agriculture, medicine).

These four general definitions have been used to formulate various more conventional definitions such as the following:

"Analytical Chemistry is a scientific discipline that develops and applies methods, instruments and strategies to obtain information on the composition and nature of matter in space and time" (Working Party on Analytical Chemistry of the European Federation of Chemical Societies) [2].

"Analytical Chemistry is a metrological discipline that develops, optimizes and applies measurement processes intended to produce quality (bio)chemical information of global and partial type from natural and artificial objects and systems in order to solve analytical problems derived from information needs" [3].

The strategic significance of Analytical Chemistry arises from the fact that it is an information discipline and, as such, essential to modern society. Analytical Chemistry as a scientific discipline has its own foundations, which materialize in keywords such as information, metrology, traceability, analytical properties, analytical problems and analytical measurement processes. Also, it shares some foundations with other scientific and technical areas such as Mathematics, Physics, Biology or Computer Science.

Aims and Objectives

To be coherent with the previous definitions, Analytical Chemistry should have the aims and objectives depicted in Figure 3.

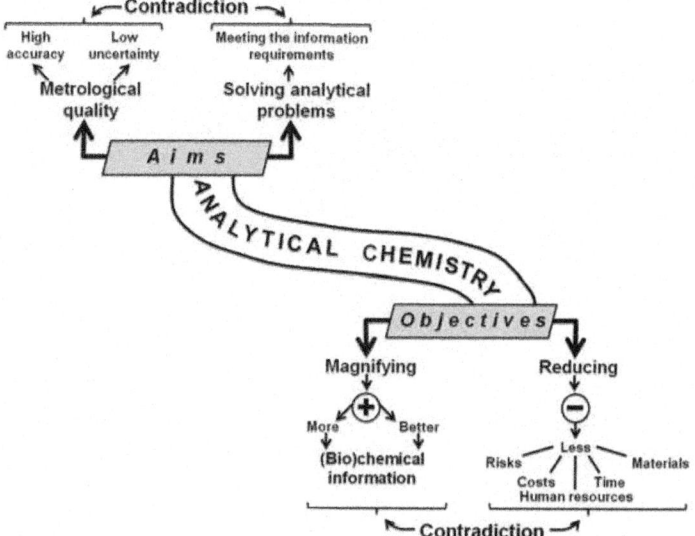

Figure 3: Primary aims and objectives of Analytical Chemistry. For details, see text.

Analytical Chemistry has two essential aims. One, which is intrinsic, is the obtainment of as high metrological quality as possible (i.e. of as true as possible analytical information with as low as possible uncertainty). The other, extrinsic aim is solving analytical problems derived from (bio)chemical information needs posed by "clients" engaged in a great variety of activities (health, general and agrifood industries, the environment).

The main magnifying objectives of Analytical Chemistry are to obtain a large amount of (bio)chemical information of a high quality, and its main reducing objectives to use less material (sample, reagents), time and human resources with minimal costs and risks for analysts and the environment.

The aims and objectives of Analytical Chemistry share its two sides (basic and applied); these are usually in contradiction and require appropriate harmonization. Thus, ensuring a high metrological quality may be incompatible with obtaining results in a rapid, economical manner. In fact, obtaining more, better (bio)chemical information usually requires spending more time, materials and human resources, as well as taking greater risks. Balancing the previous two aims and objectives requires adopting quality compromises [4] that should be clearly stated before specific analytical processes are selected and implemented.

Basic Analytical Standards

Analytical Chemistry relies on the three basic standards (milestones) shown in Figure 4 [5]. The two classical standards, which have been around for centuries, are tangible measurement standards (e.g. pure substances, certified reference materials) and written standards (e.g. the norms and guides ofFigure 1, official and standard methods). A modern approach to Analytical Chemistry requires including a third standard: (bio)chemical information and its properties it should have to facilitate correct, timely decisions. Without this reference, analytical laboratory strategies and work make no sense. In fact, it is always essential to know the level of accuracy required, how rapidly the results are to be produced, and the maximum acceptable cost per sample (or analyte), among other requirements.

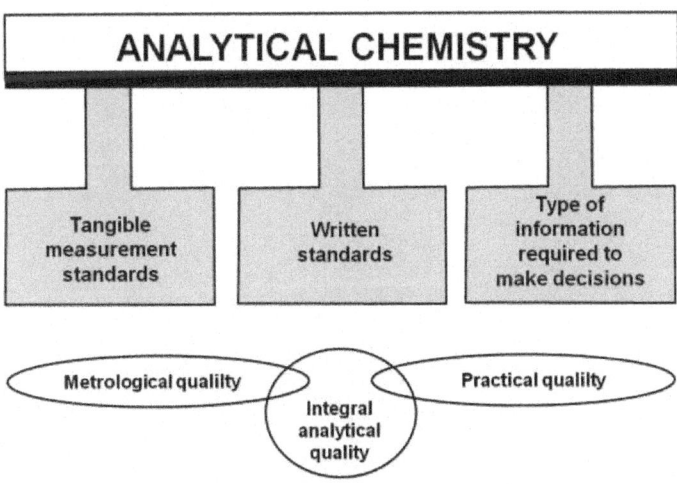

Figure 4: Basic standards supporting the Analytical Chemistry building and analytical quality related concepts. For details, see text.

As can be seen in Figure 4, conventional basic standards are related to so named "metrological quality", whereas (bio)chemical information and its required characteristics (the third basic standard) are related to "practical quality". Combining both concepts in so named "integral analytical quality" requires balancing two contradictory forces, which in turn entails the adoption of "quality compromises" (see Section 4 of this chapter).

R&D Analytical "Products"

The basic side of Analytical Chemistry encompasses a variety of R&D activities aimed at improving existing methods and/or developing new ones in response to new, challenging information needs. These activities can produce both tangible and intangible tools such as those of Figure 5 [6]. Typical tangible analytical tools include instruments, apparatus, certified reference materials, immobilized enzymes and engineering processes adapted to the laboratory scale (e.g. supercritical fluid extraction, freeze-drying). Analytical strategies, basic developments (e.g. calibration procedures) and chemometric approaches (e.g. new raw data treatments, experimental design of analytical methods) are the intangible outputs of analytical R&D activities. Transfer of technology in this context is more closely related to tangible R&D tools, whereas transfer of knowledge is mainly concerned with intangible R&D analytical tools; in any case, the two are difficult to distinguish.

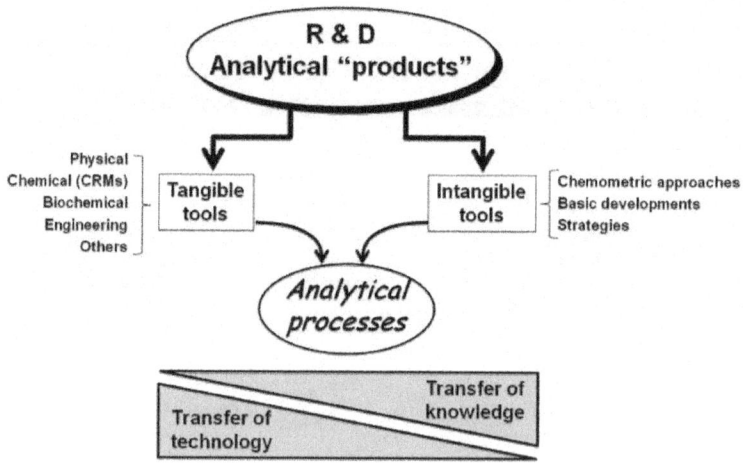

Figure 5: Main outputs of research and development (R&D) in Analytical Chemistry, transfer of knowledge and technology included. For details, see text.

Quality Indicators

Analytical properties are quality indicators for the great variety of systems, tools and outputs of (bio)chemical processes that allow one to compare and validate analytical processes and the results they provide. Traditionally, they have been dealt with separately, with disregard of the high significance of their mutual relationships. Figure 6 provides a holistic view of analytical properties [7] as classified into three groups (capital, basic and productively-related) that are assigned to analytical results and analytical processes.

Top or capital analytical properties (accuracy and representativeness) are characteristics of the quantitative results of measurement processes. Accuracy is related to two classical metrological properties: traceability and uncertainty. In qualitative analysis, this property must be replaced with "reliability", which includes precision (a basic property). Capital properties can be defined in simple terms as follows:

Accuracy is the degree of consistency between a result (or the mean of several) and the true value or that considered as true (viz. the value for a certified reference material) in quantitative analyses. Any differences between the two constitute systematic errors.

Reliability is the proportion (percentage) of right yes/no answers provided by independent tests for analyte identification in aliquots of the same sample in qualitative analyses.

Representativeness is the degree of consistency of the results with the samples received by a laboratory, the overall sample or object studied, the particular analytical problem and the information required by the client.

Basic analytical properties (precision, robustness, sensibility selectivity) are attributes of analytical processes and provide support for capital properties. Thus, it is impossible to obtain highly accurate results if the analytical process is not precise, sensitive and selective enough. These properties can be defined as follows:

Precision is the degree of consistency among a set of results obtained by separately applying the same analytical method to individual aliquots of the same sample, the mean of the results constituting the reference for assessing deviations or random errors.

Robustness in an analytical method is the resistance to change in its results when applied to individual sample aliquots under slightly different experimental conditions.

Sensitivity is the ability of an analytical method to discriminate between samples containing a similar analyte concentration or, in other words, its ability to detect (qualitative analysis) or determine (quantitative analysis) small amounts of analyte in a sample.

Selectivity is the ability of an analytical method to produce qualitative or quantitative results exclusively dependent on the analytes present in the sample.

Productivity-related properties (expeditiousness, cost-effectiveness and personnel-related factors) are attributes of analytical processes with a very high practical relevance to most analytical problems.

Expeditiousness in an analytical method is its ability to rapidly develop the analytical process from raw sample to results. Expeditiousness is often expressed as the sample frequency (i.e. in samples per hour or per day).

Cost-effectiveness is the monetary cost of analyzing a sample with a given method and is commonly expressed as the price per analyte-sample pair. This property has two basic economic components, namely: the specific costs of using the required tools and the overhead costs of the laboratory performing the analyses.

Personnel-related factors. Strictly speaking, these are not analytical properties but are occasionally essential towards selecting an appropriate analytical method. These factors include the risks associated to the use of analytical tools and the analyst's safety and comfort.

As illustrated by Figure 6, quality in the results should go hand in hand with quality in the analytical process. In other words, capital analytical properties should rely on basic properties as their supports. It is a glaring error to deal with analytical properties in isolation as it has been usual for long. In fact, these properties are mutually related in ways that can be more consequential than the properties themselves. Their relationships are discussed in detail in Section 4. Each type of analytical problem has its own hierarchy of analytical properties, which materializes in the above-described "quality compromises".

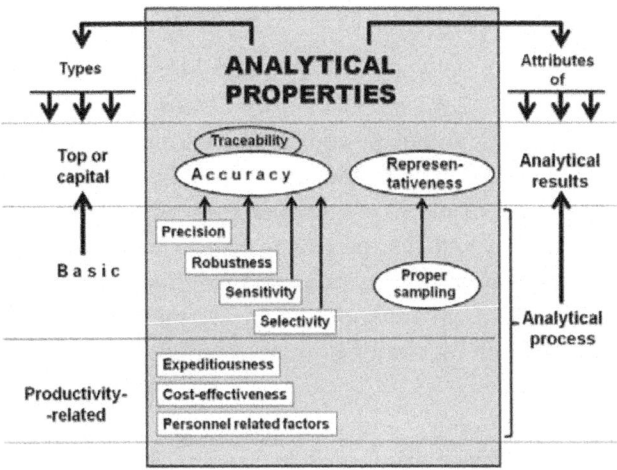

Figure 6: Holistic view of analytical properties as classified into three major groups and of their relationships with quality of the results and the analytical process. For details, see text.

(Bio)Chemical Information

The main output of (bio)chemical measurement processes is analytical or chemical/biochemical information, which is used to describe objects and systems for a variety of purposes, but especially to (*a*) understand processes and mechanisms in multidisciplinary approaches; and (*b*) provide support for grounded, efficient decision-making in a great variety of scientific, technical and economic fields. "Information" is probably the most important keyword for Analytical Chemistry, which has been aptly defined as an "information discipline" [8]. As shown below, (bio)chemical information lies in between raw data and knowledge; also, it has evolved markedly over the past few centuries and eventually become highly influential on human life and the environment by virtue of the increasing importance attached to social responsibility in Analytical Chemistry.

"(Bio)chemical information" and "analytical information" are two equivalent terms in practice. In fact, the difference between chemical and biochemical analysis is irrelevant as it depends on the nature of the analyte (e.g. sodium or proteins), sample (e.g. soil or human plasma) and tools involved (e.g. an organic reagent or immobilized enzymes).

Contextualization

Information is the link between raw data and knowledge in the hierarchical sequence of Figure 7. *Primary* or *raw data* are direct informative components of objects and/or systems, whereas *information* materializes in a detailed description of facts following compilation and processing of data, and *knowledge* is the result of contextualizing and discussing information in order to understand and interpret facts with a view to making grounded, timely decisions. Einstein [9] has proposed *imagination* as an additional step for the sequence in critical situations requiring the traditional boundaries of knowledge to be broken by establishing new paradigms.

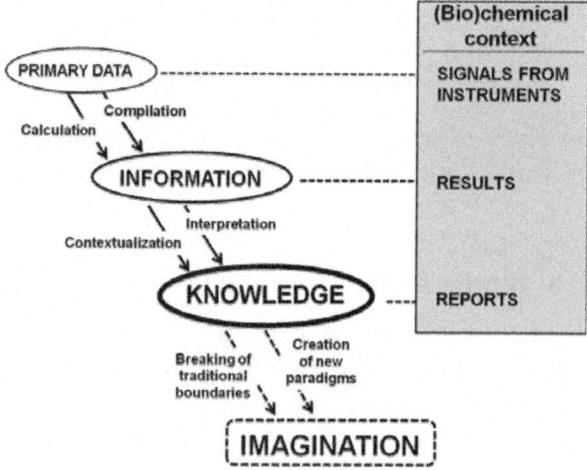

Figure 7: "Information" as an intermediate step between "raw data" and "knowledge", and their significance in the context of chemistry and biochemistry. For details, see text.

In a (bio)chemical context, "raw data" coincide with the primary "signals" provided by instruments (e.g. absorbance, fluorescence intensity, electrical potential readings). Also, "information" corresponds to the "results" of (bio) chemical measurement processes, which can be quantitative or qualitative. Finally, "knowledge" corresponds to "reports", which contextualize information, ensure consistency between the information required and that provided, and facilitate decision-making.

Types

Figure 8 shows several classifications of (bio)chemical information according to complementary criteria such as the relationship between the analyte(s) and result(s), the nature of the results, the required quality level in the results in relation to the analytical problem and the intrinsic quality of the results [10].

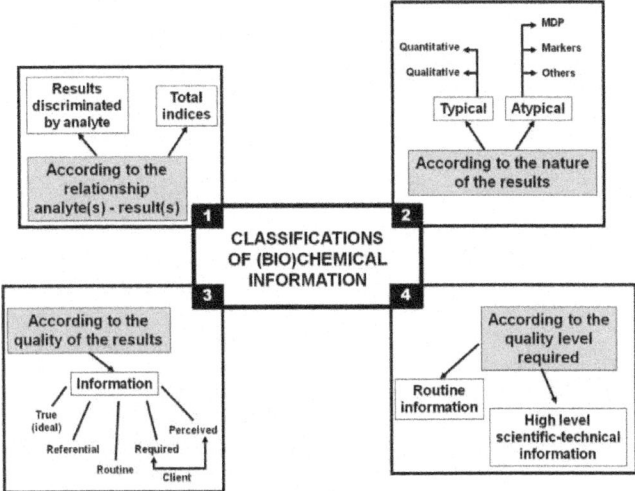

Figure 8: Four complementary classifications of (bio)chemical information based on different criteria. For details, see text.

Based on classification 1 in Figure 8, results can be discriminated by analyte (one analyte–one result), which is the most frequent situation when a separation (e.g. chromatographic, electrophoretic) is involved or when the measurement process is highly selective (e.g. immunoassays). Of increasing interest in this context are "total indices" [11], which can be defined as parameters representing a group of (bio)chemical species (analytes) having a similar structure/nature (e.g. greases, polyphenols, PAHs, PCBs) and/or exhibiting a similar operational behavior or effect (e.g. toxins, antioxidants, endocrine disruptors). More than 50% of the information required for decision-making is of this type. A large number of validated analytical methods produce this peculiar type of output. Probably, the greatest problem to be solved here is to obtain appropriate metrological support.

Classification 2 in Figure 8 establishes two types of results: typical and atypical. Typical (ordinary) results can be quantitative (viz. numerical data with an associated uncertainty range) and qualitative (e.g. yes/no binary responses); the latter have gained increasing importance in recent times. There are also atypical results requiring the use unconventional metrological approaches in response to specific social or economic problems. Thus, so named "method defined parameters" (MDPs) [12] are measurands that can only by obtained by using a specific analytical method —which, in fact, is the standard— and differ if another method is applied to the same sample to determine the same analyte. Usually, MDPs are total indices expressed in a quantitative manner (e.g. 0.4 mg/L total phenols in water; 0.02 mg/L total hydrocarbons in water). In some cases, MDPs are empirical (e.g. bitterness in beer or wine). Some can be converted into yes/no binary responses (e.g. to state whether a threshold limit imposed by legislation or the client has been exceeded). Markers [13] are especially important analytes in terms of information content (e.g. tumor markers, saliva markers to detect drug abuse).

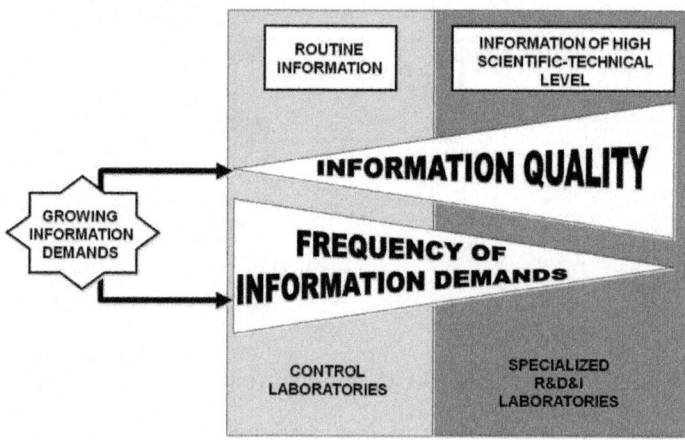

Figure 9: Contradiction between the frequency of information demands and the level of quality required in a situation of growing demands for (bio)chemical information. For details, see text.

Classification 3 in Figure 8 is based on the quality level of the results required in response to the client's information needs and comprises (*a*) routine information provided by control laboratories analyzing environmental, industrial, clinical or agrifood samples, for example; and (*b*) information of a high scientific and technical level that can only be obtained by using sophisticated instrumentation in specialized laboratories usually involved in R&D activities. The frantic recent changes in social and economic activities have promoted an impressive expansion of (bio)chemical information about objects and systems. As can be seen in Figure 9, the quality of (bio)chemical information increases from routine laboratories to specialized laboratories, whereas the frequency of information demands decreases in the same direction. A compromise must often be made between these two contradictory notions. The panoramic view of Figure 9 is essential to perceive all connotations of analytical information. Classification 4 in Figure 8 is based on the intrinsic quality of the results and is examined in detail in Section 4 of this chapter.

Evolution

The routine information provided by control laboratories has evolved dramatically in the last decades.Figure 10 summarizes the most salient general trends in this context, which are commented on briefly below.

1.   *Simplification.* Instead of delivering large amounts of high-quality information (a classical paradigm in Analytical Chemistry), there is a growing trend to delivering the information strictly required to make grounded decisions while avoiding time-consuming efforts to obtain oversized information that is useless in practice. Specially relevant here is the third basic standard supporting Analytical Chemistry (see Figure 4). The situation is quite common in routine laboratories but should be minimized or avoided altogether. Such is the case, for example, with the determination of hydrocarbons in tap water, the legal threshold limit for which is 0.1 ng/mL total hydrocarbons. Using a classical method involving several steps (e.g. filtration, cleanup, solvent changeover) and sophisticated equipment (e.g. a gas chromatograph and mass spectrometer) allows a long list of aliphatic and aromatic hydrocarbons with their concentrations —usually at the ppt or even lower level— to be produced which is utterly unnecessary to make grounded decisions, especially when a simplified method (e.g. one involving extraction into Cl_4C and FTIR measurement of the extract) can be used instead to obtain a total index totally fit for purpose.

2.   *Binary responses.* Qualitative Analysis has been revitalized [14] by the increasing demand for this type of information; in fact, clients are now more interested in yes/no binary responses than in numerical data requiring discussion and interpretation. This trend is related to the previous one because obtaining a simple response usually entails using a simple testing method. The greatest challenge here is to ensure reliability in the absence of firm metrological support. In any case, false negatives should be avoided since they lead to premature termination of tests; by contrast, false positives can always and are commonly confirmed by using more sophisticated quantitative methodologies (see Section 5.4 andFigure 14).

3.   *Total indices.* Based on classification 1 in Figure 8, a result can be a total index [11] representing a group of analytes having a common structure or behavior. This type of information is rather different from classical information, which is typically quantitative and discriminated by analyte. For example, the total antioxidant activity of a food can be easily determined with a simple, fast method using a commercially available dedicated analyzer. This avoids the usual procedure for determining antioxidants in foodstuffs, which involves time-consuming

sample treatment and the use of sophisticated instruments (e.g. a liquid chromatograph coupled to a mass spectrometer). This trend is also related to simplification and is rendering the classical paradigm of Analytical Chemistry (viz. maximizing selectivity) obsolete.

4. *Increasing importance of productivity-related properties.* The holistic approach to analytical properties of Figure 6, which considers hierarchical, complementary and contradictory relationships between them, and systematically using information needs as the third basic analytical reference (Figure 4), provide solid support for the increasingly popular productivity-related analytical properties (expeditiousness, cost-effectiveness and personnel-related factors). These properties are in contradiction with capital and basic analytical properties. Thus, achieving a high accuracy is not always the primary target and, in some cases, productivity-related properties are more important than capital properties. Such is the case with so named "point of care testing" approaches [15], the best known among which is that behind the glucose meter used to monitor the glucose level in blood at home. Glucose meter readings are inaccurate but rapid and convenient enough to control diabetes.

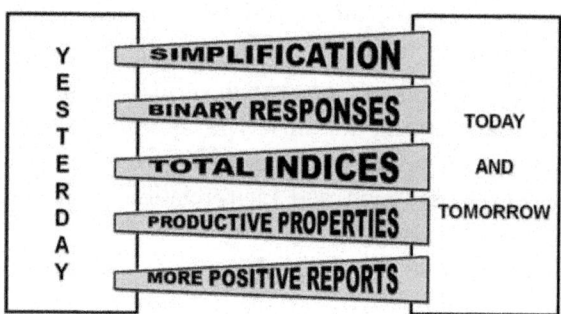

Figure 10: Major trends in the characteristics of (bio)chemical information provided by routine laboratories. For details, see text.

5. *Use of positive approaches to produce reports from results.* Analysts tend to emphasize negative aspects in delivering results and reports. A dramatic impulse of their "marketing abilities" to communicate with clients is therefore needed. One case in point is the word "uncertainty", inherited from Metrology in Physics and introduced in Metrology in Chemistry during the last few decades. This word can lead to wrong interpretations in chemistry nonmajors (e.g. politicians, economists,

managers, judges) and raise global doubts about results. Simply replacing "uncertainty" with "confidence interval", which has the same scientific and technical meaning, can facilitate interpretation and acceptance of the results [16]. One other typical case is the use of "false positives" and "false negatives" to describe errors in binary responses. There is an obvious need to revise the terms related with (bio)chemical information and find alternatives emphasizing positive aspects rather negative connotations.

Social Responsibility

Social responsibility (SR) is a concept encompassing a series of activities intended to support social well-being and help protect the environment which has extended from the corporate world to other human activities such as those involved in Science and Technology. In particular, Social Responsibility of Analytical Chemistry (SRAC) [17] is directed related to the impact of (bio) chemical information or knowledge from objects and systems to society, in general, and to human and animal health, the environment, industry and agrifoods, among others, in particular.

SRAC encompasses two basic requirements, namely: (1) producing reliable data, information and knowledge by using sustainable procedures in the framework of so named "green methods of analysis" [18]; and (2) ensuring consistency of delivered data, information and knowledge with the facts to avoid false expectations and unwarranted warnings.

Analytical Chemistry can therefore provide society with signals (data), results (information) and knowledge (reports), which can have a rather different impact. As can be seen in Figure 11, SRAC has two complementary connotations. One, intrinsic in nature, is the sustainable production of reliable data and results, and their appropriate transfer —which can be made difficult by contextualization and interpretation errors if left in the hands of nonexperts. The other, external connotation, is the appropriate delivery of reports (knowledge) to provide society with accurate information about the composition of natural and artificial objects and systems.

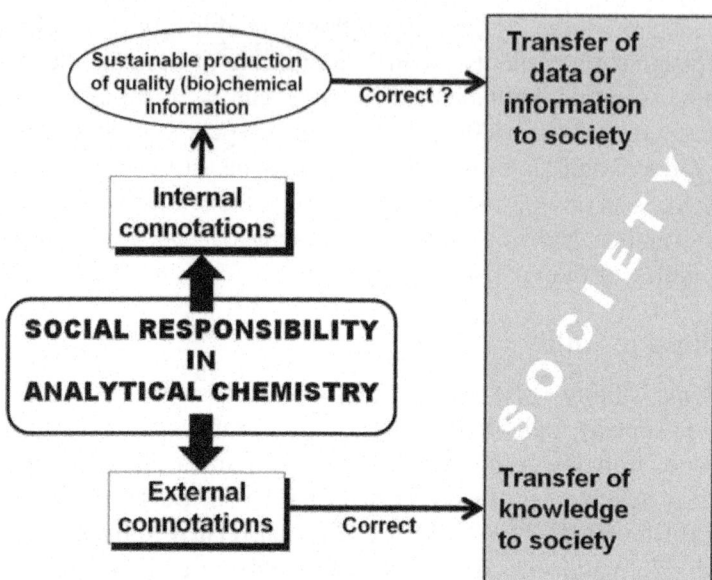

Figure 11: Connotations of Social Responsibility in Analytical Chemistry and ways to transfer data, information and knowledge to society. For details, see text.

ANALYTICAL QUALITY

An integral approach to quality should rely on the following essential components: (1) the basic connotations of the concept as related to a set of features and comparisons, which in Analytical Chemistry materialize in analytical properties (Figure 6) and the three basic standards (Figure 4); (2) the practical connotations of fulfilling the (bio)chemical information needs posed by clients, which is one of the essential aims of Analytical Chemistry (Figure 3); and (3) the measurability of quality in terms of the capital, basic and productivity-related properties for analytical methods and their results.

Classification 3 in Figure 8 allows (bio)chemical information types to be depicted as shown in Figure 12, which additionally shows their mutual relationships via a tetrahedron. The arrows in the figure represent tendencies to converge —in the ideal situation, the tetrahedron could be replaced with a single, common point. Below is briefly described each member of the tetrahedron.

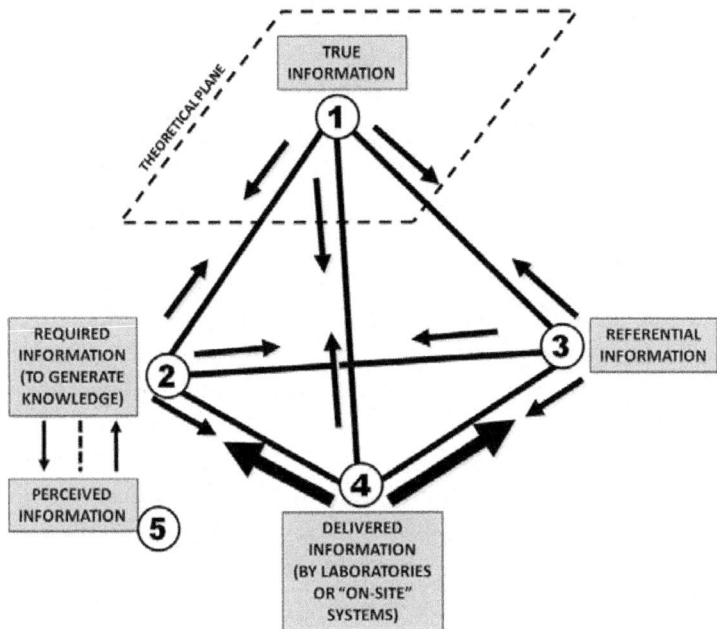

Figure 12: Types of analytical information according to quality and location in a tetrahedron. (1) denotes the ideal situation, in clear contrast with the other types (triangle 2–3–4). The triangles 1–2–4 and 1–3–4 represent problem solving and Metrology in Chemistry, respectively. For details, see text.

1. *True information* corresponds to intrinsic information about objects or systems. It is subject to no uncertainty and hence equivalent to trueness, which is unavailable to analysts. It is also known as "ideal analytical quality".

2. *Referential information* corresponds to the highest quality level that can be achieved in practice, with the information about a certified reference material (CRM) as the most typical example. Referential information is usually obtained in interlaboratory exercises where nonroutine laboratories analyze the same sample under the supervision of a renown organization (e.g. NIST in USA). Certified reference materials and their associated values are essential with a view to assuring quality in analytical methods and their results. The main problem here is their limited availability. In fact, only 3–5% of current needs for CRMs in (bio)chemical analysis have been met, in clear contrast with up to 90–95% in Metrology in Physics. Under these conditions, analysts are very often compelled to use alternative strategies to validate new analytical methods (e.g. standard addition procedures involving pure analytes).

3. *Routine information* is that produced by control laboratories or on-site systems operating outside the laboratory and largely used to control the quality of foodstuffs, industrial products or the environment.

4. *Required information* is that demanded by clients to make grounded, timely decisions and constitutes the third basic analytical standard (see Figure 4), which is frequently disregarded despite its high relevance to the major aims and objectives of Analytical Chemistry (see Figure 3).

5. *Perceived information*, which can be of a similar, higher or lower quality than that actually required by the client. Ideally, a client's perceived and required information should coincide. In some cases, the information delivered falls short of that required and can thus be deemed of low quality. Such is the case, for example, with the toxicological characterization of seawater by potential mercury contamination. The total mercury concentration is inadequate for this purpose because the toxicity of mercury species differs with their nature (inorganic, organometallic). It is therefore necessary to provide discriminate information for each potentially toxic mercury species.

The sides of the tetrahedron of Figure 12 represent the relationships between the different types of analytical information [19]. There are two contradictory relationships (forces) arising from delivered analytical information of great significance to Analytical Chemistry, namely: (1) the relationship between required and delivered information (2–4 in Figure 12), which represents problem solving and is related to the second aim of the discipline (see Figure 3); and (2) that between routinely delivered information and referential information (3–4 in Figure 12), which coincide at the highest metrological quality level —the first aim of Analytical Chemistry (Figure 3). One other significant distinction is that between required and perceived information on the client's side. Analytically, the most convenient situation is that where both types of information coincide in their level of quality —even though it is desirable that the client's perception surpass the actual requirements.

There are thus two contradictory facets of Analytical Chemistry that coincide with the its two aims, namely: a high level of metrological quality and fulfilling the client's information needs (see Figure 3). Analytical Chemistry is located at their interfaces [4]. There are some apparent conflicts, however, including (1) contradictory relationships of capital and basic analytical properties with productivity-related properties (see Figure 6); (2) failing to include required information among basic standards (seeFigure 4); and (3) conceptual differences in analytical excellence between metrology and problem solving.

MAJOR CHALLENGES

Achieving the general aims and objectives of Analytical chemistry in today's changing world requires producing tangible (reagents, sorbents, solvents, instruments, analyzers) and intangible means (strategies, calibration procedures, advances in basic science) to facilitate the development of new analytical methods or improvement of existing ones. This, however, is beyond the scope of this section, which is concerned with general trends in this context.

1. *A sound balance between metrological and problem solving approaches for each information demand.* The situation in each case depends strongly on the specific type of information and its characteristics (see Figure 8). With routine information, the challenge is to adopt well-defined quality compromises, which usually involves selecting and adapting analytical processes to fitness for purpose. Obtaining information of a higher scientific–technical level (e.g. that for materials used in R&D&I processes) calls for a high metrological quality level, as well as for exhaustive sample processing and sophisticated laboratory equipment.

2- *Information required from objects/systems far from the ordinary macroscopic dimensions.* These target objects or systems are directly inaccessible to humans because of their location or size. The size of such objects can fall at two very distant ends: nanomatter and outer space.

Analyzing the nanoworld is a real challenge for today's and tomorrow Analytical Chemistry. Extracting accurate information from nanostructured matter requires adopting a multidisciplinary approach. Nanotechnological information can be of three types according to nature; all are needed to properly describe and characterize nanomatter. Figure 13 shows the most salient types of physical, chemical and biological information that can be extracted from the nanoworld. Nanometrology, both physical and chemical, is still at an incipient stage of development. There is a current trend to using powerful hybrid instruments affording the almost simultaneous extraction of nanoinformation by using physical (e.g. atomic force microscopy, AFM) and chemical techniques (Raman and FTIR spectroscopies, electrochemistry).

The extraction of accurate information from objects and systems in outer space is a challenge at the other end of the "usual" range. This peculiar type of analysis uses miniaturized instruments requiring little maintenance and energy support. There are three different choices in this context, namely: (*a*) remote spectrometric analyses from spacecrafts with, for example, miniaturized X-ray spectrometers [20] or miniaturized mass spectrometers for the analysis of cosmic dust [21]; (*b*) analyses implemented by robots operating on the surface of other planets (e.g. to find traces of water in Mars [22], by using laser ionization-mass spectrometers [23]); or (*c*) monitoring of the inner and outer atmospheres of spacecrafts [24,25].

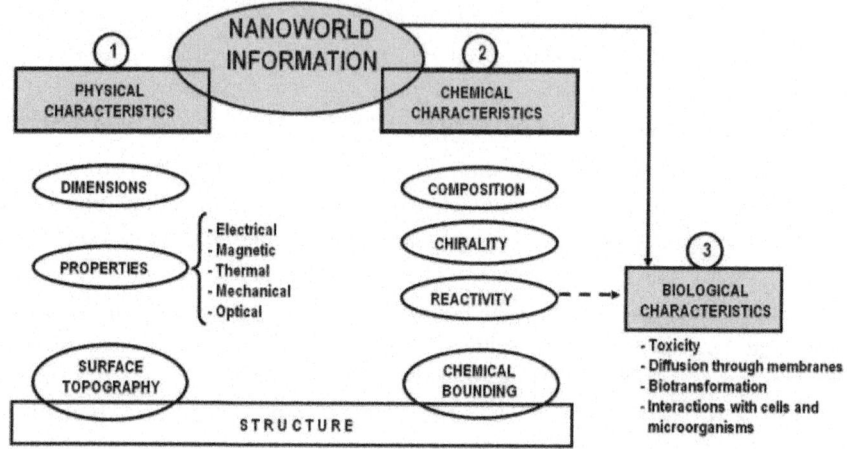

Figure 13: Types of information that can be extracted from the nanoworld. For details, see text.

3. *Breaking the traditional boundaries of the analytical laboratory.* To be consistent with its present aims and objectives (Figure 3), Analytical Chemistry cannot be exclusively confined inside the laboratory walls. In fact, it is necessary to open laboratory doors and analysts' minds in at least two complementary ways, namely:

    (*a*) Analytical Chemistry should play an active role in activities preceding and following the development of analytical processes. Analytical chemists should play a twofold external role here by participating in the design and control of sampling procedures, and also in the discussion and interpretation of analytical results with other professionals in a multidisciplinary approach to transforming information (results) into knowledge (reports).

    (*b*) Analytical Chemistry is increasingly focusing on the production of primary data from (automated) analytical processes implemented with so named "on site" systems outside the laboratory. These systems accumulate or send the requested primary data or results to a central laboratory. In the industrial field, on site monitoring can be performed "in-line" or "on-line". In clinical analysis, points of care testing systems (POCTs) [15] are extensively used for this purpose. The development of robust, reliable sensors for a broad range of analytes in a variety of sample types is a major challenge in this context, where automated calibration and quality control are the two greatest weaknesses.

4.   *Vanguard–rearguard analytical strategies* [26]. As can be seen from Figure 9, the demand for (bio)chemical information has grown dramatically in the past decade and will continue to grow in the next. As a consequence, conventional analytical laboratories have been rendered unable to accurately process large numbers of samples each day. This has raised the need for a new strategy (an intangible R&D&I analytical product according to Figure 5) intended to minimize the negative connotations of conventional sample treatment steps and facilitate the adoption of quality compromises between metrology and problem solving. This strategy uses a combination of vanguard (screening) systems and rearguard (conventional) systems as illustrated in Figure 14.

Vanguard analytical systems are in fact sample screening systems (SSS) [27,28] which are used in many activities where information is rapidly needed to make immediate decisions in relation to an analytical problem. Their most salient features are as follows: (*a*) simplicity (*viz.* the need for little or no sample treatment); (*b*) a low cost per sample–analyte pair; (*c*) a rapid response; (*d*) the production of atypical results (binary responses, total indices, method-defined parameters); and (*e*) reliability in the response. These systems act as mere sample filters or selectors and their greatest weakness is the low metrological quality of their responses —however, uncertainties up to 5–15% are usually accepted as a toll for rapidity and simplicity, which are essential and in contradiction with capital analytical properties. Sample screening systems provide a very attractive choice for solving analytical problems involving high frequency information demands. If these systems are to gain widespread, systematic use, they must overcome some barriers regarding accuracy (viz. the absence of false negatives for rapid binary responses), metrological support (traditionally, norms and guides have focused almost exclusively on quantitative data and their uncertainties) and commercial availability (e.g. in the form of dedicated instruments acting as analyzers for determining groups of analytes in a given type of sample such as antioxidants in foodstuffs or contaminants in water).

Rearguard analytical systems are those used to implement conventional analytical processes. Their most salient features are as follows: (*a*) they require conventional, preliminary operations for sample treatment and these involve intensive human participation and are difficult to automate (e.g. dissolution, solid and liquid extraction, solvent changeover); (*b*) they also usually require sophisticated instruments (e.g. GC–MS, GC–MS/MS, GC–FTIR/MS, LC–MS, LC–ICP-MS, CE–MS); (*c*) they afford high accuracy as a result of their excellent sensitivity and selectivity; (*d*) they use powerful primary data processing systems supported by massive databases easily containing 5000 to 50 000 spectra for pure substances, which ensures highly reliable results; (*e*) they usually provide information for each individual target analyte in isolation; and (*f*) they are expensive and operationally slow, but provide information of the highest possible quality level.

An appropriate combination of these two types of systems allows one to develop *vanguard–rearguard analytical strategies* (see Figure 14). With them, a large number of samples are subjected to the vanguard (screening) system to obtain binary or total index responses in a short time window. The output is named "crash results" and can be used to make immediate decisions. In fact, the vanguard system is used as a sample "filter" or selector to identify a given attribute in a reduced number of samples (e.g. a toxicity level exceeding the limit tolerated by law or by clients) which are subsequently processed systematically with the rearguard analytical system to obtain quantitative data and their uncertainty for each target analyte. The rich information thus obtained can be used for three complementary purposes, namely: (1) to confirm the crash results of vanguard systems (e.g. positives in binary responses to ensure that they are correct); (2) to amplify the simple (bio)chemical information provided by vanguard systems and convert global information about a group of analytes into discriminate information for each for purposes such as determining relative proportions; and (3) to check the quality of vanguard systems by using them to process a reduced number of randomly selected raw samples according to a systematic sampling plan.

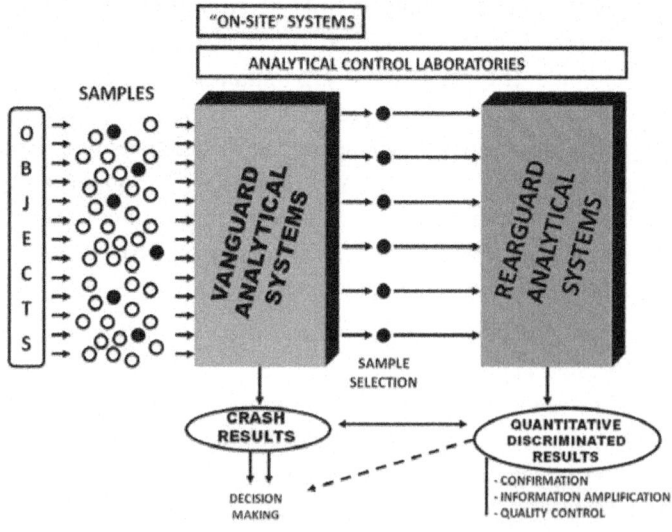

Figure 14: Vanguard–rearguard analytical strategies for the systematic analysis of large numbers of samples. For details, see text.

REFERERNCES

1. R. Murray, ". , The permanency of fading boundaries" Anal. Chem., 1996A.

2. R. Kellner, J. M. Mermet, M. Otto, H. D. Widmer, M. , Valcárcel "Analytical Chemistry" (2nd edition). 2004Wiley-VCH, Weinheim, Germany.

3. M.Valcárcel "Principles of Analytical Chemistry". 2000Springer-Verlag, Heidelberg, 135

4. M. Valcárcel, B. , Lendl "Analytical Chemistry at the interface between metrology and problem solving". Trends Anal. Chem. 2004200423527534

5. M. Valcárcel, A. , Ríos "Reliability of analytical information in the XXIst century". Anal. Chim. Acta. 19991999400425432

6. M. Valcárcel, B. M. Simonet, S. , Cárdenas "Bridging the gap between analytical R&D products and their use in practice". Analyst. 2007200713297100

7. M. Valcárcel, A. , Ríos "The hierarchy and relationships of analytical properties". Anal. Chem. 1993A-787A.

8. M. Valcárcel, E. Aguilera, Aguilera-Herrador "La información (bio) química de calidad". An. Quím. 2011

9. A. Einstein, La colección libre de citas y frases célebres en 1.4.E de http://es.wikiquote.org/wiki/Albert_Einstein.

10. M. Valcárcel, B. M. , Simonet "Types of analytical information and their mutual relationships". Trends Anal. Chem. 2008200827490495

11. J. R. Baena, M. Gallego, M. Valcárcel, ". , Total indices in analytical science". Trends Anal. Chem. 2003200322641646

12. B. M. Simonet, B. Lendl, M. Valcárcel, ". Method-defined, measurands. parameters, forgotten". sometimes, Anal. Trends, Chem, 2006200625520527

13. J. R. Baena, M. Gallego, M. , Valcárcel "Markers in Analytical Chemistry". Trends Anal. Chem., 2002200221878891

14. M. Valcárcel, S. Cárdenas, M. , Gallego "Qualitative analysis revisited". Crit. Rev. Anal. Chem. 2000200030345361

15. E. Aguilera-Herrador, M. Cruz-Vera, M. , Valcárcel "Analytical connotations of point-of-care-testing". Analyst 2010201013522202232

16. J.D.R.Thomas "Reliability versus uncertainty for analytical measurements". Analyst. 1996

17. M. Valcárcel, R. , Lucena "Social responsibility in Analytical Chemistry". Trends Anal. Chem.,201220123117

18. S. Armenta, S. Garrigues, A. de la Guardia, ". Green, Chemistry". Analytical, Trends Anal. Chem. 2008200827497511

19. M. Valcárcel, A. , Ríos "Required and delivered analytical information: the need for consistency". Trends Anal. Chem. 2000200019593598

20. C.E. Schlemm et al.The X-ray spectrometer on the Messenger spacecraft" Space Sci. Rev. 20072007131393415

21. D.E. Austin, T.J. Ahrens, J.L. Beauchamp "Dustbuster: a compact impact-ionization time-of-flight mass spectrometer for in situ analysis of cosmic dust".Rev. Sci. Instrum. 2002200273185189

22. E.K.Wilson "Mars watery mysteries". C&E News (ACS) 2008December 1, 5961

23. B. Sallé, J. L. Lacour, E. Vors, P. Fichet, S. Maurice, D. A. Cremers, R. S. , Wiens "Laser-induced breakdown spectroscopy for mass surface analysis: capabilities at stand-off distances and detection of chlorine and sulfur elements". Spectrochim. Acta B. 2004200459141131422

24. M. L. Matney, S. W. Beck, T. F. Limero, J. T. , James "Multisorbent tubes for collecting volatile organic compounds in spacecraft air". AIHAJ 20002000616975

25. G. G. Rhoderick, W. J. Thor, I. I. I. W. R. Miller, F. R. Jr Guenther, E. J. Gore, T. O. , Fish "Gas standards development in support of NASA's sensor calibration program around the space shuttle". Anal. Chem. 200920098138093815

26. M. Valcárcel, S. , Cárdenas "Vanguard-rearguard analytical strategies". Trends Anal. Chem. 20052005246774

27. M. Valcárcel, S. Cárdenas, M. , Gallego "Sample screening systems in Analytical Chemistry". Trends Anal. Chem. 1999199918685694

28. M. Valcárcel, S. , Cárdenas "Current and future screening systems". Anal. Bioanal. Chem. 200520053818183

Chapter 2

QUANTUM CHEMISTRY AND CHEMOMETRICS APPLIED TO CONFORMATIONAL ANALYSIS

Aline Thaís Bruni[1] and Vitor Barbanti Pereira Leite[2]

[1]Departamento de Química, Faculdade de Filosofia, Ciências e Letras de Ribeirão Preto, Universidade de São Paulo

[2]Departamento de Física, Instituto de Biociências, Letras e Ciências Exatas, Universidade Estadual Paulista, São José do Rio Preto Brazil

INTRODUCTION

Conformational Analysis: Early History and Importance

Molecular structure plays a special role in science. Knowledge of the atomic arrangement is essential in order to be able to elucidate chemical properties and processes. The first advances in determining molecular structure occurred in nineteenth century. Around 1812, Jean-Baptiste Biot, a French physicist, discovered optical activity by observing polarized light shifting when crossing a quartz crystal. He observed that the light was displaced to the right in some cases and to the left in others. The conclusion was that rotation of polarized light by quartz is an inherent property of the crystal. Interested in the phenomenon, Biot noticed in further studies that similar effects were found when polarized light passed through certain liquids such as natural oils (lemon extract and laurel), alcoholic solutions of camphor, some sugars and tartaric acid. (Drayer, 1993; Cintas, 2007; Gal, 2011) Biot's observations were very important in laying foundation for the concept of optical activity. In 1948, Louis Pasteur discovered molecular chirality when studying a mixture of tartaric acid crystals.(Gal, 2007) He patiently performed the manual separation of tartarate enantiomer crystals (Cintas, 2007) and observed that each solution made with them was able to displace polarized light in one direction. He concluded that compounds with nonsuperimposable molecular asymmetry have identical chemical properties despite the inverse behavior related to polarized light.

Pasteur argued that the optical activity of organic solutions is related to molecular geometry. This insight was far ahead of the organic structural theory of the time.(Drayer, 1993) Although Pasteur was the first to show a relationship between optical activity and molecular symmetry, he was not able to say exactly how a molecule could be right- or left-handed. The main advances in this idea occurred in 1874 when a theory of organic structure in three dimensions was independently and simultaneously developed by Jacobus Henricus van't Hoff in Holland, and Joseph Achille Le Bel in France. (Drayer, 1993; Cintas, 2007) In 1865, August Kekulé proposed his theory of the benzene molecular structure and proposed that the carbon atom has valence 4.(Brush, 1999) His principal idea was that the carbon atom is tetravalent and can form valence bonds with other carbon atoms yielding to chains. These carbon chains can sometimes have closed arrangements, forming rings. (Drayer, 1993) Van't Hoff and Le Bel proposed that the four valences of the carbon atom were not planar, but directed into three-dimensional space. Van't Hoff specifically proposed that the spatial arrangement was tetrahedral. Later, he used the tetrahedron as a graphic representation of the valence arrangement around the carbon atom and also used this model to explain the physical property of optical activity.(Ramberg & Somsen, 2001) A compound containing a four different substituted carbon – described by Van't Hoff as asymmetric carbon - would be capable of existing in two distinctly different nonsuperimposable forms. Finally, he stated that the asymmetric carbon atom was the cause of molecular asymmetry and optical activity.(Drayer, 1993) Le Bel, in turn, also published his stereochemical ideas in 1874, but with a different approach to the problem from that presented by Van't Hoff. His hypothesis was not based on the tetrahedral model for the carbon atom and the fixed valences between the atoms. His investigation was into the asymmetry as a whole, without evaluating the individual atoms. The full system was considered in his evaluation, and his interpretation could be inserted into the field that is currently understood as molecular asymmetry. He mentions the tetrahedral carbon atom only in special cases, and not as a general principle. Many molecules confirm Le Bel's concepts of molecular asymmetry. Allenes, spiranes, and biphenyls are some examples of asymmetric molecules that do not contain any asymmetric carbons. Van't Hoff's and Le Bel's different approaches can be explained by the origin of their formation. Van't Hoff, based on Kekulé tetrahedron models, suggested the concept of the asymmetric carbon atom. On the other hand, Le Bel based his investigations on Pasteur's considerations of the connections between optical rotation and molecular structure.(Drayer, 1993) The historical development of conformational search does not end here and has many other important aspects and particularities. Our goal was just to give a basic outline of the initial concepts and how they influence current conformational understanding. Despite the historical progress in conformational studies, the advances in structure determination has been relatively recent and have been made possible by the development of analytical instruments and computational tools. Early structural studies were applied only to small molecules or substructures that could be expressed in terms of a few settings.(Allen et al., 2010) Currently, a great evolution is occurring in mechanisms for determining and understanding molecular structures. The relationship between geometry and energy is experimentally measurable and gives an idea of the balance between energy factors involved in each structure. (Pietropaolo et al., 2011) Reactivity and other properties are directly linked to the conformational arrangement of molecules.(Hunger & Huttner, 1999) Every chemical property must be understood according to its molecular structure and atomic connections. (Pietropaolo et al., 2011) Indeed, knowledge of structural arrangement is important since it underlies studies in chemical reactions and other molecular behaviors. There are experimental techniques for the structural determination, such as X-ray, magnetic resonance, infrared, mass spectroscopy and others. In this chapter we will discuss theoretical methods for molecular conformational determination. The field that concerns ways to mimic the behavior of molecules and molecular systems is molecular modeling. It seeks a simplified or idealized description of molecular systems, making it possible to produce three-dimensional representations that provide insights into their behavior. As computer tools have enjoyed a spectacular increase in last decades, theoretical methods are invariably associated with computer modeling. This has become a powerful tool for evaluating molecular structure, from which special chemical information about molecular behavior can be inferred. (Pietropaolo et al., 2011)

STATEMENT OF THE PROBLEM

For the theoretical and computational determination of molecular properties it is necessary to previously determine the minimum energy structure of the system being studied. A central issue is to probe the equilibrium configuration of the molecular system. The way that energy varies with the coordinates is usually referred to as the potential energy surface. At the atomic level, the interaction energy between atoms is essentially ruled by quantum mechanics, which provides the basic elements and methods used in molecular modeling. However, the potential energy surface can be addressed with different degrees of approximation, i.e., ab initio, effective potentials or even more coarse-grained potentials. Irrespective of the details with which the system is considered, one usually faces the problem of a highly dimensional system with the occurrence of multiple minima. Low energy minima play an important role in determining molecular properties, and the determination of these minima conformational states is a non-trivial task, usually referred to as energy minimization method for exploring the energy surface. If four or more atoms are connected in chain by single bonds we can suppose that there is considerable flexibility in the molecule. The existence of hindered rotation about a single bond is one of the fundamental concepts in conformational analysis.(Mo & Gao, 2007) The understanding of the connections between the atoms is related to the internal coordinate parameters, i.e., bond length, bond angle and dihedral angle, and is essential in designing molecular models.

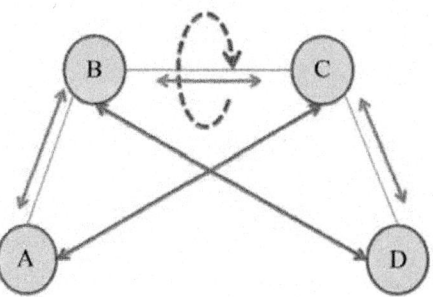

Figure. 1: Model for a generic molecule with four different atoms.

For instance, let us consider a molecule composed of four different atoms which are singlebond linked (Figure 1). The green arrows represent the bond stretch and the average value is the bond length; the red arrows correspond to the angle formed by three sequential atoms, i.e., the angle bond; the curved blue arrow indicates the free rotation around the only single bond able to perform changes in the molecule conformation, as shown in Figure 2:

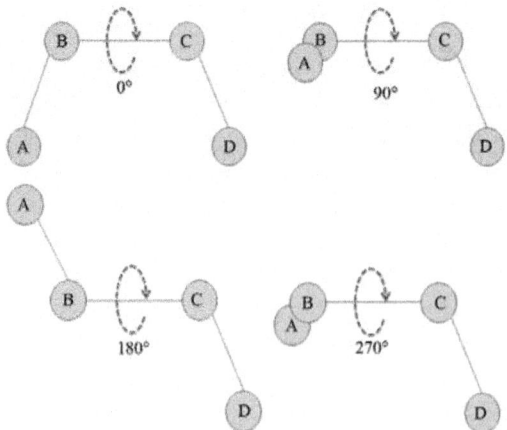

Figure. 2: 90° rotation around the single bond.

In other words, different conformations are obtained when a dihedral angle is rotated. A dihedral angle is that composed by the planes formed by the sequence of three atoms (Figure 3):

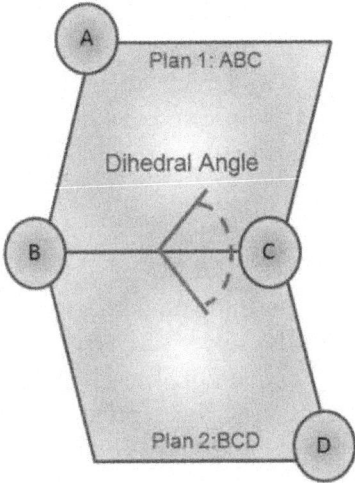

Figure. 3: Dihedral angle representation for the molecule ABCD.

The main questions on molecular modeling are concerned with a good way of finding the global minimum energy structure. Important information is also concerned with the behavior of the conformational space. It is not only necessary to know which is the global minimum, but also the whole shape of the potential energy surface (PES) The main characteristic of this conformational phase space is that it is exponentially large, and a computationally hard problem. (Fraenkel, 1993) One problem that illustrates this difficulty is the protein folding, in which one searches for the global energy minimum structure associated with its functional conformation. If a method can be used to describe the relevant potential energy surface of a given molecule, it can also accurately elucidate its behavior against many interest situations. Many techniques have been presented, and it is not the goal of this work to make a deep study on them. A good discussion of practical methods is given by Leach.(Leach, 2001) We intend to give a brief idea of the most popular techniques used for investigating the molecular structure computationally. For conformational sampling, one can imagine a hierarchy of methods with different computational costs.(Seabra et al., 2009) However, there is no sovereign truth about what is the best method for performing a conformational analysis. Each situation must be evaluated. The best method is one that has the best fit to the problem studied; in practical terms it will provide the answer as quickly as possible, using the least amount of computer resources.

Stochastic Methods of Conformational Analysis

The literature reports many methods for trying to solve multiconformational problems, and most of them are based on stochastic approaches. Put simply, stochastic methods work with random variables, such as initial conformations for the search or the steps probing the configuration phase space. The simple criterion for establishing a minimum energy conformation is that the first derivatives of the energy E with respect to each variable (x_i) is zero and the second derivatives are all positive:

$$\frac{\partial E}{\partial x_i} = 0 \quad \text{and} \quad \frac{\partial^2 E}{\partial x_i^2} > 0.$$

The algorithms that search for minimum energy states can be classified into two groups: those which use derivatives of the energy with respect to the coordinates, and those which do not. The most used derivative minimization methods are the steepest descent, line search in one dimension and conjugate gradient methods.(Leach, 2001) These algorithms are very useful for conducting local (restricted) searches of minima, or downhill searches to the nearest minimum, since they are not able to overcome energy barriers. They are often used in combination with other stochastic methods. In the remainder of this section we discuss examples of stochastic methods. Havel, Kuntz and Crippen described distance geometry algorithms in conformational analysis.(Havel et al., 1983b) Given the impossibility of examining all possible conformations, they introduced a method which is capable of finding global optima without considering all possible solutions by means of combinatorial optimization. The method is known as brunch and bound and involves logical tests that allow whole classes of solutions to be eliminated without examining them one by one. The method converts a set of distance ranges (or "bounds") into a set of Cartesian coordinates that are consistent with those bounds. (Spellmeyer et al., 1997) The efficiency of a branch and bound algorithm depends on how effective these tests are compared to the time required to perform them. (Havel et al., 1983a; Havel et al., 1983b) In another study, Havel et al presented the basic theorems of distance geometry in Euclidean space. They proposed new algorithms and described refinements to the existing ones. All these algorithms were similar because they utilize geometric principles in order to interpret structural relationships. (Havel et al., 1983b) According to Leach and Smellie,(Leach & Smellie, 1992) distance geometry is a method for searching conformational space in which a structure is initially formulated in terms of interatomic distances. Any molecular system can be described as the set of minimum and maximum interatomic distances between all pairs of atoms in the molecule. The complete conformational space of the molecule is contained within this space. In distance geometry, a matrix is defined as the set of minimum and maximum distances, and then used to create a series of conformers that are consistent with those distances.(Spellmeyer et al., 1997) Another tool for performing conformational searches is the genetic algorithm, a stochastic method first introduced by Holland in 1975. Genetic algorithm (GA) is a method applied to solve problems using a natural evolution process simulation. It is a stochastic method developed in analogy to Darwin's theory of evolution in order to perform the optimization.(Brodmeier & Pretsch, 1994; Lucasius, 1993; Nair & Goodman, 1998) Genetic algorithm is commonly used for studying a large-scale space of possible solutions. The goal is to identify the best solutions within that space without the need to evaluate all possibilities.(Yanmaz et al., 2011) The GA is the optimization of a large number of possible solutions using a randomly generated population. When applied to conformational analysis, the population of interest consists of different conformations. The biological evolution of this population is simulated. A population of trial solutions is iteratively manipulated by a series of genetic operators to satisfy an objective function. The adjustment is calculated, and a new population is generated according to operators, such as selective reproduction, recombination and mutation. The process is repeated until the minimum energy structures are obtained.(Lucasius & Kateman, 1994; Beckers et al., 1996; Beckers et al., 1997) Artificial Neural Networks (ANN) are another example of stochastic methods used in conformational analysis. This method is based on concepts of the behavior of the human brain. Although artificial neural networks are primitive compared to their biological counterparts, they exhibit some interesting properties which make them useful as multivariate tools in various fields of research. During the last decade, ANN have been successfully applied in non-linear modeling, classification, signal processing and process control.(Derks & Buydens, 1996) The properties of a molecule are intimately linked to the conformations that it adopts and so an understanding of the conformational space is important in rationalizing and predicting its behavior.(Jordan et al., 1995) Among the most popular stochastic methods for covering the conformational space are Monte Carlo (MC) and Molecular Dynamics (MD). They are similar in the sense that both procedures include the same representation of molecules and use classical force fields for the potential energy terms, under periodic boundary conditions. The main purpose of these methods is to sample the phase space and to use the force fields ability to represent the conformational space near minima and connecting transition structures.(Jorgensen & TiradoRives, 1996; Grouleff & Jensen, 2011) However, large differences are found in sampling and configuring space available to the system. For MC, a new configuration is generated by selecting a random molecule or part of it, rotating it, translating it, and performing an internal structural variation. These changes do not necessarily need to follow a realistic physical trajectory. The acceptance of the new configuration is, however, determined by the Metropolis sampling algorithm. The sampling criterion is set in a way that enhances the likelihood of probing low energy conformations. Application over enough configurations yields properly Boltzmann-weighted averages for structure and thermodynamic properties. For MD, given a set of initial conditions (position and velocities of all atoms), new configurations are generated by application of Newton's equations of motion, so that the new atomic positions and velocities of all atoms are determined simultaneously over a small time step. In both cases, the force field controls the total energy (MC) and forces (MD), which determines the evolution of the systems. (Jorgensen & TiradoRives, 1996) Examples of problems related to large systems are the interaction between drug and the receptor, and protein behavior and folding. Molecular docking procedures are capable of predicting the three-dimensional structure of macromolecular complexes and their binding affinity. The information required is simple and corresponds to the structures of the receptor and ligand and the presumable interfacing region between them. Besides the simplicity of these docking procedures, they have low computational costs. However, molecular plasticity and solvation effects are not, or are only approximately, taken into account in these approaches. Free energy simulations may be then used to investigate the molecular association process and to predict binding affinity. (Biarnés et al., 2011) It is important to realize that sometimes the probing of PES addresses singular questions, which involve association of several methods, also called hybrid methods. A particular wellknown tailored one is the quantum mechanics/molecular dynamics approach, also known as QM/MM approach. This is a molecular simulation method that combines the strength of both QM (high accuracy in specific regions) and MD fast calculations (in not so crucial regions), in such a way that it efficiently allows the study of chemical processes in solution and in proteins. When stochastic methods are used to find minimum energy conformations, asymptotic states in restricted regions of the phase space are probed. This means that there is no end point in the search, and the convergence cannot be assured.

SYSTEMATIC SEARCH IN CONFORMATIONAL ANALYSIS

As seen before, stochastic techniques use different heuristics to randomly cover the conformational space. These algorithms apply a perturbation to the initial conformer and minimum energy conformation is associated with the lowest energy state that is found through out this procedure. They provide a sampling of energy minima structures and the shape of the PES is obtained in an indirect way. Beyond the stochastic methods there are procedures that do not work with random choice to cover conformational space. These classes of methods are described as deterministic and are capable of searching the conformational map in a systematic way, providing a direct knowledge of PES shape. These searches divide conformational space into quantized units and apply algorithms to search this discrete space or define a set of heuristic rules that are used to drive the search.(Smellie et al., 2003) Systematic methods are those that explore all conformational space at some fixed degree of resolution. To perform the systematic search, a molecule must be numerically described by its atoms' internal coordinates. The internal coordinates are bond length, angle bond and dihedral (torsion) angle. For a given initial structure the systematic conformational search is conducted by regular variation in dihedral angles (Figure 2). Although a systematic search can obtain the morphology of a molecule's energetic behavior directly, this method is not feasible for evaluating complex systems. (Beusen et al., 1996) Systematic search is most usefully applied for molecules with few degrees of freedom.(Li Manni et al., 2009) According to literature (Beusen et al., 1996), to cover the PES corresponding to the conformational space, different molecular structures must be systematically generated by rotating the torsion angles around the single bonds between 0° and 360°. The number of conformations is given by:

$$\text{Number of conformations} = s^N \tag{1}$$

where N is the number of free rotation angles, and s is the number of defining steps according to the angle increment:

$$s = \frac{360°}{\theta_i} \tag{2}$$

with θ_i being the dihedral increment of angle i. An examination of equation (1) reveals that the number of conformations generated will exponentially increase in proportion to the number of bonds with free rotation in the molecule under study. A problem arises if the number of steps is large, i.e., when a very refined surface is required by small angle increments. This problematic behavior of the systematic study of PES, described as combinatorial explosion, is the major restriction involved in this kind of search. Figures 4 and 5 illustrate how combinatorial explosion works. In Figure 4, we have a representation of the system growth where many single bonds can be rotated. The combinatorial explosion problem is represented by Figure 5. The number of branches to be considered is shown by the ramification achieved according the number of angles (A, B, C, D...) and will depend on the dihedral increment chosen.

Due to the problem involved in combinatorial explosion, systematic search becomes nonviable for studying large molecules, since the number of degrees of freedom increases. A useful strategy for reducing the dimensionality of the conformational space is to perform systematic conformational searches on small portions of the molecule (either as isolated fragments or in situ). Using these optimal parts, one builds the conformation of the whole molecule with only limited additional searching of the relative conformations of the fragments. Approaches that incorporate this principle are known as "build-up" methods. (Beusen et al., 1996; Izgorodina et al., 2007) There are some strategies for overcoming the combinatorial explosion. We will focus our discussion on procedures that involve chemometrical approaches.

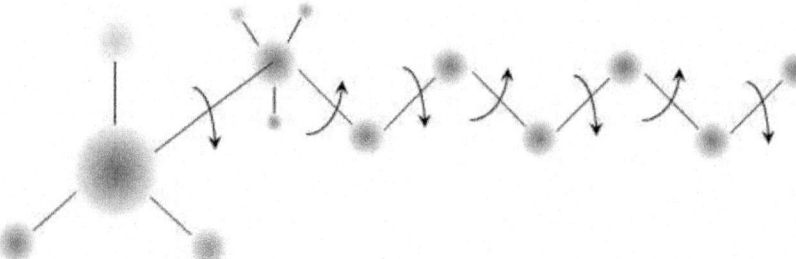

Figure. 4: A general structure with many single bonds.

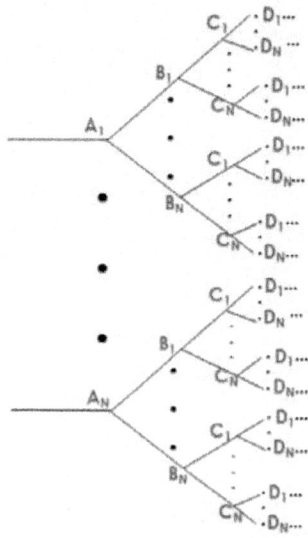

Figure. 5: Branches generated by the dihedral angles A, B, C and D.

CHEMOMETRICS AND STRUCTURE DETERMINATION

A conformational search, independently of the method chosen, usually involves large amounts of data. Sometimes, data achieved from a given methodology must be explored by an additional technique. According to Geladi (2003) "data exploration means taking a look at the data to find interesting phenomena, often without prior expectations. As a result, outliers, clustering of objects and gradients between clusters may be detected."(Geladi, 2003) Chemometrics has been used extensively in recent years for exploring chemical problems by means of computer tools and statistical observations. The literature presents many definitions for chemometrics. For our purposes, this field of knowledge is better defined as a combination of two definitions found in the literature:

a. According to Wold (1995), chemometrics can be understood as a way "to get chemically relevant information out of measured chemical data, how to represent and display this information, and how to get such information into data";(Wold, 1995)

b. For Beeb (1998), chemometrics corresponds to "the entire process whereby data (e.g., numbers in a table) are transformed into information used for decision making." (Beebe, 1998)

The above definitions indicate that chemometrics offers a broad approach to chemical measurement sciences. It is not restricted to the actual experimental analysis but also considers what happens before and after it. (Massart et al., 2004) It is the goal of chemometrics to extract the information from the data. (Ramos et al., 1986) Chemometrical approaches have been applied to conformational analysis for handling special difficulties of large amounts of data generated both by stochastic and by systematic searches. Among the various chemometrical techniques, Principal Component Analysis (PCA) is the most commonly used for conformational problems. In many ways, it forms the basis for multivariate data analysis. PCA is a multivariate method of analysis whose main concern is to reduce the dimensions needed to portray accurately the characteristics of a large dimensional data matrix.(Beebe, 1998; Wold et al., 1987) This mathematical procedure consists of eliminating a large number of correlated variables without changing the characteristics of the original data-set that contribute most to its variance. For an easy graphical representation, consider a two-dimensional set of variables as shown in Figure 6 (a). PCA can be performed on the original variables as shown in Figure 6 (b) and new axes, called Principal Components, arise to account for the maximum variation. A subsequent rotation (Figure 6(c)) is made on these new PC axes in order to rewrite the original variables in terms of this new axes-system. Each PC is constructed as a linear combination of variables:

$$P_i = \sum_{j=1}^{v} c_{i,j} x_j$$

(3)

where P_i is the ith principal component and $c_{i,j}$ is the coefficient of the variable $x_{i,j}$. (Leach, 2001) There are v such variables. The first principal component PC1 is chosen in order to maximize the data variance of the axis. The second and subsequent ones are chosen to be orthogonal to each other and account for the maximum variance in the data not yet described by previous principal components. A variety of algorithms can be used to calculate the principal components. The most commonly employed approach is singular value decomposition SVD. (Golub & Loan, 1996)

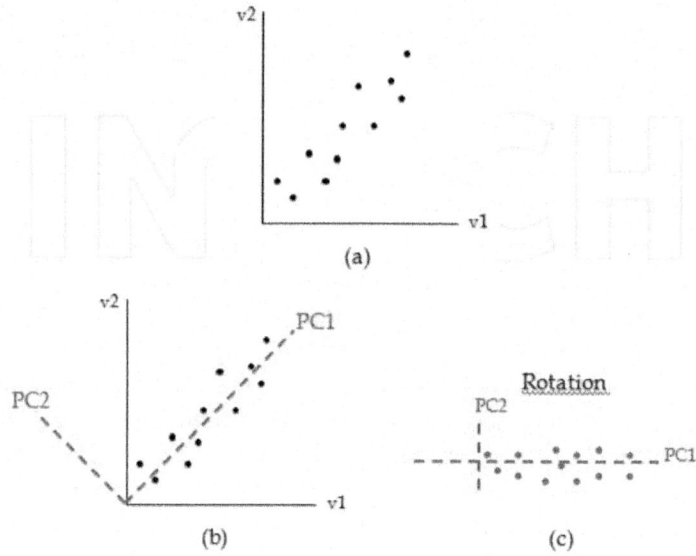

Figure. 6: PCA procedure: (a) original data set; (b) PCA on original data set and (c) Variables according to the new PC coordinates.(Beebe, 1998)

A matrix of arbitrary size can be decomposed into the product of three matrices in such a way that:

X = USV$_t$

(4)

where U and V are square orthogonal matrices. The matrix U (whose columns are the eigenvectors of XXt) contains the coordinates of samples along the PC axes. The V matrix (which contains the eigenvectors of the correlation matrix XtX) contains the information about how the original variables were used to make the new axes[$c_{i,j}$ coefficients in eq. (3)]. The S matrix is a diagonal matrix that contains the eigenvalues of the correlation matrix (standard deviations) or singular values of each of the new PCs. The diagonalization of symmetric matrices (such as XXt and XtX) and SVD are fundamental problems in linear algebra (Golub & Loan, 1996), for which computationally efficient software has been developed and can be used on a routine basis (Hanselma et al., 1997) for very large-size matrices.

In chemistry, PCA was introduced by Malinowski around 1960 under the name Principal Factor Analysis, and further developed after 1970.(Malinowski, 2003) Principal Component Analysis can be used for crystallographic structure data; in its general form, conformational analysis is applied to multivariate numerical problems. (Allen et al., 2010) Many studies report on the use of PCA for handling Molecular Dynamics data. Among them, we highlight the application that uses PCA for mapping potential energy surfaces, by the quantitative visualization of a macromolecular energy funnel. (Becker, 1998) Other examples where PCA can be applied in molecular structure determination can also be found in recent studies. (Das et al., 2011; AraujoAndrade et al., 2010; Kiralj et al. 2007; Oblinsky et al., 2009; Silva et al., 2011)

PAIRS OF DIHEDRAL ANGLES-SYSTEMATIC ANALYSIS

There is a variety of theoretical methods that are capable of locating minimum energy structures in the potential energy surface. The problem of stochastic methods is that there is no natural end point for the conformational search. In some cases, only a small subset of conformational space is explored and the convergence of the system is not guaranteed. Only Systematic Conformational Analysis maps the conformational space completely. We stress the principal difficulty inherent in this method is the combinatorial explosion. In a previous study (Bruni et al., 2002), a new methodology was introduced for controlling the combinatorial explosion through a systematic reduction in the size of the system by means of chemometrics. This method consists of a small systematic conformational analysis, in which the conformational space is studied by rotation of the important free rotation in pairs, described as Pairs of Dihedral Angles-Systematic Analysis – PDA-SA. The main objective is to reduce the dimension of the investigated system. The idea is to address the conformational space in small portions, evaluating PES in combinations of angles in pairs. If the problem of combinatorial explosion is controlled, the conformational space can be sufficiently refined in the regions of minimum energy, taking care to minimize the information lost. The energy surfaces are obtained for each pair of angles and the number of conformations is given by Equation 5:

$$\text{Number of conformations} = s^2 \frac{N(N-1)}{2}$$

$$(5)$$

where s and N have the same meaning as in Equation 1. The number of conformations, in this case, is given by the combinatory arrangement of the N dihedrals in pairs. The main observation of the comparison between equations (1) and (5) is that the number of conformations as given by Eq. (1) increases exponentially with the number of bonds with free rotation, while from Eq. (5), the number of studied conformations increases quadratically with N. As the number of free rotation angles increases, the difference in the number of conformers generated by these two equations becomes more evident.

The computational procedure for PDA-SA can be organized in five basic steps:

1. Molecular Building: The interest molecule must be defined in terms of its internal coordinates: bond length, angle bond and dihedral angles. There are many softwares able to define this molecular initial structure. A quantum chemistry optimization is required at this step in order to adjust internal parameters. The best method must be chosen according to the system under study.

2. Dihedral Pair Rotation: The PDA-SA conformational search begins and the combination of the existing pairs of angles is taking account. Sometimes it is only possible to choose a dihedral increment with a less refined value. A rough PES is obtained in this case. The matrix to be analyzed consists of energy values from potential surfaces for angle combinations, and they are grouped according to Figure 7 for N angles. Appendix A shows the energy values for omprazole basic structure. The idea is to perform a cyclical permutation on the data, and this matrix form ensures that no information about the total PES is lost. The energy values obtained for each angle rotation as a function of the others allow the conformational space to be completely mapped. The major advantage is that the shape of these small portions can be visually observed, since we have a 3-D fitting. (see Figures 9 and 10)

3. PCA application on data matrix: After the energy matrix statement, PCA is performed on the data. The regions with minimum energy points on the grid search can be easily selected. The number of selected regions will depend on the nature of the studied system.

4. Refinement with a short dihedral increment: The regions initially obtained in step 3 can be refined with a small angle increment. It is important to emphasize that this step is not obligatory, since a small dihedral increment can be used in step 3, depending on the studied

system. However, previous experience in this methodology (Bruni et al., 2002; Bruni & Ferreira, 2008) shows that this is the easiest procedure, i.e., firstly make rotations with a large dihedral increment and subsequently refine the minimum energy regions selected by PCA with small dihedral increments.

5. Optimization of the final structure: the procedure described above provides angle values for the conformational search with a good level of accuracy. When these values are combined, we obtain all the possible minimum structures. Those structures constrained by the angle values obtained by PCA analysis are submitted to final optimization and the resulting structures are considered to be those of minimum energy.

In the study that introduced this method, the approach was successfully tested in the analysis of omeprazole and its derivatives, in which the results were in agreement with the experimental ones.(Bruni et al., 2002) In a second study, the technique was used to find minimum energy conformations of omeprazole derivative molecules in a QSAR study. (Bruni & Ferreira, 2002) It was shown that conformational analysis is crucial when establishing SAR/QSAR models using theoretically calculated descriptors, and they are strongly dependent on the details of molecular structure. Though all minima conformation have similar energetic values, some calculated properties are very sensitive to the structural variation, which is understandable since electronic properties are intrinsically dependent on molecular conformation.(Bruni & Ferreira, 2002) Omeprazole's racemization barrier and decomposition reaction was also studied. Quantum chemistry coupled to PDA-SA chemometric method was used to find all omeprazole minimum energy structures. To obtain the racemization barriers it was essential that the starting structure was in a global energy minimum. In that work, for all the studied structures, there was no change in the values of the racemization barriers, which confirmed the identification of the most stable structures for omeprazole.(Bruni & Ferreira, 2008)

$i\backslash j$	1	2	3	$\ldots$	N
1	-	E_{12}	E_{13}	$\ldots$	E_{1N}
2	E_{21}	-	E_{23}	$\ldots$	E_{2N}
3	E_{31}	E_{32}	-	$\ldots$	E_{3N}
$\vdots$	$\vdots$	$\vdots$	$\vdots$	$\ddots$	$\vdots$
N	E_{N1}	E_{N2}	E_{N3}	$\ldots$	-

Figure. 7: Matrix scheme for N angles: the discrete energy values for each rotation angle must be evaluated. E_{ij} are the energy matrices with elements E_{ij}^{km}, in which k and m are the angle increment indices for the angles i and j, respectively.

This approach is straightforward and in principle would have no size limits for its application. However, it presents limitation due to some initial condition dependence. Given a system with N degrees of freedom, for each pair of angles there are N-2 parameters that can interfere in the method. For example, in Figure 4 the potential energy surface for first and last dihedral angle combinations depends on the dihedral angles conformation between them. When the dihedral angles are too far from each other along the chain of atoms, the method may not become feasible. In this case the method may need to be repeated with different initial conditions to improve the sampling of the configurational phase space, and moreover we cannot be sure that we have reached the global minimum. When the correlations between the pairs of angles do not depend strongly on these initial conditions the method is very useful. Such system corresponds to small molecules, not so flexible, in which there are few large potential basins, such as omeprazol and its derivatives. (Bruni et al., 2002; Bruni & Ferreira, 2002; Bruni & Ferreira, 2008) The limit of validity for this method is under investigation. We are applying this method to study the IAN peptide, which is a tetrapeptide isobutyryl-(ala)3-NH-methyl. (Nascimento et al., 2009) This is the smallest polypeptide that can have secondary-like structure (an helix) (Becker & Karplus, 1997) and it has 11 free rotation bonds. For a flexible system, such as this, the initial condition dependence in the calculation of the minimum energy conformations is expected to increase with the size of the system. Since the system is more flexible and expected to be more rugged, we partially overcome this problem by using small angle increments steps, in order to probe all local minima of the system and compare them.

NUMERICAL RESULTS

Study of Basic Structure for Omeprazole and Derivatives

Initially, the basic structure of omeprazole and derivatives was evaluated. This structure has three bonds with free rotation. To validate the proposed methodology, two different approaches were performed. In the first approach, pairs of angles were taken account ((1,2), (1,3) e (2,3) in Fig. 8) and the number of conformations is given according to Equation 5. The resulting matrix analyzed was composed by the energy values from the potential energy surface for each angle combination (see matrix example in Figure 7). A matrix with discrete energy values for the basic structures with 30° angle increment in

Equation 2 is showed in Appendix A.

Three PES were obtained for a 30° dihedral increment and are showed in Figures 9 and 10. Figure 9 shows the original energy values and Figure 10 shows the same surfaces, but with a 0,12 hartrees cut off for better visualization. PCA was performed on autoscaled original data and the results are shown in Figure 11. 64% of the whole information is cumulated in first and second Factors (or Principal Components-PCs). The convergence of the points for one region is observed. Figure 12 shows the PCA results for the leveled data in 0,12 hartrees. Factor 1 and Factor 2 now cumulate 73% of the entire information.

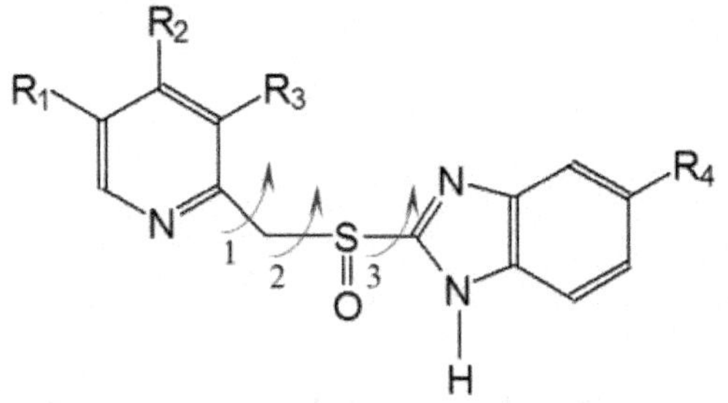

Figure. 8: Basic structure for omeprazole and derivatives.

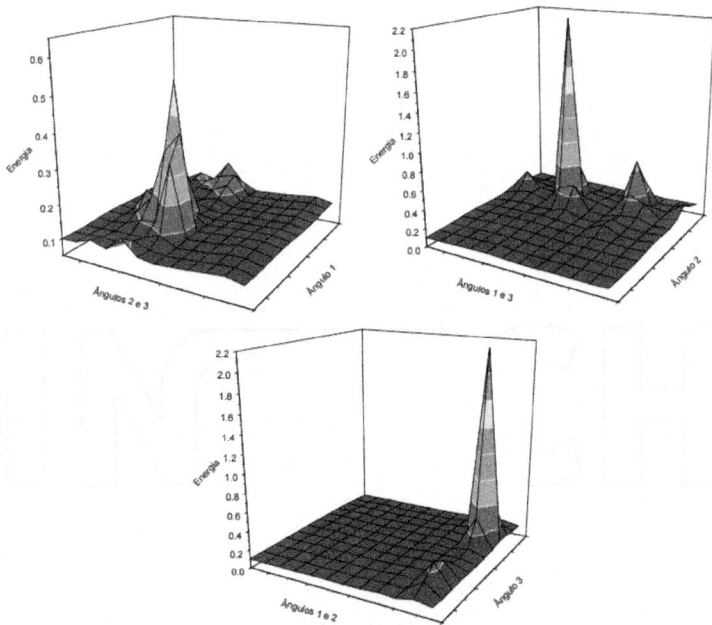

Figure. 9: Orignal PES obtained from PDA-SA method for structure from Fig. 8.

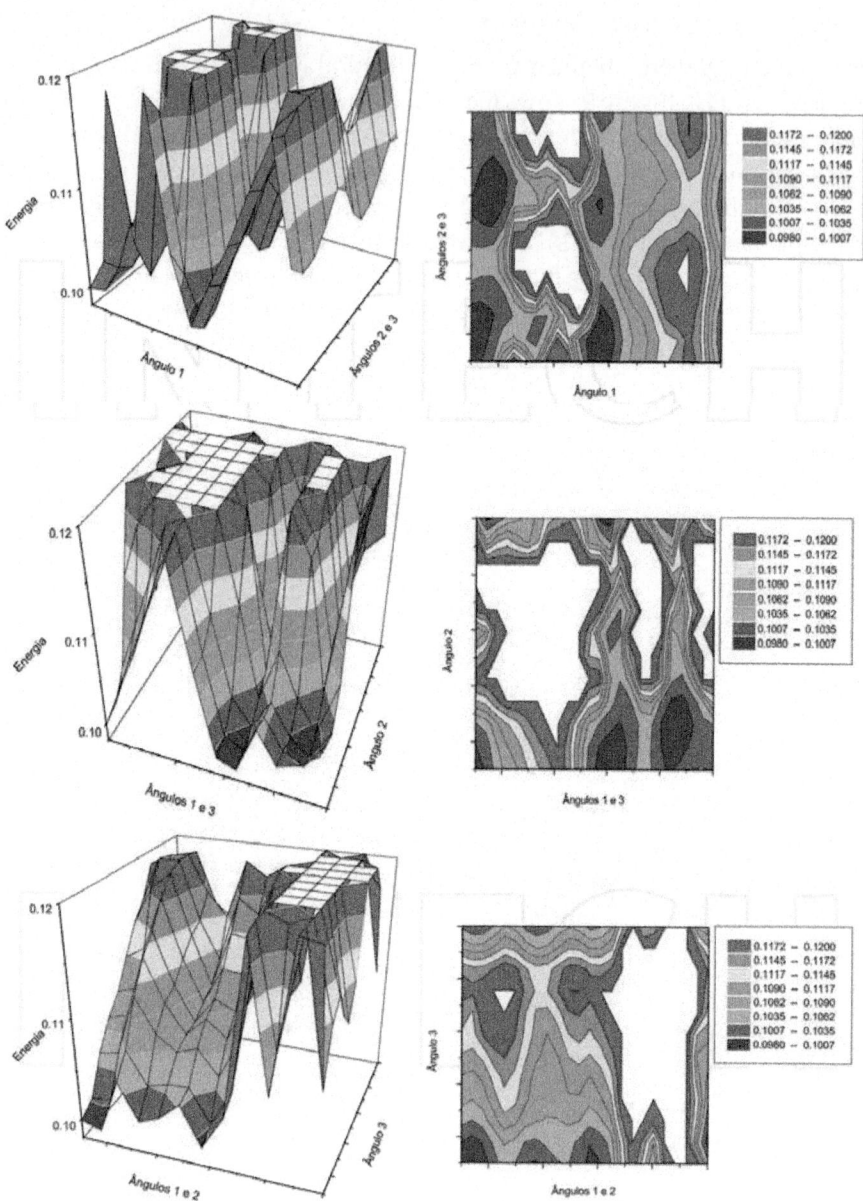

Figure. 10: PES obtained from PDA-SA method for structure from Fig. 8, with a 0,12 hartress cutoff.

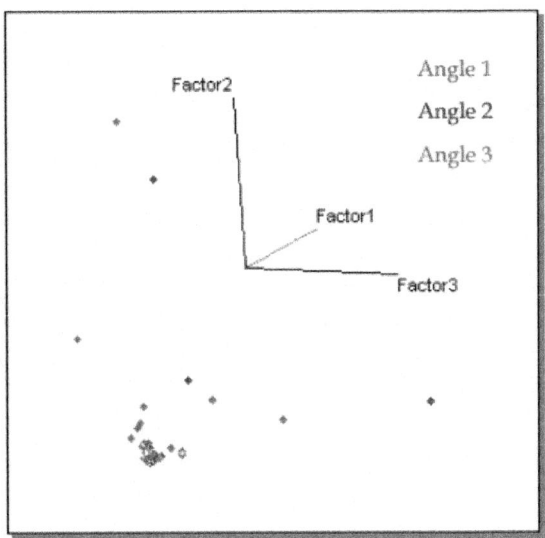

Figure. 11: PCA for data from original PES (Fig.9).

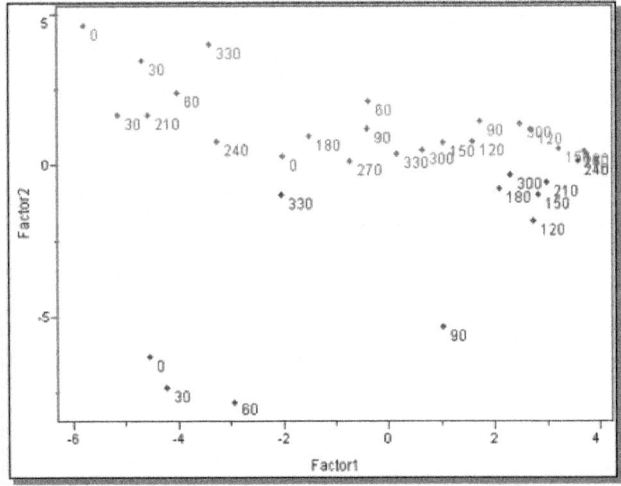

Figure. 12: Principal Component Analysis forPES from Fig. 10, with a 0,12 cutoff.

Factor 1 accounts to the minimum region in each case and Factor 2 accounts for the energy range for the different combinations. Table 1 shows the selected minimum energy for each angle. The first column shows that two different regions were chosen for Angle 1 and only one region for Angles 2 and 3. Second column shows the rotation over the initial angle value (third column) resulting in the fourth column.

Table 1: Regions separated by PCA

Angle	Rotation	Initial Value	Value obtained by PCA
1 (a)	0° - 60°	48,48°	48,48° - 108,48°
1 (b)	180° - 240°	48,48°	228,48° - 288,48°
2	0° - 60°	209,79°	209,79° - 269,79°
3	330° - 30°	289,09°	259,09° - 319,09°

Once minima energy regions were defined, a small angle increment (5°) was used on them. Results for PCA are in Figure 13. In all cases a parabolic behavior was observed. When data variation decreases, curves are more easily observed and the minimum point is detectable. The amount of information accounted for both first and second PC´s (Factors) is around 90%. Table 2 shows the final values for each angle. When these values are combined, two different geometries were obtained with similar energy values (Table 3). These conformations are shown in Figure 14.

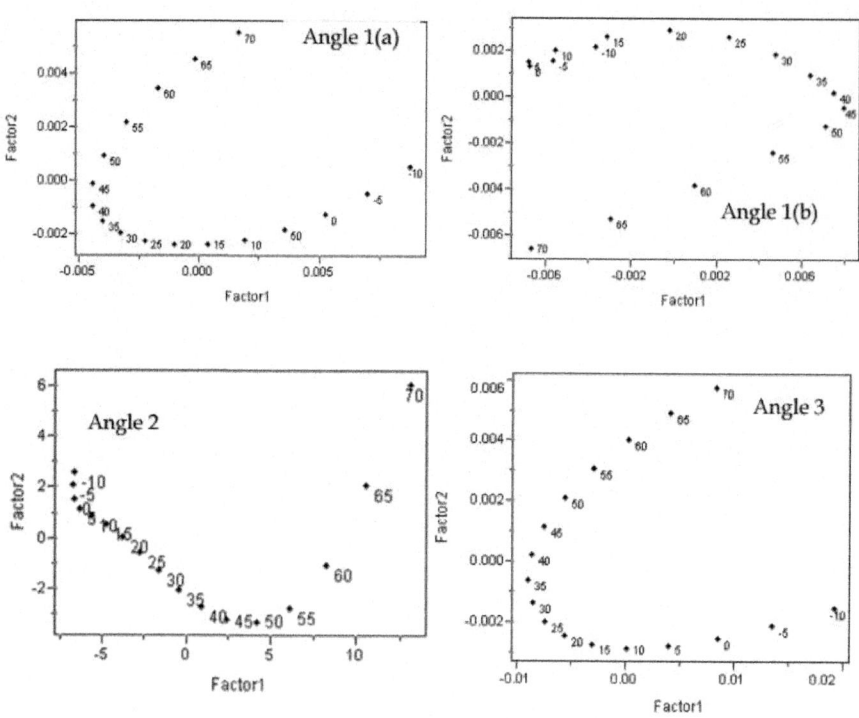

Figure. 13: PCA results for 5° angle increment refinement.

Table 2: Regions obtained through PCA

Angle	Rotation	Initial Value	Value obtained by PCA
1 (a)	45°	48,48° - 108,48°	93,48°
1 (b)	45°	228,48° - 288,48°	273,48°
2	45°	209,79° - 269,79°	254,79°
3	35°	259,09° - 319,09°	294,09°

Table 3: Minimum conformation characteristics (basic structure)

Conformation	Angle	Obtained value	$\Delta H_{f\,(PM3)}$/kcal mol^{-1}	$E_{e(6\text{-}31G^{**})}$/hartree
A	1	92.29	54.96	-1134.33
	2	107.10		
	3	298.32		
B	1	266.45	54.71	-1134.35
	2	175.70		
	3	296.45		

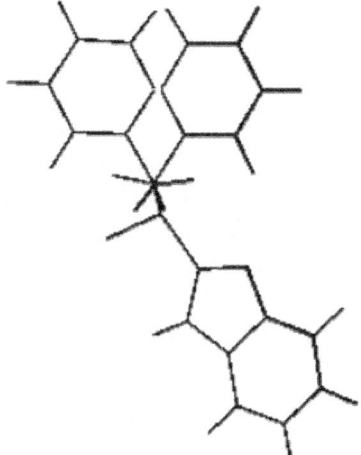

Figure. 14: Optimized superposed conformations for structure from Fig. 8.

In the second approach, the conformational analysis was made according to Equation 1 and took into account all possible conformations. PCA was performed on data matrix and minimum energy regions were selected. The next step a lower dihedral increment of 5° was used to refine those selected regions. PCA was performed again, and the same structures and energy, shown in Table 3, were obtained. This indicates that the two approaches are equivalent. The details of this complete systematic search can be found in (Bruni et al., 2002).

IAN Preliminary Studies

IAN (isobutyryl-Ala3-NH-methyl) tetrapeptide has also been studied to validate PDA-SA methodology. IAN has 11 consecutives dihedrals and its main characteristic is to be the shorter peptide able to make a complete helix turn. Figure 15 shows the IAN 2D structure (Becker, 1998). Red arrows indicate the ψ, Φ e ω dihedrals. The dihedral angles ψ, ω and Φ are related to the rotations of single bonds between atoms in the main chain C (i)-C, OC-NH and N-C(i+1), respectively, where C (i) is the ith alpha carbon of the polypeptide chain. Angles τ and Φ are connected to two arrays of functional protein chain: alpha-helix or beta-sheet.

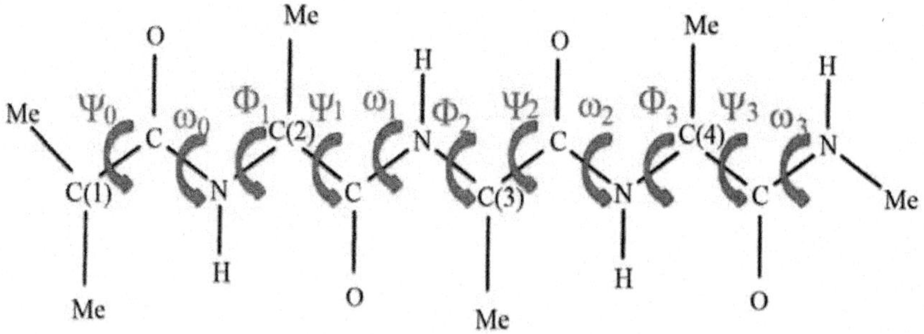

Figure. 15: 2D IAN peptide structure.

Ten random different starting conformations were studied. Table 4 shows the angles and energy values corresponding to these initial conformations. The red values indicate dihedrals that were changed in comparison to initial conformation number 1. The starting conformation 2 is close to an alpha-helix. Energy values correspond to single point AM1 semi-empirical calculation, in kcal mol^{-1}.

IAN was analyzed using the PDA-SA procedure. The eleven dihedral angles provide 55 different conformations according to all possible combinations. Conformational analysis was performed with a 20° increment. PCA was carried out and Figure 16 shows that all points converge to specific regions of the phase space. Each selected region for each angle was refined with a 5° angle increment. PCA was performed again and the final structures characteristics are shown in Table 5.

Table 4: Energy(kcal mol^{-1}) and dihedrals values (degrees) for each starting IAN structure.

Number	Energy	ψ_0	ω_0	ϕ_0	ψ_1	ω_1	ϕ_1	ψ_2	ω_2	ϕ_2	ω_3	ϕ_3
1	-179.66	79.45	-169.01	-64.73	-44.75	171.78	-84.62	44.47	179.30	-144.22	-59.37	178.07
2	-122.24	-60.55	179.00	-64.73	-64.75	-180.00	-64.62	-65.55	180.00	-64.22	-59.37	178.07
3	-156.02	79.45	170.99	-64.73	-34.75	171.78	-84.62	74.47	179.30	-124.22	-59.37	178.07
4	195.46	79.45	-169.01	-14.73	-44.75	171.78	-134.62	44.47	179.30	-144.22	-39.37	178.07
5	-80.48	49.45	-169.01	-64.73	-44.75	151.78	-84.62	44.47	179.30	-144.22	-39.37	178.07
6	-169.24	79.45	-149.01	-64.73	-44.75	171.78	-84.62	44.47	159.30	-144.22	-79.37	178.07
7	-149.11	59.45	-169.01	-84.73	-44.75	-168.22	-84.62	74.47	179.30	-144.22	-59.37	178.07
8	-147.67	79.45	-169.01	-64.73	-24.75	171.78	-104.62	44.47	-160.70	-144.22	-39.37	178.07
9	-29.05	109.45	-169.01	-64.73	-44.75	171.78	-84.62	24.47	179.30	-174.22	-59.37	178.07
10	-167.76	79.45	-169.01	-44.73	-44.75	-178.22	-84.62	44.47	179.30	-144.22	-59.37	178.07

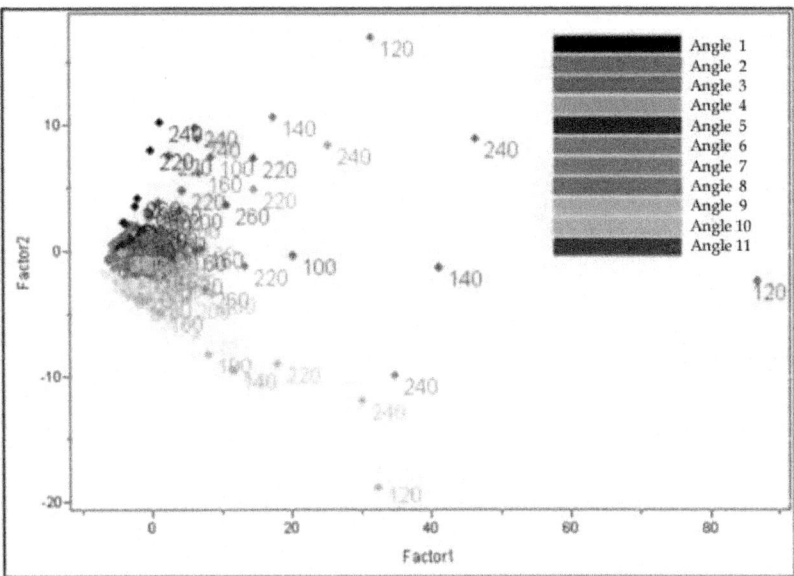

Figure. 16: PCA results for IAN peptide.

Table 5 shows the obtained energy values for the final structures and they indicate that some correspond to identical conformations. Three different groups were identified. Figure 17 shows the group that corresponds to structures 1, 5, 6, 9 and 10 superposed (blue ones in Table 5). Structure 9 shows a slightly different value on ψ_0 but it does not change the energy value. These five structures have two stabilizing hydrogen bonds which are indicated by the red circle and the resulting conformations for them resemble a beta-sheet.

Table 5: Energy (kcal mol-1) and dihedrals values (degrees) for each obtained IAN structure

Number	Energy	ψ_0	ω_0	ϕ_0	ψ_1	ω_1	ϕ_1	ψ_2	ω_2	ϕ_2	ω_3	ϕ_3
1	- 181.498	70.50	176.87	85.57	66.93	-177.14	-85.01	67.39	179.60	-112.24	-46.78	177.92
2	- 180.70	-61.83	-172.66	-78.26	25.05	157.24	-90.68	-29.19	177.15	-110.70	-45.10	178.62
3	- 179.661	79.45	-168.99	-64.80	-44.71	171.76	-84.61	44.32	179.26	143.99	-59.33	178.07
4	-179.276	72.93	-174.15	-104.24	-32.54	-179.9	-82.66	70.38	-179.69	-115.57	-51.50	179.42
5	-181.498	70.49	-176.89	-85.56	66.89	-177.17	-84.96	67.52	179.62	-112.34	-46.80	177.93
6	-181.498	70.49	-176.88	-85.57	66.93	-177.15	-85.01	67.40	179.61	-112.26	-46.78	177.92
7	-179.276	72.93	-174.15	-104.24	-32.52	-179.9	-82.65	70.39	-179.69	-115.58	-51.50	179.44
8	-179.276	72.94	-174.14	-104.27	-32.45	-179.89	-82.66	70.40	-179.67	-115.55	-51.48	179.36
9	-182.385	-62.43	-177.31	-84.72	67.46	-177.22	-84.95	67.38	179.58	-112.28	-46.85	177.94
10	-181.498	70.49	-176.89	-85.56	66.93	-177.15	-85.00	67.43	179.62	-112.32	-46.82	177.91

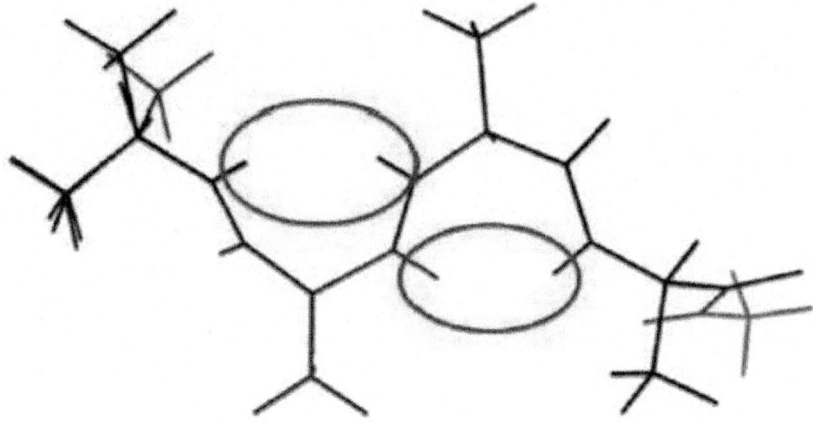

Figure. 17: Superposed final conformations for 1, 5, 6, 9 and 10 structures (blue ones in Table 5).

The second group is composed by final conformations 4, 7 and 8 (green ones in Table 5). The superposed conformations can be observed in Figure 18. These conformations are more open and have only one hydrogen-bond (red circle in Fig.18). The last group, the black ones in Table 5, are superposed in Figure 19. The resulting structures show an alpha-helix like behavior, with two stabilizing hydrogen bond (red circles, Fig. 19).

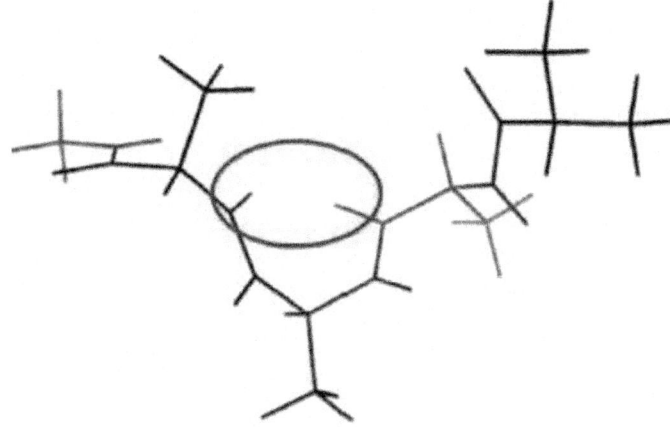

Figure. 18: Superposed final conformations for 4,7and 8 structures (green ones in Table 5).

Figure. 19: Superposed final conformations for structures 2 and 3 (black ones in Table 5)

Results presented for IAN peptide are partial and were only performed for one minimum region for each starting structure. Other minimum energy regions of this system are being investigated. A gradual increase in the size of the chain is also been explored.

CONCLUSION

The arrangement of atoms in a molecule or its structure determination has intrigued scientists through history. However only with recent experimental and computational advances the discussions on this theme became more effective and elucidative. The nature of PES is intrinsically multidimensional, usually has a very complex landscape. The global minima search, like the one encountered in the protein folding problem, is a NP-hard problem. This means that this task belongs to a large set of computational problems, assumed to be very hard ("conditionally intractable") (Fraenkel, 1993). The search for its relevant minima in molecular modeling has motivated the development of methods with very specific applications, as discussed in this chapter. For each particular problem one finds a variety of methods that allows feasible solutions, and most likely a combination of methods provides the optimum solution. In this chapter, we discussed some aspects of conformational search that controls the combinatorial explosion. In particular, Principal Component Analysis was associated with a systematic search method to find structures with low energy in PES. The methodology can be useful to handle small- and medium-size molecules. The maximum size which the method can efficiently handle is being investigated (Nascimento et al., 2009). Due to the PCA dimension reduction, the method's efficiency is highly increased, allowing it to be of practical use in the study of more complex molecules.

ACKNOWLEDGMENT

We thank Prof. Márcia M.C. Ferreira (Unicamp) for the helpful discussions. We were supported by Fundação de Amparo à Pesquisa do Estado de São Paulo (FAPESP) and Conselho Nacional de Desenvolvimento Científico e Tecnológico (CNPq), Brazil. Computational resources were provided by Centro Nacional de Processamento e Alto Desempenho em São Paulo (CENAPAD-SP), Brazil.

Apendix A

Matrix with the discrete values for each rotation angle and its corresponding energy value for the first rotation for basic structure in Figure 8 Labels in bold were not used in PCA analysis, they are shown to help the matrix notation and visualization.

Table (left panel)

Angle 1

Rotation	0	30	60	90	120	150	180	210	240	270	300	330
0	0.0998	0.100	0.103	0.111	0.108	0.111	0.114	0.115	0.118	0.116	0.108	0.102
30	0.0984	0.099	0.101	0.105	0.108	0.110	0.113	0.114	0.118	0.115	0.107	0.101
60	0.0992	0.099	0.100	0.102	0.109	0.112	0.117	0.125	0.119	0.115	0.107	0.101
90	0.1020	0.101	0.101	0.103	0.112	0.116	0.120	0.119	0.120	0.117	0.109	0.103
120	0.1046	0.106	0.104	0.107	0.115	0.117	0.119	0.120	0.122	0.119	0.111	0.106
150	0.1037	0.105	0.104	0.109	0.113	0.112	0.113	0.118	0.120	0.118	0.110	0.105
180	0.1010	0.103	0.100	0.107	0.110	0.109	0.110	0.114	0.116	0.114	0.107	0.102
210	0.1003	0.103	0.098	0.106	0.108	0.108	0.110	0.111	0.114	0.112	0.106	0.101
240	0.1019	0.099	0.098	0.107	0.107	0.108	0.109	0.110	0.114	0.113	0.106	0.102
270	0.1033	0.101	0.100	0.106	0.107	0.108	0.110	0.113	0.117	0.115	0.109	0.104
300	0.1036	0.102	0.102	0.106	0.110	0.111	0.114	0.117	0.120	0.118	0.111	0.106
330	0.1023	0.102	0.103	0.108	0.110	0.112	0.115	0.117	0.120	0.118	0.111	0.105

Angle 2

Rotation	0	30	60	90	120	150	180	210	240	270	300	330
0	0.0998	0.100	0.104	0.107	0.108	0.111	0.114	0.118	0.146	0.133	0.106	0.099
30	0.1006	0.101	0.105	0.108	0.110	0.113	0.115	0.106	0.117	0.111	0.103	0.099
60	0.1032	0.104	0.108	0.111	0.113	0.116	0.118	0.120	0.107	0.115	0.123	0.108
90	0.1115	0.113	0.117	0.120	0.122	0.125	0.127	0.128	0.132	0.125	0.120	0.116
120	0.1194	0.121	0.126	0.129	0.132	0.134	0.136	0.138	0.141	0.138	0.127	0.121
150	0.1136	0.116	0.119	0.127	0.130	0.131	0.132	0.134	0.139	0.150	0.171	0.114
180	0.1062	0.109	0.119	0.134	0.141	0.133	0.125	0.147	0.167	0.126	0.111	0.116
210	0.1182	0.133	0.172	0.238	0.189	0.147	0.166	0.319	0.434	1.241	0.376	0.126
240	0.1467	0.225	0.339	0.279	0.174	0.146	0.243	0.178	2.168	0.552	0.181	0.135
270	0.1467	0.133	0.166	0.167	0.140	0.121	0.145	0.123	0.533	0.368	0.167	0.120
300	0.1196	0.106	0.111	0.113	0.110	0.108	0.107	0.104	0.146	0.140	0.117	0.112
330	0.1047	0.099	0.101	0.104	0.105	0.107	0.100	0.101	0.115	0.117	0.112	0.105

Angle 3

Rotation	0	30	60	90	120	150	180	210	240	270	300	330
0	0.0998	0.100	0.103	0.102	0.104	0.103	0.101	0.100	0.101	0.103	0.103	0.102
30	0.1009	0.099	0.105	0.101	0.104	0.104	0.100	0.099	0.099	0.101	0.102	0.102
60	0.1045	0.101	0.100	0.101	0.104	0.103	0.103	0.098	0.098	0.100	0.102	0.103
90	0.1071	0.105	0.102	0.107	0.107	0.109	0.103	0.100	0.099	0.101	0.106	0.112
120	0.1088	0.106	0.103	0.113	0.115	0.117	0.116	0.104	0.103	0.107	0.111	0.128
150	0.1115	0.105	0.104	0.112	0.113	0.112	0.121	0.104	0.104	0.112	0.116	0.128
180	0.1140	0.103	0.100	0.109	0.110	0.109	0.110	0.112	0.104	0.104	0.121	0.117
210	0.1157	0.101	0.098	0.099	0.102	0.108	0.109	0.113	0.107	0.113	0.113	0.110
240	0.1187	0.104	0.098	0.101	0.107	0.108	0.108	0.114	0.114	0.114	0.113	0.114
270	0.1166	0.101	0.100	0.106	0.107	0.108	0.111	0.119	0.117	0.120	0.117	0.116
300	0.1086	0.102	0.102	0.106	0.108	0.111	0.114	0.117	0.120	0.118	0.120	0.112
330	0.1025	0.101	0.103	0.108	0.113	0.114	0.115	0.116	0.117	0.117	0.117	0.107

Table (right panel)

Angle 1

Rotation	0	30	60	90	120	150	180	210	240	270	300	330
0	0.099	0.100	0.104	0.107	0.108	0.111	0.114	0.115	0.118	0.116	0.108	0.102
30	0.098	0.099	0.103	0.106	0.108	0.110	0.113	0.114	0.118	0.115	0.107	0.101
60	0.099	0.101	0.104	0.107	0.109	0.112	0.117	0.125	0.119	0.115	0.107	0.101
90	0.102	0.103	0.110	0.113	0.115	0.117	0.119	0.120	0.122	0.117	0.109	0.103
120	0.104	0.105	0.110	0.113	0.115	0.117	0.119	0.120	0.122	0.119	0.111	0.106
150	0.103	0.103	0.107	0.112	0.113	0.112	0.113	0.118	0.120	0.118	0.110	0.105
180	0.101	0.103	0.106	0.109	0.110	0.109	0.110	0.111	0.114	0.112	0.106	0.102
210	0.100	0.103	0.106	0.107	0.108	0.108	0.109	0.110	0.114	0.113	0.106	0.101
240	0.101	0.104	0.106	0.107	0.107	0.108	0.109	0.110	0.114	0.113	0.106	0.102
270	0.103	0.104	0.106	0.107	0.108	0.108	0.110	0.113	0.117	0.115	0.109	0.104
300	0.103	0.104	0.106	0.108	0.110	0.111	0.114	0.117	0.120	0.118	0.111	0.106
330	0.102	0.103	0.106	0.108	0.110	0.112	0.115	0.117	0.120	0.118	0.111	0.105

Angle 2

Rotation	0	30	60	90	120	150	180	210	240	270	300	330
0	0.099	0.098	0.099	0.102	0.104	0.103	0.101	0.100	0.101	0.103	0.103	0.102
30	0.100	0.099	0.099	0.101	0.104	0.103	0.100	0.099	0.099	0.101	0.102	0.102
60	0.103	0.101	0.100	0.101	0.104	0.104	0.103	0.098	0.098	0.100	0.102	0.103
90	0.111	0.105	0.102	0.107	0.107	0.109	0.103	0.100	0.102	0.101	0.106	0.112
120	0.119	0.106	0.103	0.115	0.124	0.173	0.116	0.104	0.105	0.107	0.132	0.128
150	0.113	0.105	0.104	0.109	0.196	0.186	0.136	0.252	0.174	0.132	0.127	0.117
180	0.106	0.103	0.103	0.108	0.145	0.252	0.166	0.516	0.145	0.322	0.152	0.180
210	0.118	0.117	0.115	0.201	0.322	0.516	0.243	0.459	0.634	0.150	0.249	0.216
240	0.146	0.111	0.123	0.310	0.634	0.137	0.145	0.241	2.168	0.146	0.163	0.126
270	0.133	0.111	0.123	0.201	0.383	0.137	0.107	0.104	0.118	0.146	0.163	0.126
300	0.106	0.103	0.108	0.151	0.116	0.105	0.100	0.101	0.108	0.117	0.111	0.103
330	0.099	0.099	0.108	0.108	0.116	0.105	0.100	0.104	0.108	0.117	0.111	0.105

Angle 3

Rotation	0	30	60	90	120	150	180	210	240	270	300	330
0	0.099	0.100	0.103	0.111	0.119	0.113	0.106	0.118	0.146	0.133	0.106	0.099
30	0.100	0.101	0.104	0.113	0.121	0.116	0.109	0.133	0.225	0.133	0.106	0.099
60	0.104	0.105	0.108	0.117	0.126	0.122	0.119	0.172	0.339	0.166	0.111	0.101
90	0.107	0.108	0.111	0.120	0.129	0.127	0.134	0.238	0.279	0.167	0.113	0.104
120	0.108	0.110	0.113	0.122	0.132	0.130	0.141	0.189	0.174	0.140	0.110	0.105
150	0.111	0.113	0.116	0.125	0.134	0.131	0.133	0.147	0.167	0.121	0.123	0.107
180	0.114	0.115	0.118	0.127	0.136	0.132	0.125	0.147	0.167	0.178	0.146	0.115
210	0.115	0.118	0.120	0.128	0.138	0.134	0.123	0.434	0.319	0.533	0.178	0.120
240	0.118	0.123	0.120	0.129	0.141	0.139	0.127	0.319	2.168	0.552	0.146	0.115
270	0.133	0.111	0.115	0.125	0.138	0.150	0.167	1.241	0.552	0.368	0.140	0.116
300	0.106	0.103	0.111	0.120	0.127	0.138	0.376	0.181	0.167	0.117	0.112	0.112
330	0.099	0.099	0.105	0.113	0.121	0.114	0.116	0.140	0.135	0.120	0.108	0.105

REFERENCES

1. Allen, F. H., Galek, P. T. a, & Wood, P. a. (2010). Energy matters! Crystallography Reviews, 16(3), 169-195. doi:10.1080/08893110903476919

2. Araujo-Andrade, C., Lopes, S., Fausto, R., & Gómez-Zavaglia, A. (2010). Conformational study of arbutin by quantum chemical calculations and multivariate analysis.

3. Journal of Molecular Structure, 975(1-3), 100-109. doi:10.1016/j.molstruc.2010.04.002 Becker, O M. (1998). Principal coordinate maps of molecular potential energy surfaces.

4. Journal of Computational Chemistry, 19(11), 1255-1267. 605 Third Ave, New York, NY 10158-0012 Usa: John Wiley & Sons Inc. doi:10.1002/(SICI)1096- 987X(199808)19:11<1255::AID-JCC5>3.3.CO;2-H

5. Becker, Oren M., & Karplus, M. (1997). The topology of multidimensional potential energy surfaces: Theory and application to peptide structure and kinetics. The Journal of Chemical Physics, 106(4), 1495. doi:10.1063/1.473299

6. Beckers, M. L. M., Derks, E. P. P. A., Melssen, W. J., & Buydens, L. M. C. (1996). Pergamon oo!n-8485(%~6-0. Science, 20(4), 449-457.

7. Beckers, M. L., Buydens, L. M., Pikkemaat, J. a, & Altona, C. (1997). Application of a genetic algorithm in the conformational analysis of methylene-acetal-linked thymine dimers in DNA: comparison with distance geometry calculations. Journal of biomolecular NMR, 9(1), 25-34. Retrieved from http://www.ncbi.nlm.nih.gov/pubmed/9081542

8. Beebe, K. R. (1998). Chemometrics: A Practical Guide (p. 360). Wiley-Blackwell. Retrieved from erscienceLaboratoryAutomation/dp/0471124516/ref=sr_1_1?ie=UTF8&qid=1320084260&sr=8-1

9. Beusen, D. D., Shands, E. F. B., Karasek, S. F., Marshall, G. R., & Dammkoehler, R. A. (1996). Systematic search in conformational analysis. Theochem-Journal of Molecular Structure, 370(2-3), 157-171. Po Box 211, 1000 Ae Amsterdam, Netherlands: Elsevier Science Bv.

10. Biarnés, X., Bongarzone, S., Vargiu, A. V., Carloni, P., & Ruggerone, P. (2011). Molecular motions in drug design: the coming age of the metadynamics method. Journal of computer-aided molecular design, 25(5), 395-402. doi:10.1007/s10822-011-9415-3

11. Brodmeier, T., & Pretsch, E. (1994). Application of genetic algorithms in molecular modeling. Journal of Computational Chemistry, 15(6), 588-595. doi:10.1002/jcc.540150604

12. Bruni, A. T., Leite, V. B. P., & Ferreira, M. M. C. (2002). Conformational analysis: A new approach by means of chemometrics. Journal Of Computational Chemistry, 23(2), 222- 236. Commerce Place, 350 Main St, Malden 02148, MA USA: Wiley-Blackwell. doi:10.1002/jcc.10004

13. Bruni, A. T., & Ferreira, M. M. C. (2008). Theoretical study of omeprazole behavior: Racemizatin barrier and decomposition reaction. International Journal of Quantum Chemistry, 108(6), 1097-1106. doi:10.1002/qua.21597

14. Bruni, A. T., & Ferreira, M. M. C. (2002). Omeprazole and analogue compounds: a QSAR study of activity againstHelicobacter pylori using theoretical descriptors. Journal of Chemometrics, 16(8-10), 510-520. doi:10.1002/cem.737

15. Brush, S. G. (1999). Dynamics of Theory Change in Chemistry : Part 1 . The Benzene Problem 1865 – 1945. Science, 30(1), 21-79.

16. Cintas, P. (2007). Tracing the origins and evolution of chirality and handedness in chemical language. Angewandte Chemie (International ed. in English), 46(22), 4016-24. doi:10.1002/anie.200603714

17. Das, G., Gentile, F., Coluccio, M. L., Perri, a M., Nicastri, a, Mecarini, F., Cojoc, G., et al. (2011). Principal component analysis based methodology to distinguish protein SERS spectra. Journal of Molecular Structure, 993(1-3), 500-505. Elsevier B.V. doi:10.1016/j.molstruc.2010.12.044

18. Derks, E. P. P. A., & Buydens, L. M. C. (1996). E. P. P. A DERKS,* M. L. M. BECKER& W. J. MELSSEN and L. M. C. BUYDENS, 20(4), 439-448.

19. Drayer, D. (1993). The Early History of Stereochemistry: From the Discovery of Molecular Asymmetry and the First Resolution of a Racemate by Pasteur to the Asymmetrical Chiral Carbon of van't Hoff and Le Bel. Clinical Pharmacology-New York-Marcel Dekker Incorporated-, 18(3), 1–1. Marcel Dekker Ag. Retrieved from http://scholar.google.com/schol ar?hl=en&btnG=Search&q=intitle:The+Early+His tory+of+Stereochem istry+:+From+the+Discovery+of+Molecular+Asymmetry+and +the+Fi rst+Resolution+of+A+Racemate+By+Pasteur+To+The+Asymmetrical+ Chir al+Carbon+Of+Van+?+T+Hoff+And+Le+Bel+*#0

20. Fraenkel, a S. (1993). Complexity of protein folding. Bulletin of mathematical biology, 55(6), 1199-210. Retrieved from http://www. pubmedcentral.nih.gov/articlerender.fcgi?artid=3042729&tool=pmce ntrez&rendertype=abstract

21. Gal, J. (2007). Review Article Carl Friedrich Naumann and the Introduction of Enantio Terminology : A Review and Analysis on the

150th Anniversary. Chirality, 98(May 2006), 89-98. doi:10.1002/chir

22. Gal, J. (2011). Review Article Louis Pasteur , Language , and Molecular Chirality . I . Background and Dissymmetry. Clinical Laboratory, 16(March 2010), 1-16. doi:10.1002/chir

23. Geladi, P. (2003). Chemometrics in spectroscopy. Part 1. Classical chemometrics. Spectrochimica Acta Part B Atomic Spectroscopy, 58(5), 767-782. doi:10.1016/S0584- 8547(03)00037-5

24. Golub, G. H., & Loan, C. F. van V. (1996). Matrix Computations (Johns Hopkins Studies in Mathematical Sciences)(3rd Edition) (p. 728). The Johns Hopkins University Press.

25. Retrieved   from   http://www.amazon.com/Computations-Hopkins-StudiesMathematical- ciences/dp/0801854148

26. Grouleff, J., & Jensen, F. (2011). Searching Peptide Conformational Space. Journal of Chemical Theory and Computation, 1783-1790.

27. Hanselman, D., Littlefield, B., Inc., M., & Mathworks. (1997). The Student Edition of Matlab Version 5 User's Guide (p. 429). Prentice Hall College Div. Retrieved from http://www.amazon.com/Student-Matlab-Version-Users-Guide/dp/0132725509

28. Havel, T. F., Crippen, G. M., Kuntz, I. D., & Blaney, J. M. (1983). The combinatorial distance geometry method for the calculation of molecular conformation. II. Sample problems and computational statistics. Journal of theoretical biology, 104(3), 383-400.

29. Retrieved from http://www.ncbi.nlm.nih.gov/pubmed/6197591 Havel, T. F., Kuntz, I. D., & Crippen, G. M. (1983). The combinatorial distance geometry method for the calculation of molecular conformation. I. A new approach to an old problem. Journal of theoretical biology, 104(3), 359-81. Retrieved from

30. Hunger, J., & Huttner, G. (1999). Optimization and analysis of force field parameters by combination of genetic algorithms and neural networks. Journal of Computational Chemistry, 20(4), 455-471. doi:10.1002/(SICI)1096-987X(199903)20:4<455::AIDJCC6>3.0.CO;2-1

31. Izgorodina, E. I., Lin, C. Y., & Coote, M. L. (2007). Energy-directed tree search: an efficientsystematic algorithm for finding the lowest energy conformation of molecules.

32. Physical chemistry chemical physics : PCCP, 9(20), 2507-16. doi:10.1039/b700938k Jordan, S. N., Leach, A. R., & Bradshaw, J. (1995). The Application of Neural Networks in Conformational Analysis. 1. Prediction of Minimum and Maximum Interatomic Distances. Journal

of Chemical Information and Modeling, 35(3), 640-650. doi:10.1021/ci00025a035

33. Jorgensen, W. L., & TiradoRives, J. (1996). Monte Carlo vs molecular dynamics for conformational sampling. Journal of Physical Chemistry, 100(34), 14508-14513. 1155 16th St, Nw, Washington, Dc 20036: Amer Chemical Soc. doi:10.1021/jp960880x

34. Kiralj, R., Ferreira, M. C., Donate, P. M., & Silva, R. (2007). Combined Computational, Database Mining, NMR, and Chemometric Approaches. Analysis, 6316-6333.

35. Lucasius, C. B., & Kateman, G. (1994). Understanding and Using Genetic Algorithms.2. Representation, Configuration and Hybridization. Chemometrics and Intelligent Laboratory Systems, 25(2), 99-145. Po Box 211, 1000 Ae Amsterdam, Netherlands:

36. Elsevier Science Bv. doi:10.1016/0169-7439(94)85038-0 Leach, A. (2001). Molecular Modelling: Principles and Applications (2nd Edition). Prentice Hall. Retrieved from http://www.amazon.ca/exec/obidos/redirect?tag=citeulike09- 20&path=ASIN/0582382106

37. Leach, A. R., & Smellie, A. S. (1992). A combined model-building and distance-geometry approach to automated conformational analysis and search. Journal of Chemical Information and Modeling, 32(4), 379-385. doi:10.1021/ci00008a019

38. Lucasius, C. (1993). Understanding and using genetic algorithms Part 1. Concepts, properties and context. Chemometrics and Intelligent Laboratory Systems, 19(1), 1-33. doi:10.1016/0169-7439(93)80079-W

39. Malinowski, E. R. (2003). Factor Analysis in Chemistry. Technometrics (Vol. 45, pp. 180-181). Wiley. doi:10.1198/tech.2003.s145

40. Li Manni, G., Barone, G., Duca, D., & Murzin, D. Y. (2009). Systematic conformational search analysis of the SRR and RRR epimers of 7-hydroxymatairesinol. Journal of Physical Organic Chemistry, (June 2009), n/a-n/a. doi:10.1002/poc.1595

41. Massart, D. L., Heyden, Y. V., & Brussel, V. U. (2004). What Can Chemometrics Do for Separation Science ? Europe, 17(9).

42. Mo, Y., & Gao, J. (2007). Theoretical analysis of the rotational barrier of ethane. Accounts of chemical research, 40(2), 113-9. doi:10.1021/ar068073w

43. Nair, N., & Goodman, J. M. (1998). Genetic Algorithms in Conformational Analysis. Journal of Chemical Information and Modeling, 38(2), 317-320. doi:10.1021/ci970433u

44. Nascimento, R. R., Bruni, A. T. , & Leite, V. B. P. (2009). Estudo conformacional do peptide IAN e seus fragmentos pelo método de análise sistemática reduzida. 07/10/09. Retrieved November 1, 2011, from http://www.athena.biblioteca.unesp.br/exlibris/bd/brp/33004153068P9/2009/na scimento_rr_me_sjrp_parcial.pdf Oblinsky, D. G., Vanschouwen, B. M. B., Gordon, H. L., & Rothstein, S. M. (2009).

45. Procrustean rotation in concert with principal component analysis of molecular dynamics trajectories: Quantifying global and local differences between conformational samples. The Journal of chemical physics, 131(22), 225102. doi:10.1063/1.3268625

46. Pietropaolo, A., Branduardi, D., Bonomi, M., & Parrinello, M. (2011). A Chirality-Based Metrics for Free-Energy Calculations in Biomolecular Systems. Journal of Computational Chemistry. doi:10.1002/jcc

47. Ramberg, P. J., & Somsen, G. J. (2001). Annals of Science The Young J . H . van ' t Hoff : The Background to the Publication of his 1874 Pamphlet on the Tetrahedral Carbon Atom , Together with a New English Translation. Annals of Science, (September 2011), 51-74.

48. Ramos, L. S., Beebe, K. R., Carey, W. P., M, E. S., Erickson, B. C., Wilson, B. E., Wangen, L. E., et al. (1986). L. Scott Ramos, Kenneth R. Beebe, W. Patrick Carey, Eugenio Sfinchez M., Brice C. Erickson, Bruce E. Wilson, Lawrence E. Wangen,' and Bruce R. Kowalski* Laboratory for Chemometrics, Department. Education, (300), 31-49.

49. Seabra, G. D. M., Walker, R. C., & Roitberg, A. E. (2009). Are current semiempirical methods better than force fields? A study from the thermodynamics perspective. The journal of physical chemistry. A, 113(43), 11938-48. doi:10.1021/jp903474v

50. Silva, D.-A., Domínguez-Ramírez, L., Rojo-Domínguez, A., & Sosa-Peinado, A. (2011). Conformational dynamics of L-lysine, L-arginine, L-ornithine binding protein reveals ligand-dependent plasticity. Proteins, 79(7), 2097-108. doi:10.1002/prot.23030

51. Smellie, A., Stanton, R., Henne, R., & Teig, S. (2003). Conformational analysis by intersection: Conan. Journal of Computational Chemistry, 24(1), 10-20. 111 River St, Hoboken, Nj 07030 Usa: John Wiley & Sons Inc. doi:10.1002/jcc.10175

52. Spellmeyer, D. C., Wong, a K., Bower, M. J., & Blaney, J. M. (1997). Conformational analysis using distance geometry methods. Journal of molecular graphics & modelling, 15(1), 18-36. Retrieved from http://www.ncbi.nlm.nih.gov/pubmed/9346820

53. Wold, S., Esbensen, K., & Geladi, P. (1987). Principal Component Analysis. Chemometrics And Intelligent Laboratory Systems, 2(1-3), 37-52. Po Box 211, 1000 Ae Amsterdam, Netherlands: Elsevier Science Bv. doi:10.1016/0169-7439(87)80084-9

54. Wold, S. (1995). Chemometrics; what do we mean with it, and what do we want from it? Chemometrics and Intelligent Laboratory Systems, 30(1), 109-115. doi:10.1016/0169- 7439(95)00042-9

55. Yanmaz, E., Sarıpınar, E., Şahin, K., Geçen, N., & Çopur, F. (2011). 4D-QSAR analysis and pharmacophore modeling: electron conformational-genetic algorithm approach for penicillins. Bioorganic & medicinal chemistry, 19(7), 2199-210. doi:10.1016/j.bmc.2011.02.035

Chapter 3

AUTOCLICKCHEM: CLICK CHEMISTRY IN SILICO

Jacob D. Durrant[1], J. Andrew McCammon[2,3,4]

[1] Department of Chemistry & Biochemistry, University of California San Diego, La Jolla, California, United States of America

[2] Department of Chemistry & Biochemistry, NSF Center for Theoretical Biological Physics, National Biomedical Computation Resource, University of California San Diego, La Jolla, California, United States of America

[3] Department of Pharmacology, University of California San Diego, La Jolla, California, United States of America

[4] Howard Hughes Medical Institute, University of California San Diego, La Jolla, California, United States of America

ABSTRACT

Academic researchers and many in industry often lack the financial resources available to scientists working in "big pharma." High costs include those associated with high-throughput screening and chemical synthesis. In order to address these challenges, many researchers have in part turned to alternate methodologies. Virtual screening, for example, often substitutes for high-throughput screening, and click chemistry ensures that chemical synthesis is fast, cheap, and comparatively easy. Though both *in silico* screening and click chemistry seek to make drug discovery more feasible, it is not yet routine to couple these two methodologies. We here present a novel computer algorithm, called AutoClickChem, capable of performing many click-chemistry reactions *in silico*. AutoClickChem can be used to produce large combinatorial libraries of compound models for use in virtual screens. As the compounds of these libraries are constructed according to the reactions of click chemistry, they can be easily synthesized for subsequent testing in biochemical assays. Additionally, *in silico* modeling of click-chemistry products may prove useful in rational drug design and drug optimization. AutoClickChem is based on the *pymolecule* toolbox, a framework that may facilitate the development of future python-based programs that require the manipulation of molecular models. Both the *pymolecule* toolbox and AutoClickChem are released under the GNU General Public License version 3 and are available for download from http://autoclickchem.ucsd.edu.

INTRODUCTION

Though the pharmaceutical industry has been the traditional steward of drug development, in recent years academic institutions have played an increasingly important role as well. Formal academic drug-discovery centers established at universities in Belgium, Sweden, the United Kingdom, and the United States have already made great contributions towards the development of novel treatments for neglected and orphan diseases, projects that are generally not financially appealing to industry [1]. Academia may be particularly well suited for the earliest stages of drug discovery, such as target and lead identification [2]. Fruitful collaborations between academia and industry are also becoming more commonplace.

Despite their growing interest in drug discovery, academic researchers, as well as some in industry, often lack the financial resources available to scientists working in "big pharma." High costs include those associated with high-throughput screening and chemical synthesis. Fortunately, limited financial resources have spurred innovation. Virtual screening, a computational technique that can, in part, mimic high-throughput screening *in silico*, is one example of this kind of innovation. Traditionally, high-throughput biochemical screens have constituted and continue to constitute a critical but expensive step in the earliest stages of drug development. Vast and costly libraries of chemical compounds, often in excess of 100,000 molecules, are screened against identified targets of known pharmacological importance in an attempt to identify potent ligands. Robotics and miniaturized/parallelized biochemical assays make such large-scale screening efforts possible. However, with some notable exceptions, the high cost and man-power demands of high-throughput screens make them inaccessible to many researchers.

Virtual screening aims to make high-throughput projects more feasible. Computer docking programs attempt to position candidate ligands within the binding pockets of crystallographic, NMR, or theoretical protein structures in order to predict binding affinity. While docking programs are powerful tools, they do have shortcomings that limit applicability [3], [4]. The programs depend on accurate, atomistic, small-molecule and receptor models (including important bound waters) that can be laborious to prepare; they employ scoring functions that are optimized for speed at the expense of accuracy, often making it difficult to distinguish between nanomolar and micromolar inhibitors; and they often ignore aspects of molecular flexibility that doubtless play important roles in receptor-ligand binding.

Consequently, docking algorithms are not yet accurate enough to assess the binding of a single ligand with certainty, but they can in many circumstances be used to enrich a pool of candidate ligands for true binders [3], [5], [6]. The compounds of this enriched pool of potential ligands, in number far fewer than the total number of compounds in the original library, are then experimentally validated to identify true binders. Virtual screening methodologies have already been used to identify many ligands [7], [8]. A few examples include inhibitors of *Trypanosoma brucei* RNA editing ligase 1 [9], [10], *Trypanosoma brucei* UDP-galactose 4'-epimerase [11], and *Homo sapiens* stromelysin-1 [12].

The high costs associated with high-throughput screens are not the only impediments to drug design. Chemical synthesis can also be very costly and time consuming. The libraries of hundreds of thousands of compounds required for high-throughput screens are expensive to synthesize and/or to purchase commercially. Additionally, following the identification of true ligands, drug optimization requires chemical synthesis in order to improve potency and other pharmacological and toxicological properties.

Dr. Barry Sharpless recently proposed a new chemistry paradigm called "click chemistry" [13]that can help overcome the financial impediments associated with chemical synthesis. There are approximately 10^{60} possible drug-like compounds [2]. Any hopes of thoroughly exploring so large a chemical space must be abandoned from the outset. Given that only an infinitesimally small portion of all possible molecules can ever be synthesized, the chemical reactions used to synthesize potential ligands might as well be limited to those reactions that are ideal; only "click" reactions that are comparatively easy to perform, safe, and cheap need be considered[14]. Using these ideal click-chemistry reactions, academic researchers have produced inhibitors of α-1,3-fucosyltransferase [15], HIV protease [16], acetylcholine esterase [17], [18],[19], carbonic anhydrase II [20], influenza neuraminidase [21], and protein tyrosine phosphatase 1B [22].

Both virtual screening and click chemistry have, in part, the same objective: to make drug discovery practical even when financial resources are limited. Given their philosophical similarities, it is curious that these two methods have not been coupled. We here present a novel algorithm called AutoClickChem that can simulate many click-chemistry reactions *in silico*. Like some other freely available [23], [24], [25] and commercial software packages (e.g., CambridgeSoft's ChemOffice Ultra [26], Tripos' CombiLibMaker [27], [28], ChemAxon's Reactor[29], etc.), AutoClickChem can be used to generate combinatorial libraries for virtual screening. However, AutoClickChem is unique in that it simultaneously satisfies the following criteria: 1) the program is freely available under an open-source license; 2) a web-server application has been implemented that permits use without requiring installation; 3) the generated compounds can be easily synthesized for subsequent testing in biochemical assays because they are constructed according to the reactions of click chemistry; 4) there is no need to specify linker atoms *a priori* because reacting functional groups are automatically detected; and 5) all structures are automatically generated in three dimensions (Table 1). Additionally, AutoClickChem is based on the *pymolecule* toolbox, a framework that may facilitate the development of other python-based programs that require the manipulation of molecular models.

Table 1: A comparison of several computer programs for virtual combinatorial-library generation

	Reference	Free	Open Source	Server Application	Synthesizability of Products	Auto-Identification of Reactive Atoms/Groups	3D Products Produced
AutoClickChem[1]		+	+	+	+ (click chemistry)	+	+
SmiLib[2]	[26]	+	+	–	–	–	–
SLF_Libmaker[3]	[24]	–	–	–	–	–	?
ChemOffice Ultra[4]	[26]	–	–	–	–	–	–
CombiLibMaker[5]	[27,28]	–	–	–	?	?	+
ChemAxon Reactor[6]	[29]	+ (for academics only)	–	+ (restricted)	+ (user-specified reactions)	+	–

[1] autoclickchem.ucsd.edu.
[2] gecco.org.chemie.uni-frankfurt.de/smilib/.
[3] www.idealp-pharma.com.
[4] cambridgesoft.com.
[5] tripos.com.
[6] chemaxon.com.
"Free" means the software is available free of charge, "Open Source" means the source code can be freely modified, "Server Application" means the software is available for use remotely over the internet (without installation), "Synthesizability of Products" means the software takes into account actual chemical reactions when generating compounds *in silico*, "Auto-Identification of Reactive Atoms/Groups" means the program automatically identifies reactive atoms or chemical groups so that the user need not manually annotate, and "3D Products Produced" means the program automatically generates models with 3D coordinates.
doi:10.1371/journal.pcbi.1002397.t001

DESIGN AND IMPLEMENTATION

AutoClickChem.

As input, AutoClickChem accepts PDB models of two small molecules, the two desired reactants. The program begins by automatically identifying functional groups such as alkynes, azides, and epoxides that are known to participate in any of a number of predefined chemical reactions, described in detail Text S1. Once the relevant functional groups have been identified, the program determines which reactions are possible and begins to assemble models of the appropriate products.

The steps required to assemble the products associated with each predefined chemical reaction are unique. As AutoClickChem has been implemented in python and is open source, interested readers can examine the source code to determine how each reaction is programmed. Additional details can also be found in Text S1. To illustrate the general procedure, we here describe how AutoClickChem mimics the azide-alkyne Huisgen cycloaddition, a representative reaction that has been called the "cream of the crop" of click chemistry [13].

The azide-alkyne Huisgen cycloaddition combines an alkyne and an azide (Figure 1A) into a 1,2,3-triazole product. As a first step, AutoClickChem fragments the alkyne along its triple bond and the azide along the bond connecting its proximal and medial azide nitrogen atoms (Figure 1B). Note that the resulting fragments have atomic "handles" comprised of what were the alkyne carbon atoms and the proximal azide nitrogen atom. The fragments are then translated so that these handles are superimposed on top of the corresponding atoms of a 1,2,3-triazole model (Figure 1C). Next, the fragments are rotated about the handle atoms in order to minimize the distance between the handle-adjacent atoms and the corresponding atoms on the 1,2,3-triazole model (Figure 1D). The positioned fragments are then rotated in order to reduce steric hindrance (Figure 1E). Finally, redundant atoms are deleted, and the fragment and 1,2,3-triazole model atoms are merged into a single final structure (Figure 1F). For non-symmetric alkynes, AutoClickChem generates both regioisomers.

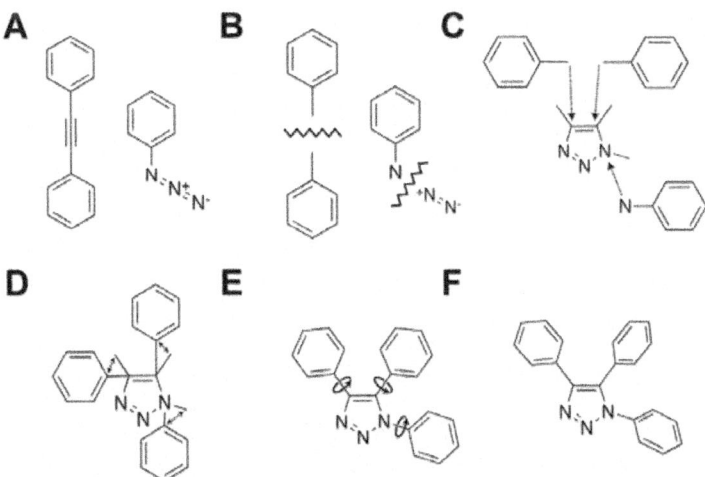

Figure 1: A schematic showing how AutoClickChem mimics the azide-alkyne Huisgen cycloaddition.

A) This cycloaddition combines an alkyne and an azide into a 1,2,3-triazole product. B) As a first step, AutoClickChem fragments the alkyne along its triple bond and the azide along the bond connecting its proximal and medial azide nitrogen atom. C) The fragments are then translated so that atomic "handles" are superimposed on top of the corresponding atoms of a 1,2,3-triazole model. D) Next, the fragments are rotated about the handle atoms in order to minimize the distance between the handle-adjacent atoms and the corresponding atoms on the 1,2,3-triazole model. E) The positioned fragments are then rotated in order to reduce steric hindrance. F) Finally, redundant atoms are deleted, and the fragment and 1,2,3-triazole model atoms are merged into a single final structure.

The *pymolecule* toolbox.

AutoClickChem is based in part on the open-source *pymolecule* toolbox, a framework that facilitates the manipulation of molecular models. We have used beta versions of this toolbox to develop a number of other applications, including HBonanza [30], BINANA [31], POVME [32], and NNScore [33]. With AutoClickChem, the *pymolecule* toolbox has matured. All supporting functions are now contained within a single python file (pymolecule.py) that can be easily included in other projects. Additionally, full documentation is available describing each*pymolecule* definition.

The *pymolecule* toolbox contains three python classes: Point, Atom, and Molecule. The Point class is used to create and manipulate objects with three coordinates, x, y, and z, be they points or vectors in three-dimensional space, and the Atom class stores and manipulates atomic information. The details of these classes are well documented in the source code.

However, the Molecule class, a useful class for manipulating entire molecular structures, merits a more detailed description because it is likely the class that will be most frequently accessed by those developing *pymolecule*-based applications. First, the Molecule class contains two python definitions, *load_pdb* and *save_pdb*, for loading and saving PDB information from/to files.

Six additional Molecule definitions can be used to manipulate the atomic coordinates of a molecular model. Two definitions are used for model translation: *translate_molecule* translates all atomic coordinates by a specified vector, and *set_atom_location* translates all atomic coordinates such that a specified atom resides at a desired coordinate. Three additional definitions rotate the molecular model: *rotate_molecule_around_pivot* rotates all atomic coordinates about a specified point, and *rotate_molecule_around_a_line* and*rotate_molecule_around_a_line_use_atom_indicies* rotate all atomic coordinates about a line segment defined by two terminal Point objects or by the coordinates of two Molecule atoms, respectively. Finally, the *align_another_molecule_to_this_one* definition aligns a second molecule (*molecule_to_align*) to the current one. "Tethers" are defined connecting pairs of atoms, where each of the constituent atoms belong to a separate molecular model. The *molecule_to_align* model is then translated and rotated as necessary to minimize the summed length of the defined tethers.

Several definitions return information about bond connectivity. The *number_of_neighors_of_element* definition counts the total number of atoms of a specified element bound to an atom of interest; *index_of_neighbor_of_element* considers all the atoms bound to a specified atom and returns the index of the first atom of the specified element; *hybridization* determines the orbital hybridization of a specified atom, based in large part on its connectivity; *in_same_ring* determines if two specified atoms are contained in the same ring system; and *get_branch* partitions a molecular model into two by essentially "cutting" along a specified bond.

Finally, two definitions are used to manipulate multiple Molecule objects. The *merge_with_another_molecule* definition merges a second Molecule object with the current one, and the *distance_to_another_molecule* function calculates the minimum distance between the atoms of the current Molecule object and a second one.

Examples illustrating how the *pymolecule* toolbox is used to simulate click-chemistry reactions *in silico* can be found in Text S1.

Results

We here present a novel computer algorithm, called AutoClickChem, capable of performing click-chemistry reactions *in silico*. AutoClickChem can be used to produce large combinatorial libraries of compound models for use in virtual screens. As the compounds of these libraries are constructed according to the reactions of click chemistry, predicted ligands can be easily synthesized for subsequent testing in biochemical assays. AutoClickChem is based in part on the *pymolecule* toolbox, an open-source framework that may facilitate the creation of other python-based applications requiring the manipulation of molecular models.

Click Chemistry Reactions

Though the azide-alkyne Huisgen cycloaddition [34] is the quintessential click-chemistry reaction, there are in fact many reactions with high chemical yields, inoffensive byproducts, simple reaction conditions, and physiologically stable/easily purified products [13], [14]. A description of the "click" reactions that AutoClickChem can simulate *in silico* is given in Text S1; a useful summarizing graphic is also provided (Figure S1).

By generating molecular models based on the reactions of click chemistry, AutoClickChem facilitates interactions between computational and synthetic chemists. When pursuing *de-novo*drug-design projects, many computational chemists (ourselves included!) are notorious for generating compounds that, while predicted to be potent, are nevertheless difficult to synthesize. AutoClickChem helps computational chemists stay within the realm of synthesizability, thus facilitating the transition from *in silico* to *ex silico* testing.

Generating a Virtual Library of Easily Synthesizable Compounds

To demonstrate how AutoClickChem can be used to generate a large virtual library of easily synthesizable compound models for virtual-screening projects, we constructed a library from models of compounds available commercially through hit2lead.com. In all, 939 suitable alkyne models and 1,220 suitable bromide models were ultimately generated from selected hit2lead compounds. AutoClickChem was first used to convert the 1,220 bromides into 1,215 azides. Next, these azide products were reacted with the 939 alkynes *in silico* to produce 2,281,770 1,2,3-triazole products. Any of these products could in theory be easily synthesized *in vitro via*the azide-alkyne Huisgen cycloaddition reaction [34]. When only those models that satisfied all of Lipinski's rule-of-five criteria were considered [35], approximately 800,000 drug-like models remained. Additional details describing the generation of this virtual library can be found in Text S1.

When creating large virtual libraries, the ability to generate products in three dimensions is particularly useful. While programs certainly do exist for converting dimensionless molecular representations (*e.g.*, SMILES strings) into 3D structures, converting hundreds of thousands of models is computationally intensive. With AutoClickChem, this extra step is unnecessary.

To demonstrate the diversity of the compounds generated, we randomly selected fifty azide and fifty alkyne models from the libraries described above. OpenBabel [36] was subsequently used to characterize the corresponding 1,2,3-triazole products according to molecular weight, the number of atoms, the partition coefficient (logP), the polar surface area, and the molar refractivity (Table 2). This characterization confirmed that the compounds are diverse despite having been generated from a limited set of reactants.

Table 2. To demonstrate the diversity of the compounds generated, fifty azides and fifty alkynes were selected at random and reacted *in silico* using AutoClickChem

	Molecular Weight	Number of Atoms	logP	PSA	MR
Minimum	395.5	41	0.9	69.0	103.3
Maximum	593.6	92	6.5	219.0	168.8
Mean ± Stan. Dev.	502.8±29.2	74.6±9.6	3.8±1.1	117.0±23.5	146.4±13.5

"logP" refers to the estimated partition coefficient, "PSA" refers to the polar surface area, and "MR" refers to the molar refractivity.
doi:10.1371/journal.pcbi.1002397.t002

Though we recommend creating custom libraries specifically designed for target proteins of interest, this large, diverse virtual library may nevertheless serve as a useful starting point for any virtual-screening project. A fast docking program like AutoDock Vina [37] running on a 100-processor cluster should be able to screen the whole library against a single protein structure in a matter of days. The AutoClickChem-generated virtual library herein described is freely available for download in several formats on the AutoClickChem website athttp://autoclickchem.ucsd.edu.

Optimization of Tacrine, a Known Acetylcholinesterase Inhibitor

Having demonstrated how AutoClickChem can be used to generate a large virtual library of easily synthesizable compound models, we next show how the program can be used for ligand optimization. To this end, we replicate *in silico* a recent study conducted by Krasinski et al. [18]that sought to optimize the binding affinity of tacrine, a known inhibitor of acetylcholinesterase (AChE). AChE inhibitors are among the approved pharmacological treatments of Alzheimer's disease, myasthenia gravis, and glaucoma. Krasinski et al. started by creating an azide analogue of tacrine. This azide was then mixed in the presence of the enzyme with 23 acetylene reagents not known to bind AChE. Remarkably, of the 46 possible 1,2,3-triazole products, only two formed *in situ*. These two ligands were subsequently identified by HPLC-mass spectrometry. The *syn* compounds (R)-TZ2PIQ-A5, TZ2PIQ-A6, and (S)-TZ2PIQ-A5 were ultimately found to inhibit mouse AChE with K_d values of 100, 410, and 500 fM, respectively.

To replicate this study *in silico*, AutoClickChem was used to generate the same 46 compounds synthesized by Krasinski et al. When alternate charged, tautomeric, ring-conformational, and stereoisomeric states were considered, 1,416 small-molecule models were ultimately produced. These were docked into a crystal structure of mouse AChE (PDB ID: 1Q83) [38] using AutoDock Vina [37], and subsequently rescored with the AutoDock 4.0 scoring function [39], without redocking. Details describing the docking protocol used can be found in Text S1.

AutoDock predicted that the binding affinities of the *syn* compounds (R)-TZ2PIQ-A5, TZ2PIQ-A6, and (S)-TZ2PIQ-A5, the three most potent inhibitors, would be −17.56, −18.43, and −17.74 kcal/mol, respectively. Remarkably, these three compounds were among the four best ranked compounds of the virtual screen. Additionally, compounds in the *syn* conformation tended to be favored, in harmony with experiment.

Optimization of Analogues of a Known Protein Tyrosine Phosphatase 1B Inhibitor

As a second demonstration of drug optimization, AutoClickChem was used to replicate a recent study conducted by Srinivasan et al. [22] wherein analogues of a known protein tyrosine phosphatase 1B (PTP1B) inhibitor, a potential treatment for type 2 diabetes, were optimized to improve binding affinity. Srinivasan et al. began by attaching alkynes to 5 of the analogues. Additionally, 14 aromatic azides were synthesized that were thought likely to bind to a nearby secondary site. Copper (I) was used to catalyze the azide-alkyne Huisgen cycloaddition so that only the 1,4 regioisomers were produced [40]. Of the roughly 70 1,2,3-triazole compounds synthesized, one, called A13, was particularly potent, with an IC_{50} of 4.7 μM against PTP1B.

To replicate this study *in silico*, we used AutoClickChem to generate the same 70 compounds. When alternate charged, tautomeric, ring-conformational, and stereoisomeric states were considered, there were 108 small-molecule models. These were docked into a crystal structure of PTP1B (PDB ID: 2F71) [41] using AutoDock Vina [37], and subsequently rescored with the AutoDock scoring function [39], without redocking. The best inhibitor identified experimentally ranked 5th in our virtual screen, placing it in the top 5% of all models docked.

As the inhibitors identified by Srinivasan et al. [22] were only potent in the low micromolar regime, we next used AutoClickChem to identify ligands with even higher predicted binding energies. The same five alkyne analogues used previously were reacted *in silico* with the 1,215 azides used to generate the large virtual library. The 14,580 resulting products were again docked with Vina and rescored with the AutoDock 4.0 scoring function. In all, 214 compounds scored better than A13 (−11.07 kcal/mol). The best ligand (Figure 2) had a predicted binding energy of −13.33 kcal/mol.

Protein residues that participate in electrostatic interactions are highlighted in yellow. Atoms that participate in receptor-ligand hydrogen bonds are shown in ball-and-stick representation. The aromatic ring of the receptor tyrosine residue that participates in π-π stacking and T-stacking interactions with the ligand is shown in thick licorice representation. The crystallographic pose of a known inhibitor is shown in purple, with key sulfonate moieties shown colored by element in licorice representation. Portions of the protein have been removed to facilitate visualization.

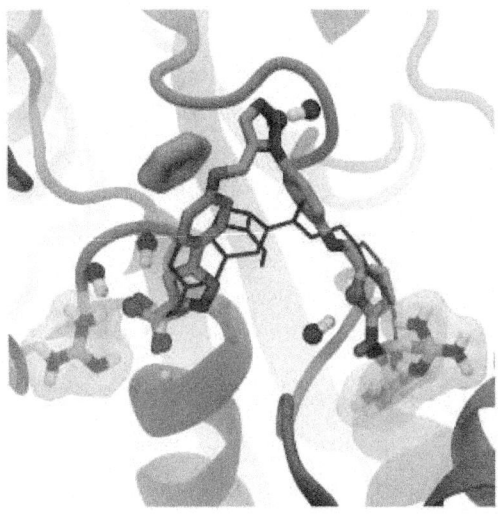

Figure 2: The top-scoring predicted PTP1B ligand (in licorice representation), docked into the receptor active site.

The predicted binding pose of the best-scoring ligand is plausible (Figure 2). The 5-phenylisoxazole-3-carboxylic-acid portion of the ligand, first identified as a PTP1B inhibitor by researchers at Abbott Laboratories, was correctly positioned in the appropriate pocket as judged by x-ray crystallography [42]. This molecular fragment is predicted to participate in electrostatic, hydrogen-bond, and π-π stacking interactions with the protein receptor (Figure 2). The 1,2,3-azole ring is likewise predicted to participate in a hydrogen-bond interaction, as well as in a T-stacking interaction. Finally, the 2-nitrofuran azide fragment extends a nitro group near two arginine side chains, potentially facilitating additional receptor-ligand electrostatic interactions. A hydrogen bond with the furan oxygen atom is also predicted, further improving molecular recognition.

Interestingly, the top predicted ligand identified using AutoClickChem is similar to another ligand whose binding pose was recently characterized by x-ray crystallography (Figure 2, shown in purple) [41]. Both ligands span the same two pockets, and both position functional groups with negative charges (carboxylate, nitro, and sulfonate groups) at the same two locations.

In summary, we herein presented a computer algorithm called AutoClickChem that can simulate the reactions of click chemistry *in silico*. AutoClickChem can be used to generate large combinatorial libraries of easily synthesizable compound models for use in virtual screening. Additionally, the algorithm may prove useful in rational drug design and drug optimization. To demonstrate its utility, we used AutoClickChem to generate a large virtual library of easily synthesizable, drug-like, 1,2,3-azole compounds for use in virtual screens. Additionally, we reproduced two experimental applications of click-chemistry inhibitor optimization *in silico*.

We have also described the *pymolecule* toolbox, a python-based framework that facilitates the development of programs that require the manipulation of molecular models. Beta versions of *pymolecule* have been used to create a number of other useful python scripts; we are hopeful that the *pymolecule* toolbox, now well documented and consolidated into a single file (pymolecule.py), will be helpful to other computational chemists as well.

Availability and Future Directions

While implementations of AutoClickChem and the *pymolecule* toolbox are available from the PLoS Computational Biology website, we recommend visiting http://autoclickchem.ucsd.edu to obtain the latest versions. Additionally, AutoClickChem has been implemented as an opal web service [43] and a server application at http://autoclickchem.ucsd.edu, enabling use without requiring installation.

The authors have plans to incorporate AutoClickChem into future projects as well. For example, the next generation of the AutoGrow algorithm [44] is currently being developed; among many improvements, the program will be extended using AutoClickChem. The original AutoGrow algorithm generated novel ligands by swapping hydrogen atoms with new molecular fragments. Unfortunately, this often produced molecular models of compounds that are difficult to synthesize. Newer versions of AutoGrow will add molecular fragments *via* the reactions of click chemistry, facilitating subsequent synthesis.

In time, we expect to add new features to *pymolecule* as well. Beta versions of the *pymolecule* toolbox have already been used in several projects; as new needs arise in the context of future projects, appropriate additions will be made to the public version of *pymolecule* as well.

We encourage others to modify the AutoClickChem and *pymolecule* source code. As both these resources are python implemented, extending the source code is not difficult. For example, users could extend AutoClickChem to include additional reactions. Some may also wish to expand the *pymolecule* toolbox by adding new functionality (*e.g.*, rmsd-alignment definitions, the ability to read formats other than PDB, etc.) as needs arise in their own projects. We encourage users to contact the authors with any significant modifications so they can be included in future versions of the software.

Acknowledgments

We would like to thank Drs. Barry Sharpless, Suresh Pitram, and Jason Hein for helpful discussions; Daniel Dadon and Henrik Keränen for help with programming AutoClickChem; and Dr. Wilfred W. Li, Nadya Williams, and Jane Ren for help with programming the AutoClickChem Rocks roll.

Author Contributions

Conceived and designed the experiments: JDD JAM. Performed the experiments: JDD. Analyzed the data: JDD. Contributed reagents/materials/ analysis tools: JDD JAM. Wrote the paper: JDD. Helped with editing the paper: JAM.

REFERENCES

1. Wyatt PG (2009) The emerging academic drug-discovery sector. Future Med Chem 1: 1013–1017.

2. .Ohlmeyer M, Zhou MM (2010) Integration of Small-Molecule Discovery in Academic Biomedical Research. Mt Sinai J Med 77: 350–357.

3. Kitchen DB, Decornez H, Furr JR, Bajorath J (2004) Docking and scoring in virtual screening for drug discovery: methods and applications. Nat Rev Drug Discov 3: 935–949.

4. Waszkowycz B, Clark DE, Gancia E (2011) Outstanding challenges in protein–ligand docking and structure-based virtual screening. Wiley Interdiscip Rev Comput Mol Sci 1: 229–259.

5. Kruger DM, Evers A (2010) Comparison of structure- and ligand-based virtual screening protocols considering hit list complementarity and enrichment factors. ChemMedChem 5: 148–158.

6. Cross JB, Thompson DC, Rai BK, Baber JC, Fan KY, et al. (2009) Comparison of Several Molecular Docking Programs: Pose Prediction and Virtual Screening Accuracy. J Chem Inf Model 49: 1455–1474.

7. Talele TT, Khedkar SA, Rigby AC (2010) Successful applications of computer aided drug discovery: moving drugs from concept to the clinic. Curr Top Med Chem 10: 127–141.

8. Schneider G, Fechner U (2005) Computer-based de novo design of drug-like molecules. Nat Rev Drug Discov 4: 649.

9. Amaro RE, Schnaufer A, Interthal H, Hol W, Stuart KD, et al. (2008) Discovery of drug-like inhibitors of an essential RNA-editing ligase in Trypanosoma brucei. Proc Natl Acad Sci U S A105: 17278–17283.

10. Durrant JD, Hall L, Swift RV, Landon M, Schnaufer A, et al. (2010) Novel Naphthalene-Based Inhibitors of Trypanosoma brucei RNA Editing Ligase 1. PLoS Negl Trop Dis 4: e803.

11. Durrant JD, Urbaniak MD, Ferguson MA, McCammon JA (2010) Computer-Aided Identification of Trypanosoma brucei Uridine Diphosphate Galactose 4'-Epimerase Inhibitors: Toward the Development of Novel Therapies for African Sleeping Sickness. J Med Chem 53: 5025–5032.

12. Puerta DT, Mongan J, Tran BL, McCammon JA, Cohen SM (2005) Potent, selective pyrone-based inhibitors of stromelysin-1. J Am Chem Soc 127: 14148–14149.

13. Kolb HC, Finn MG, Sharpless KB (2001) Click Chemistry: Diverse Chemical Function from a Few Good Reactions. Angew Chem Int Ed Engl 40: 2004–2021.

14. Kolb HC, Sharpless KB (2003) The growing impact of click chemistry on drug discovery. Drug Discov Today 8: 1128–1137.

15. Lee LV, Mitchell ML, Huang SJ, Fokin VV, Sharpless KB, et al. (2003) A potent and highly selective inhibitor of human alpha-1,3-fucosyltransferase via click chemistry. J Am Chem Soc 125: 9588–9589.

16. Brik A, Muldoon J, Lin YC, Elder JH, Goodsell DS, et al. (2003) Rapid diversity-oriented synthesis in microtiter plates for in situ screening of HIV protease inhibitors. Chembiochem 4: 1246–1248.

17. Lewis WG, Green LG, Grynszpan F, Radic Z, Carlier PR, et al. (2002) Click chemistry in situ: acetylcholinesterase as a reaction vessel for the selective assembly of a femtomolar inhibitor from an array of building blocks. Angew Chem Int Ed Engl 41: 1053–1057.

18. Krasinski A, Radic Z, Manetsch R, Raushel J, Taylor P, et al. (2005) In situ selection of lead compounds by click chemistry: target-guided optimization of acetylcholinesterase inhibitors. J Am Chem Soc 127: 6686–6692.

19. Manetsch R, Krasinski A, Radic Z, Raushel J, Taylor P, et al. (2004) In situ click chemistry: enzyme inhibitors made to their own specifications. J Am Chem Soc 126: 12809–12818.

20. Mocharla VP, Colasson B, Lee LV, Roper S, Sharpless KB, et al. (2004) In situ click chemistry: enzyme-generated inhibitors of carbonic anhydrase II. Angew Chem Int Ed Engl 44: 116–120.

21. Li J, Zheng M, Tang W, He PL, Zhu W, et al. (2006) Syntheses of triazole-modified zanamivir analogues via click chemistry and anti-AIV activities. Bioorg Med Chem Lett 16: 5009–5013.

22. Srinivasan R, Uttamchandani M, Yao SQ (2006) Rapid assembly and in situ screening of bidentate inhibitors of protein tyrosine phosphatases. Org Lett 8: 713–716.

23. Schuller A, Hahnke V, Schneider G (2006) SmiLib v2.0: A Java-Based Tool for Rapid Combinatorial Library Enumeration. QSAR Comb Sci 26: 407–410.

24. Krier M, Araujo-Junior JX, Schmitt M, Duranton J, Justiano-Basaran H, et al. (2005) Design of small-sized libraries by combinatorial assembly of linkers and functional groups to a given scaffold: application to the structure-based optimization of a phosphodiesterase 4 inhibitor. J Med Chem 48: 3816–3822.

25. Melnikov AA, Palyulin VA, Zefirov NS (2007) Generation of molecular graphs for QSAR studies: an approach based on supergraphs. J Chem Inf Model 47: 2077–2088.

26. CambridgeSoft Corporation. ChemOffice Ultra 7.0.1. Cambridge, MA.

27. Cramer RD, Patterson DE, Clark RD, Soltanshahi F, Lawless MS (1998) Virtual compound libraries: A new approach to decision making in molecular discovery research. J Chem Inf Model 38: 1010–1023.

28. Pearlman RS, Smith KM (1998) Novel software tools for chemical diversity. Perspect Drug Discovery Des 9–11: 339–353.

29. ChemAxon (2011) Reactor. Budapest (Hungary): ChemAxon.

30. Durrant JD, McCammon JA (2011) HBonanza: A Computer Algorithm for Molecular-Dynamics-Trajectory Hydrogen-Bond Analysis. J Mol Graphics Modell 31: 5–9.

31. Durrant JD, McCammon JA (2011) BINANA: A novel algorithm for ligand-binding characterization. J Mol Graphics Modell 29: 888–893.

32. Durrant JD, de Oliveira CA, McCammon JA (2011) POVME: An algorithm for measuring binding-pocket volumes. J Mol Graphics Modell 29: 773–776.

33. Durrant JD, McCammon JA (2010) NNScore: A Neural-Network-Based Scoring Function for the Characterization of Protein-Ligand Complexes. J Chem Inf Model 50: 1865–1871.

34. Huisgen R (1961) pp. 357–396. Centenary Lecture - 1,3-Dipolar Cycloadditions.

35. Lipinski CA, Lombardo F, Dominy BW, Feeney PJ (2001) Experimental and computational approaches to estimate solubility and permeability in drug discovery and development settings. Adv Drug Deliv Rev 46: 3–26.

36. Guha R, Howard MT, Hutchison GR, Murray-Rust P, Rzepa H, et al. (2006) The Blue Obelisk-interoperability in chemical informatics. J Chem Inf Model 46: 991–998.

37. Trott O, Olson AJ (2009) AutoDock Vina: Improving the speed and accuracy of docking with a new scoring function, efficient optimization, and multithreading. J Comput Chem 31: 455–461.

38. Bourne Y, Kolb HC, Radic Z, Sharpless KB, Taylor P, et al. (2004) Freeze-frame inhibitor captures acetylcholinesterase in a unique conformation. Proc Natl Acad Sci U S A 101: 1449–1454.

39. Morris GM, Goodsell DS, Halliday RS, Huey R, Hart WE, et al. (1998) Automated docking using a Lamarckian genetic algorithm and an empirical binding free energy function. J Comput Chem 19: 1639–1662.

40. Tornøe CW, Christensen C, Meldal M (2002) Peptidotriazoles on Solid Phase: [1,2,3]-Triazoles by Regiospecific Copper(I)-Catalyzed 1,3-Dipolar Cycloadditions of Terminal Alkynes to Azides. J Org Chem 67: 3057–3064.

41. Klopfenstein SR, Evdokimov AG, Colson AO, Fairweather NT, Neuman JJ, et al. (2006) 1,2,3,4-Tetrahydroisoquinolinyl sulfamic acids as phosphatase PTP1B inhibitors. Bioorg Med Chem Lett 16: 1574–1578.

42. Liu G, Xin Z, Pei Z, Hajduk PJ, Abad-Zapatero C, et al. (2003) Fragment screening and assembly: a highly efficient approach to a selective and cell active protein tyrosine phosphatase 1B inhibitor. J Med Chem 46: 4232–4235.

43. Ren J, Williams N, Clementi L, Krishnan S, Li WW (2010) Opal web services for biomedical applications. Nucleic Acids Res 38: W724–731.

44. Durrant JD, Amaro RE, McCammon JA (2009) AutoGrow: A Novel Algorithm for Protein Inhibitor Design. Chem Biol Drug Des 73: 168–178.

Chapter 4

USING WORKFLOWS TO EXPLORE AND OPTIMISE NAMED ENTITY RECOGNITION FOR CHEMISTRY

BalaKrishna Kolluru[1], Lezan Hawizy[2], Peter Murray-Rust[2], Junichi Tsujii[1], Sophia Ananiadou[1]

[1] National Centre for Text Mining, Manchester Interdisciplinary Biocentre, University of Manchester, Manchester, United Kingdom

[2] Unilever Centre for Molecular Informatics, University of Cambridge, Cambridge, United Kingdom

ABSTRACT

Chemistry text mining tools should be interoperable and adaptable regardless of system-level implementation, installation or even programming issues. We aim to abstract the functionality of these tools from the underlying implementation via reconfigurable workflows for automatically identifying chemical names. To achieve this, we refactored an established named entity recogniser (in the chemistry domain), OSCAR and studied the impact of each component on the net performance. We developed two reconfigurable workflows from OSCAR using an interoperable text mining framework, U-Compare. These workflows can be altered using the *drag-&-drop* mechanism of the graphical user interface of U-Compare. These workflows also provide a platform to study the relationship between text mining components such as tokenisation and named entity recognition (using maximum entropy Markov model (MEMM) and pattern recognition based classifiers). Results indicate that, for chemistry in particular, eliminating noise generated by tokenisation techniques lead to a slightly better performance than others, in terms of named entity recognition (NER) accuracy. Poor tokenisation translates into poorer input to the classifier components which in turn leads to an increase in Type I or Type II errors, thus, lowering the overall performance. On the Sciborg corpus, the workflow based system, which uses a new tokeniser whilst retaining the same MEMM component, increases the F-score from 82.35% to 84.44%. On the PubMed corpus, it recorded an F-score of 84.84% as against 84.23% by OSCAR.

INTRODUCTION

Text mining for the domain of chemistry is a very challenging task because of the several semantic and syntactic styles in which domain texts are usually expressed. Different aspects such as named entity recognition (NER), tokenisation and acronym detection require bespoke approaches because the complex nature of such texts [1]–[5]. Chemical compounds such as:

17-α-hydroxy-16-α-methyl-3,20-dioxopregna-1,4-dien-21-yl acetate

P(Cy)3

1-cyclopropyl-6-fluoro-4-oxo-7-(piperazin-1-yl)-1,

4-dihydroquinoline-3-carboxylic acid hydrochloride

illustrate the complexity of the mining task. Typical word delimiters such as spaces, brackets, hyphens and commas cease to bear the same meaning as in a natural language. As a consequence, the normal text mining approaches such as tokenisers, part-of-speech (POS) taggers and parsers will need to be re-calibrated for this domain as already done for other domains such as biochemistry, biomedicine *etc.*, [6], [7].

In the chemistry domain, researchers have presented a few successful approaches to handle some tasks such as named entity recognition [8]–[13]. However, these approaches usually require reconfiguring and sometimes rewriting everytime a new training corpus or dictionary is released [14], [15]; typically this could be due to different data format or additional information in the new resource. For example, if the new resource is in a different format, the whole system or at least a part of it may need to be rewritten. With the growing number of freely available resources such as Chemspider (http://www.chemspider.com/), Chemlist [14] and [5], [16]–[18]*etc.*, the ability to reconfigure the systems becomes more acute. Such reconfiguring takes time and the subtle changes in the throughputs of these components, which may seem innocuous, could result in the lowering of the net performance of a system; this could be a direct consequence of a suboptimal composition of the workflow. Therefore, it is imperative to configure the optimal set by exploring the various manifestations of the different components[19]. To be able to arrive at an optimum combination of components, one has to substitute one component for another in a workflow and then assess if the performance has indeed improved. This warrants an understanding of inter-component relations working together as a system. It would also be desirable if components using different machine learning techniques could easily be replaced to observe differences in performance. This ability to reconfigure an approach has the advantage of allowing scientists to concentrate more on science rather than format conversion and code refactoring. Usage of workflows for chemistry and its related disciplines has been pursued very actively in the community [20]–[23]. Thus, there is already good familiarity, if not expectation, of this methodology. For the experiments discussed in this paper, we implement reconfigurable workflows that are interchangeable by *drag-&-drop* on the graphical user interface. To do this we employ U-Compare [24]: an open UIMA-based [25]framework (http://incubator.apache.org/uima/[26]) which allows shareable components, using a common type system, to be used together to form different workflows. In doing so, we also design an interoperable type system for UIMA-compliant systems.

Figure 1 (b) illustrates the composition of a reconfigurable workflow system, wherein one component can be substituted by another component from a repository.

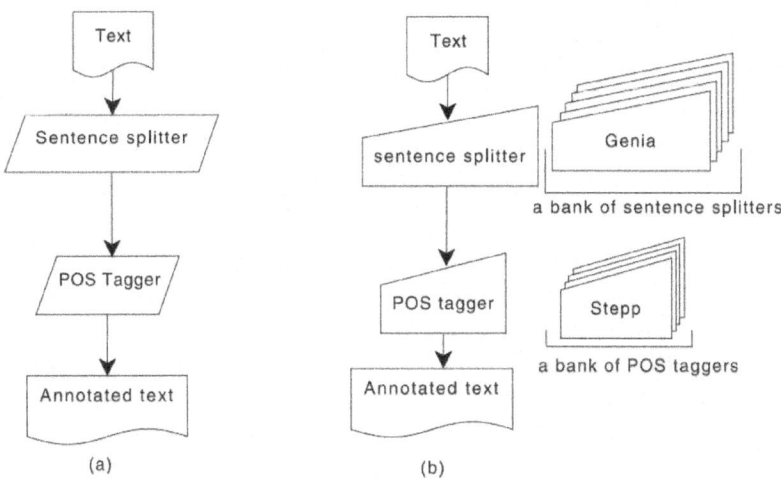

Figure 1: Showing a normal workflow and a reconfigurable workflow as can be built by using U-Compare.

U-Compare framework provides a platform for reconfigurable workflow experiments. Its UIMA-based framework provides the necessary component repository, consisting of several shareable components such as Genia tagger [7], Stepp tagger [27] and OpenNLP [28]sentence splitters, for other UIMA-based components. This extensive repository readily allows for several combinations of components as workflows.

As a consequence, we use a set of individual components to handle the different aspects of text mining, which together form a workflow.

Related work for workflows

The use of several individual components to construct workflows is quite prevalent amongst the scientific community with interdisciplinary sciences [29]. Bespoke workflows have been employed in several domains such as bio-informatics and earth sciences. Some studies have introduced workflow tools as a Lego®-like setup [30], wherein several simple components form a complex workflow which can be easily deployed, modified and tested without the overhead of implementing it into a monolithic application. Taverna [31] is such a workflow management suite for building scientific workflows which offers *loosely-coupled* services. Kepler [32] is another workflow management system for designing, executing, processing and sharing scientific workflows. The workflows in this context are directed graphs where the nodes represent components, the edges represent data paths along which data and results can flow between components.

There are also commercial products which provide environments to create and manage workflows. Pipeline Pilot [33] is an example of a commercial application that combines workflows with data analysis to represent information visually for informatics and scientific business intelligence needs. Pipeline Pilot has been extended to track bibliography in chemistry literature using a web-based graphical user interface [34].

The UIMA platform introduced a new framework for developing shareable components into a repository. Mellebeek *et al.* show the usage of UIMA and text mining applications for curation purposes in the domain of bio-informatics [35]. In doing so, they demonstrate the possible synergy from a combination of diverse expertise in biology, computer science and linguistics. Their application was fundamental to the development of a successful curation tool. The U-Compare [36], based on the UIMA Framework, is an integrated text mining system which provides a graphical user interface for easy *drag-&-drop* workflow creation. It has built-in tools for evaluation and visualisations of components and also has a number of syntactic and semantic tools to generate workflows. Kano *et al.* showed the advantages of using workflows in U-Compare framework by developing a protein-protein interaction extraction system [37].

Our paper presents a similar workflow to [37] but in the domain of chemistry. As a first step, we used Oscar3 [38] to extract chemical named entities from the literature. Subsequently, we segregated Oscar3 into separate components. Townsend *et al.* have developed a methodology and a workflow (CHIC) for the automatic semantic enrichment and structuring of legacy scientific documents by using Oscar3[38]. U-Compare has a plug-in for Taverna [37] which implicitly means the workflows discussed here can be ported to Taverna, which increases the audience and applicability of our workflows.

In the area of chemistry and text mining, Wilbur *et al.* employed two approaches with an aim of separating chemical terms from non-chemical terms [39]: thesaurus-based lexical text analysis using chemical patterns

Bayesian Classification Using n-Grams

They found that the Bayesian approach had an overall classification accuracy of 97%, while the thesaurus-based method had an accuracy of 84%. While the work by Wilbur *et al.* operates on individual words (or entities) based on thesaurus-style lists [39], the work described here processes full papers (and abstracts), tokenizes them for analysis and classifies chemical compounds found in the text.

Materials and Methods

In this paper, our principal task was to elicit chemical compounds from free-flowing text in the chemistry literature. We have used the Sciborg [40] and PubMed [41] corpora for this task.

Sciborg Corpus

This corpus was compiled as part of the Sciborg project [42]. It consists of 42 articles (full papers) published in the chemical literature which were provided by the Royal Society for Chemistry (RSC). It was curated for linguistic analysis by [41]. This corpus was split randomly into two groups of 14 and 28 papers, such that they form two disjoint testing and training sets respectively. MEMM models (discussed later in paper) were trained on the set of 28 papers having 4102 manually annotated chemical compounds and the 14 papers were used as a test set. The test set was hand-annotated by three chemistry experts and an inter-annotator agreement (κ) of 0.91 was observed on this set.

PubMed Corpus

This corpus was compiled for linguistic analysis by Corbett *et al.* [41]. It had 500 abstracts from the PubMed [43] collection. This corpus was randomly split into 400 and 100 abstracts for training and test sets respectively. MEMM models were trained on the 400 abstracts consisting of 4048 annotations of chemical compounds. The test set was hand-annotated by one expert.

Models

For the MEMM-based component of our approach we experimented with two models, *chempaper-M*: trained on Sciborg training data (28 papers)

pubmed-M: trained on 400 PubMed abstracts

Overview of Oscar3

Oscar3 is an open extensible system for the automated annotation of chemical entities in scientific articles [9]; it was created as part of the Sciborg project [40]. The overall architecture of Oscar3 is shown in Figure 2 and the individual components are discussed below:

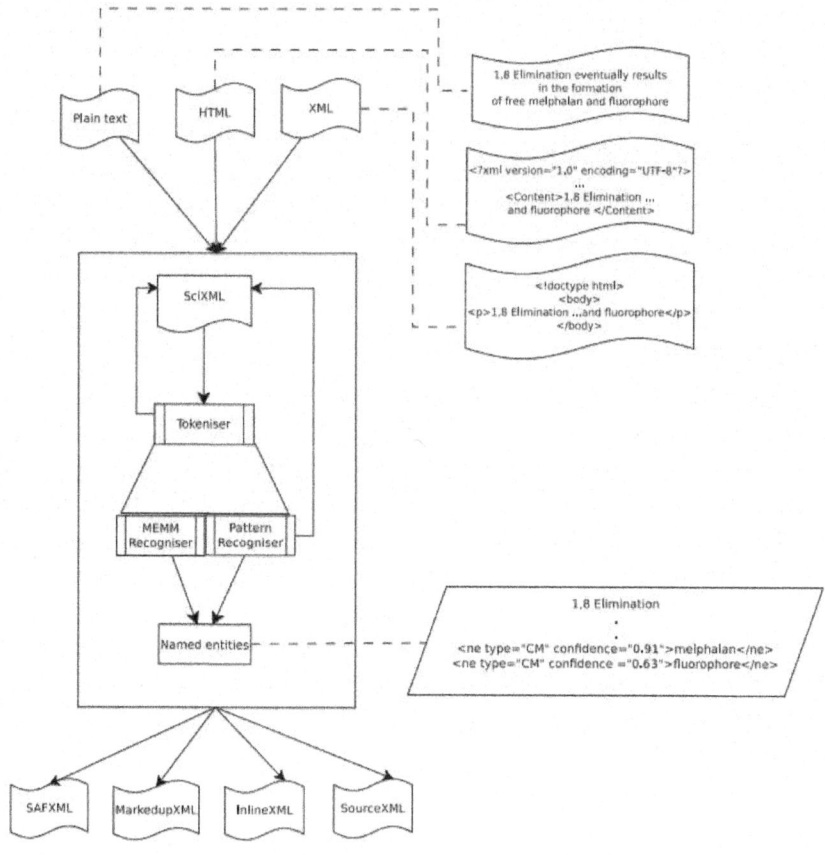

Figure 2: The original architecture of Oscar3.

SciXML.

SciXML is the interface used when working with Oscar3, all forms of input (such as XML, HTML and plain text) are converted into this format before any processing is done. It is a form of XML markup used for providing logical structure to scientific papers. Further information about SciXML and its schema can be found in [40].

Tokeniser.

The tokenisation with Oscar3 is chemistry specific; chemical names are fragile to common methods of tokenisation as they contain potential inter token and intra token characters such as space, hyphens, brackets and comma. The tokeniser here also refers back to the SciXML document to store information about the start and end points of a token as well as its content. For example, some of the tokens in data are:

aztreonam__Metallo-β-lactamase

Cu2+

C2(MONO)

C__O/C__N

Zn...O3S(monobactam)

Chemical Entity Recognisers.

Oscar3 contains two types of chemical entity recognisers, each producing a list of named entities as an output containing chemical annotations, such as token, types and likelihood scores (where applicable).

Pattern Recogniser

This recogniser was initially used before the machine learning component was introduced. It uses deterministic finite state automata alongside ontologies (such as CHEBI [44]), dictionaries and n-gram models to recognise the named entities. As it relies on the regular expression based rules, it does not use any mathematical models for classification.

MEMM Recogniser

This recogniser uses MEMM and character level n-grams to recognise chemical entities based on their likelihoods. The MEMM was trained using the annotated corpora discussed earlier. Corbett *et al.* reported an F-score of 80.7% for a model trained on Sciborg and PubMed training sets at a confidence threshold of 0.3 [41].

Why the new Oscar?

Oscar3 is an efficient annotation tool and is widely used within the chemistry domain. However, the architecture is rigid and, due to its dependency on the SciXML format and the interdependency within the different components, it is difficult to modularise and it does not readily adapt to new and emerging trends in annotation and corpora. This puts a limitation on enabling and refactoring reusable components.

Oscar3 as a Workflow of Reconfigurable Components

Conceptually, Oscar3 [9] is a named entity recogniser which classifies tokens into chemical entities based on either likelihoods or a pattern match. Therefore, Oscar3 was divided into the following components as shown in Figure 3. This is just one of the many possible manifestations of the workflows; other configurations such as different tokenisers and components implementing machine learning techniques can be easily accommodated to make a new workflow.

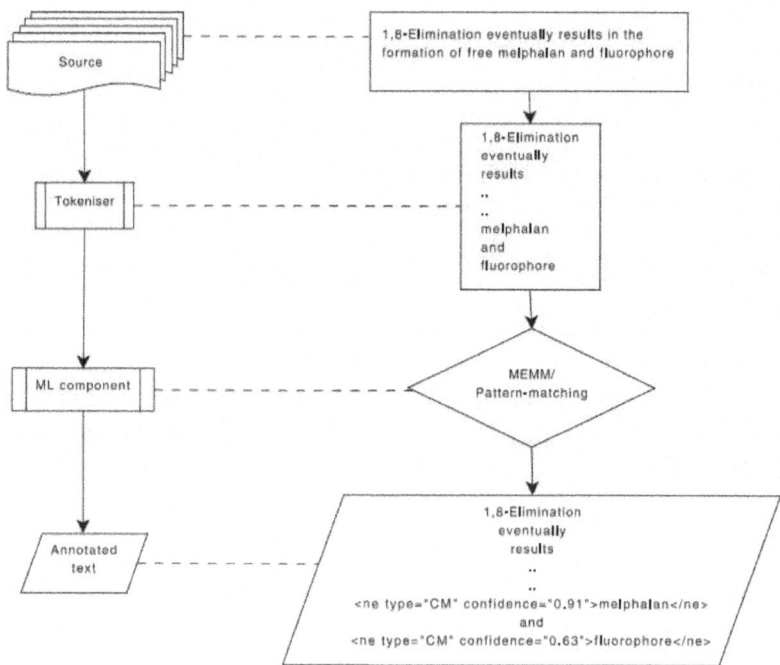

Figure 3: Oscar3 refactored as a workflow of different components.

A tokeniser: This tokeniser is a white-space delimited, word eliciting component which reads content from files in text, XML or HTML and yields tokens similar to the syntactic token of the U-Compare type system [45]. It must be noted here that any tokeniser that yields the syntactic tokens for a given source file can be used as the first stage of the Oscar workflow.

A MEMM Component: trained on two chemistry-specific corpora (as mentioned in the data section).

A Pattern matching Component: based on a finite state automaton driven regular expression matcher; the rules of which were designed after several observations of the training data. As a consequence, this component does not use any statistical models for classification purposes.

Shown in Figure 4 is one of the workflows (left in the figure) using three components (right in the figure): a file system component to read the files which are then split into individual tokens by the OscarTokeniser component which subsequently feeds into the OscarMER component to classify the tokens into chemical names.

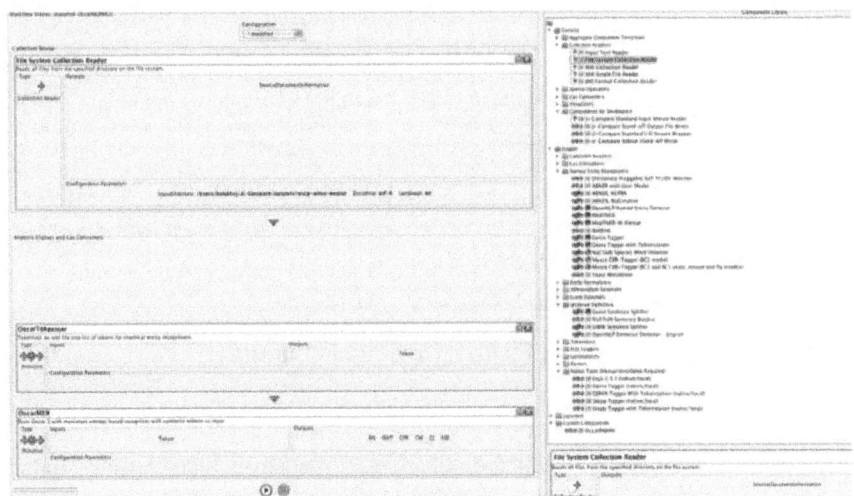

Figure 4: U-Compare view of Oscar workflow.

Right side of the figure shows a workflow made from the Oscar components shown on the left.

Results

The experiments described earlier with Oscar3 and its refactored version were designed to study the effect of tokenisation on chemical element identification. We present the results of modularising Oscar3 and compare it with the existing version. Also, to present the robustness of the workflow, we compare the performance of Oscar3 on the corpora described in the data section.

Reconfiguring Oscar3: a confidence-driven approach

The machine learning components used by the two variations of Oscar3 (Oscar3 stand-alone version and Oscar workflow) yield a confidence score, which is a likelihood estimate (see [41]for more details), to show the confidence in that annotation. In order to arrive at an optimum threshold for each of the corpora, we have plotted the ROC (A ROC curve is a receiver operating characteristic which plots the rate of true positives against the rate of false positives.) curves for each of the data sets. Shown in Figure 5 are the ROC curves for different combinations of data sets and Oscar3 variants.

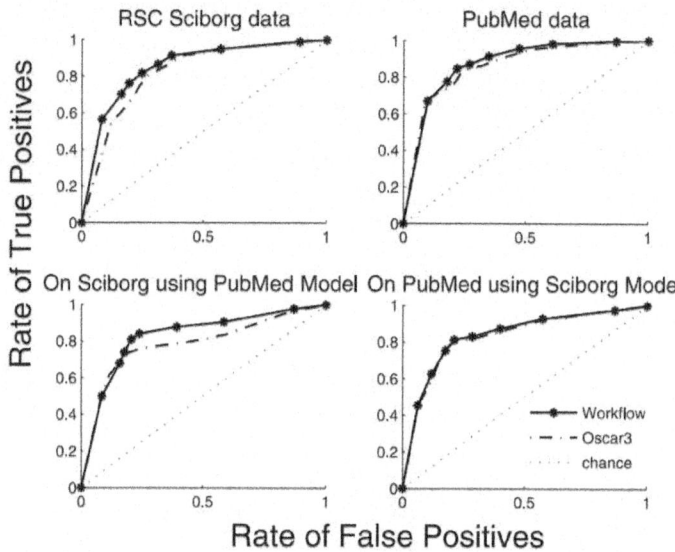

Figure 5: ROC curves comparing the performance of various Oscar variants.

In all the four different experiments, Oscar workflow has a slight edge over the Oscar 3 variant.

Oscar vs. Oscar: One Variant against Another

As described earlier, currently there are two types of named entity recognisers, a MEMM-based and a pattern matching one. The MEMM-based versions were tested with two models;*chempaper-M* and *pubmed-M*.

The results are presented in terms of percentages of precision (P), recall (R) and F score (F).Table 1. shows the overall performance of both the variants of Oscar3 on Sciborg test data. The MEMM-driven systems were tested using both models (*chempaper-M* and *pubmed-M*) which are trained on Sciborg and PubMed training data respectively.

Table 1: Performance (%) of different variants of Oscar on Sciborg test data using the models trained on Sciborg data and PubMed data

Variants on Sciborg	Model used	
	chempaper-M	pubmed-M
Oscar3 (MEMM)	P 88.24	74.76
	R 77.19	65.18
	F 82.35	69.64
Oscar workflow (MEMM)	P 90.31	80.19
	R 79.29	71.22
	F 84.44	75.44

doi:10.1371/journal.pone.0020181.t001

Table 2. shows the performances of Oscar3 with pattern recogniser (Oscar3 (PAT)) and as a workflow with pattern recogniser (Oscar workflow (PAT)) on the Sciborg test data.

Table 2: Performance of different Oscar pattern recogniser versions on Sciborg

Variants on Sciborg	Scores (%)
Oscar3 (PAT)	P 70.43
	R 67.42
	F 68.89
Oscar workflow (PAT)	P 74.11
	R 73.68
	F 73.90

doi:10.1371/journal.pone.0020181.t002

As described in Corbett *et al.* [41], Oscar3 can be tuned to filter out some false positives (Type I) errors based on a confidence score derived from the logit scores (see [41] for more details). At a confidence score of 0.42, Oscar3 as a workflow with MEMM recorded an F-score of 84.84% while Oscar3 with MEMM recorded 82.35% on the Sciborg data. It is also noteworthy that although the pattern recognition variants were less accurate than their MEMM counterparts, the workflow variant still outperforms its monolithic parent.

Table 3 shows the performance of the Oscar3 variants when used on the PubMed test set against models trained on Sciborg and PubMed training data sets. It can be observed that a different tokeniser gives an extra boost of 0.61% (84.84 by the workflow MEMM variant as against 84.23 by the Oscar3 variant) whilst retaining the same machine learning component and the model.

Table 3: Performance of different variants of Oscar on PubMed test data using the models trained on Sciborg data and PubMed data

Variants on PubMed	Model used	
	chempaper-M	*pubmed-M*
Oscar3 (MEMM)	P 75.28	89.04
	R 63.42	79.91
	F 68.84	84.23
Oscar workflow (MEMM)	P 75.06	85.66
	R 64.58	84.03
	F 69.43	84.84

doi:10.1371/journal.pone.0020181.t003

Table 4 shows the performance of pattern-recognition based variants of Oscar3 and Oscar3 workflow on the Pubmed data. Although having lower scores than the MEMM variants, Oscar workflow (PAT) outperforms the Oscar3 (PAT) by 1.7%.

Table 4: Performance of different Oscar pattern recogniser versions on Pubmed

Variants on Pubmed	Scores (%)
Oscar3 (PAT)	P 44.22
	R 58.24
	F 50.27
Oscar workflow (PAT)	P 45.64
	R 60.35
	F 51.97

doi:10.1371/journal.pone.0020181.t004

Wren used the single-order Markov models to distinguish between chemical and non-chemical terms on Medline [46] corpus with an average precision of about 82.7% [10]. The work described here uses maximum entropy Markov models on 2 different corpora: Sciborg and PubMed. For this corpus, our approach recorded a precision of 90.31% and a recall of 85.66%. As the work by Wren ([10]) and our approach vary on the corpora and methodology, we do not see it fair to compare head-to-head; however, our system does perform as well, if not better.

Discussion

To observe the efficacy of workflows, we have used two sets of workflows, one each in pattern recognition based variant and the MEMM variant. As shown in Tables 1 and 2, the workflow variants of Oscar3 achieve better performance the two variants.

On the Sciborg, the workflow-based MEMM model achieved an F-score of 84.44% as opposed to 82.35% for Oscar3. We observed that this increase was due to removal of the dependency on SciXML conversions within the workflow.

A reconfigurable approach enabled us to identify the erroneous (or underperforming) component and relate some of the errors to severe dependency on SciXML conversions, when using *chempaper-M*. We infer that there was a net increase in false positives due to the noise in several inter-conversions of formats in SciXML. It should be noted here that the dependence of Oscar3 on SciXML was due to them both being part of the Sciborg project. This dependency could make it difficult to adapt to newer corpora. It was observed that the new tokenisation identified more chemical words such as

β-lactam_zn2+

bis-monodentate

gold_sulfur

which automatically led to a decrease in false positives (Type I errors). It also avoided wrongly tokenizing a few words such as,

diimine

mono-bidentate

which were subsequently omitted as non-chemistry words by the Oscar3 but accurately identified by the workflow variant. In this example, the complete chemical was ruthenium (ii) diimine, but Oscar3 returned only ruthenium(ii) as CM, whilst the workflow version got the complete entity. by the Oscar3 but accurately identified by the workflow variant. This led to fewer false negatives (Type II errors) and hence a better recall.

As the machine learning classifiers and the models they used were exactly the same for all experiments, we infer that tokenisation on the Sciborg test avoided partial entities for recognition and this helped reduce both Type I and Type II errors. Figure 6 shows the chemical names as annotated by the Oscar workflow in the U-Compare framework. When these entities (which are underlined) are clicked, more information about the entity such as confidence scores, metadata *etc.* is available to the user.

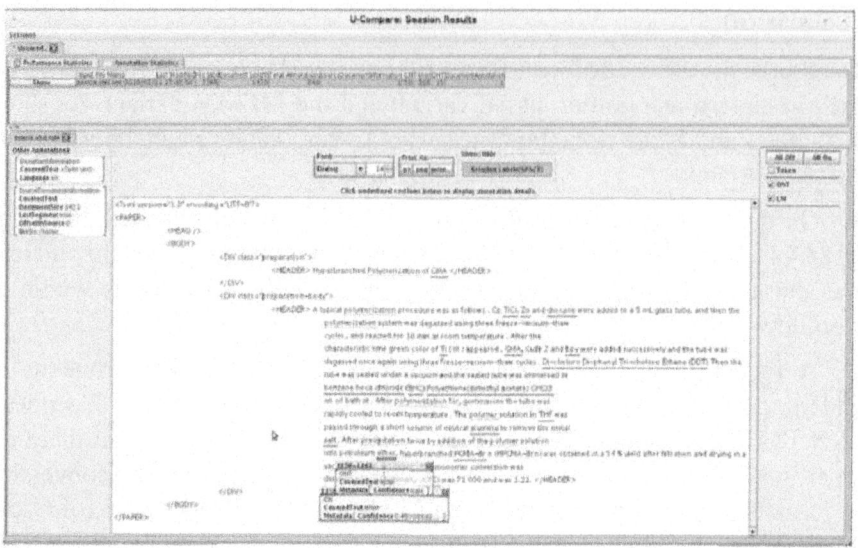

Figure 6: U-Compare output for a test document.

Chemical Names (underlined) as Identified by the MEMM-based Workflow.

Table 3 shows a decrease in precision of ~3% with an increase in recall of ~ 4% for the PubMed test data. Perhaps, this could be attributed to the possible shortcomings in the ability of the new tokeniser to adapt to the biochemical entities; we are working on enhancing the tokeniser to suit multiple domains where chemistry plays an important role.

The current version of Oscar3, which can be downloaded fromhttp:// sourceforge.net/projects/oscar3-chem/, had an F-score of 82.35% on Sciborg as against 80.7% achieved by [41]. This could be due to the fact that [41] used a 3-fold cross validation, whilst we used only 1 combination. The usage of one training set and one test set, instead of multi-fold cross validation, was guided by the focus of our paper, namely: the advantages of workflows for text mining in chemistry. Also, as described in the Data section, the test set comprised of 14 full papers, manually annotated by three experts, whilst, the training set was annotated by a single expert. We conjecture that this test set had enough data points to support our inferences.

On the Sciborg (Table 1), the pattern-recognition based workflow achieved a precision of 66.32% while the Oscar3 using the pattern recognition module achieved a precision of 44.65%. Again, it seems the only difference between the two variants was the tokenisation which stems from issues relating to

SciXML conversions. This could be perceived as an example of having an optimal combination in a workflow to derive a better performance.

The results indicate the success of workflows described in our experiments discussed earlier. Currently, we are in the process of converting implementations of other machine learning algorithms such as Conditional Random Fields (CRF) into the U-Compare framework. This will enable us to compare the performance of different algorithms on data sets. Every time a new annotation scheme is announced, it obliges the existing applications to adapt, sometimes subtly and at times extensively. We have shown that a reconfigurable system (or application) is better for such adaptation.

Conclusions

We have shown that, using a reconfigurable workflow, it is possible to assess different components in a system to elicit the best combination. As a consequence, it helps users to focus less on system implementation issues. Using these workflows, we studied the impact of using different tokenisation techniques on the task of named entity recognition in chemistry. The potential for expanding the scope of inter-component analysis is immense and more so, with complex systems involving several components. We have demonstrated the impact of tokenisation in recognising complex named entities in chemistry, wherein a named entity may contain two, three or even four words with numerals, Greek letters, punctuation marks, *etc.* Work is currently underway to make a CRF component so that one can freely replace MEMM models with a CRF model and thus benefit from a pool of machine learning algorithms for various tasks, named entity recognition being one of them. We are also working on workflows to combine a set of taggers and named entity recognisers for application in the domain of chemistry, biochemistry and biological sciences [47].

ACKNOWLEDGMENTS

We would like to thank Richard Kidd and Colin Batchelor of the Royal Society for Chemistry for providing the Sciborg corpus, as well as their insight into the annotation guidelines. Also, we would like to thank Dr. Paul Dobson for valuable comments, and Dr. John McNaught for the numerous edits of our paper. Last, but not least, we would like to thank Dr. Yoshinobu Kano for his help with U-Compare.

AUTHOR CONTRIBUTIONS

Conceived and designed the experiments: BK LH. Performed the experiments: BK LH. Analyzed the data: BK LH SA PM JT. Wrote the paper: BK LH. Supervised the experiments: PM SA JT.

REFERENCES

1.   Kemp N, Lynch M (1998) Extraction of information from the text of chemical patents. 1. identification of specific chemical names. Journal of Chemical Information and Computer Sciences 4: 544–551.

2.   Murray-Rust P, Rzepa H (1999) Chemical markup, xml, and the worldwide web. 1. basic principles. Journal of Chemical Information and Computer Sciences 39: 928–942.

3.   Murray-Rust P, Mitchell J, Rzepa H (2005) Chemistry in bioinformatics. BMC Bioinformatics 6: 141.

4.   Banville D (2006) Mining chemical structural information from the drug literature. Drug Discovery Today 11: 35–42.

5.   Kolrik C, Hofmann-Apitius M, Zimmermann M, Fluck J (2007) Identification of new drug classification terms in textual resources. Bioinformatics 13: 264–272.

6.   Miyao Y, Tsujii J (2005) Probabilistic disambiguation models for wide-coverage hpsg parsing. ACL-2005. pp. 83–90.

7.   Tsuruoka Y, Tateishi Y, Kim J, Ohta T, McNaught J, et al. (2005) Developing a robust part-ofspeech tagger for biomedical text. In: Bozanis P, Houstis EN, editors. volume 3746, chapter 36. Berlin Heidelberg: Advances in Informatics, Springer. pp. 382–392. doi:http://dx.doi.org/10.1007/11573036.

8.   Steinbeck C, Hoppe C, Kuhn S, Floris M, Guha R, et al. (2006) Recent developments of thechemistry development kit (cdk) - an open-source java library for chemo- and bioinformatics. Current Pharmaceutical Design. pp. 2111–2120.

9.   Corbett P, Murray-Rust P (2006) High-throughput identification of chemistry in life science texts.In: Computational Life Sciences II.107–118. doi: http://dx.doi.org/10.1007/11875741_11.

10. Wren J (2006) A scalable machine learning approach to recognize chemical names within large textdatabases. BMC Bioinformatics 7: S3.

11. Florian B, Juan MTM, El-Bèze M (2008) Mixing statistical and symbolic approaches for chemicalnames recognition. In: CICLing. pp. 334–343.

12. Klinger R, Kolarik C, Fluck J, Hofmann-Apitius M, Friedrich C (2008) Detection of IUPAC and IUPAC-like Chemical Names. Bioinformatics 24: i268–276.

13. Jiao D, Wild DJJ (2009) Extraction of cyp chemical interactions from biomedical literature using natural language processing methods. Journal of chemical information and modeling 49: 263–269.

14. Hettne K, Stierum R, Schuemie M, Hendriksen P, Schijvenaars B, et al. (2009) A dictionary to identify small molecules and drugs in free text. Bioinformatics 25: 2983–2991.

15. Hettne K, Williams A, van Mulligen E, Kleinjans J, Tkachenko V, et al. (2010) Automatic vs. manual curation of a multi-source chemical dictionary: The impact on text mining. Journal of Cheminformatics 2: 4.

16. Kolarik C, Klinger R, Friedrich C, Hofmann-Apitius M, Fluck J (2008) Chemical names: Terminological resources and corpora annotation. In: Workshop on Building and evaluating resources for biomedical text mining, 6th edition LREC.

17. Klinger R, Friedrich C, Hofmann-Apitius M, Fluck J, Birlinghoven S (2009) Chemical names: Terminological resources and corpora annotation.

18. Muller B, Klinger R, Gurulingappa H, Mevissen H, Hofmann-Apitius M, et al. (2010) Abstractsversus full texts and patents: A quantitative analysis of biomedical entities. Advances in Multidisciplinary Retrieval. volume 6107, chapter 12. Berlin, Heidelberg: Springer Berlin Heidelberg,. pp. 152–165. doi: http://dx.doi.org/10.1007/978-3-642-13084-7_12.

19. Rupp CJ, Copestake A, Corbett P, Waldron B (2007) Integrating general-purpose and domainspecific components in the analysis of scientific text.

20. Hassan M, Brown R, Varma-O'brien S, Rogers D (2006) Cheminformatics analysis and learning in a data pipelining environment. Molecular diversity 10: 283–299.

21. Tiwari A, Sekhar A (2008) Workflow based framework for life science informatics. Computational Biology and Chemistry 31: 306–319.

22. Shon J, Ohkawa H, Hammer J (2008) Scientific workflows as productivity tools for drug discovery. Current opinion in drug discovery and development 11: 381–388.

23. Kuhn T, Willighagen E, Zielesny1 A, Steinbeck C (2010) Cdk-taverna: An open workflow environment for cheminformatics. Bioinformatics 11:

24. Kano Y, Baumgartner W, McCrohon L, Ananiadou S, Cohen K, et al. (2009) U-compare: Share and compare text mining tools with uima. Bioinformatics 25: 1997–1998.

25. Ferrucci D, Lally A, Gruhl D, Epstein E, Schor M, et al. (2006) Towards an interoperabilitystandard for text and multi-modal analytics. Technical report, IBM.

26. Apache (2010) Unstructured information management. URL http://uima. apache.org/. [last accessed on 02-May-2011].

27. Tsuruoka Y, Tsujii J, Ananiadou S (2009) Fast full parsing by linear-chain conditional random fields. In: EACL '09: Proceedings of the 12th Conference of the European Chapter of the Association for Computational Linguistics. Morristown, NJ, USA: Association for Computational Linguistics. pp. 790–798.

28. OpenNLP (2010) Opennlp. URL http://incubator.apache.org/opennlp/. [last accessed on 02-May-2011].

29. Taylor I, Deelman E, Gannon D, Shields ME (2007) Workflows for e-Science: Scientific Workflows for Grids. Springer.

30. Kuhn T, Zielesny A, Steinbeck C (2009) Creating chemo- and bioinformatics workflows, further developments within the cdk-taverna project. Chemistry Central Journal 3: 42.

31. Oinn T, Addis M, Ferris J, Marvin D, Senger M, et al. (2004) Taverna: A tool for the composition and enactment of bioinformatics workflows. Bioinformatics 20: 3045–3054.

32. Ludascher B, Altintas I, Berkley C, Higgins D, Jaeger-Frank E, et al. (2006) Scientific workflow management and the kepler system. Special Issue: Workflow in Grid Systems Concurrency and Computation: Practice & Experience 18: 1039–1065.

33. Accelrys' (2010) Pipeline pilot. URL http://accelrys.com/products/pipeline-pilot/. [last accessed on 02-May-2011].

34. Vellay SG, Latimer ME, Paillard G (2009) Interactive text mining with pipeline pilot: A bibliographic web-based tool for pubmed. Infectious disorders drug targets 9: 366–374.

35. Mellebeek B, Rodriguez-Penagos C, Furlong LI (2009) Uima in the biocuration workflow: A coherent framework for cooperation between biologists and computational linguists. Nature Precedings.

36. Kano Y, N N, R S, Fukamachi K, Kazuhiro Y, et al. (2008) Sharable type system design for tool inter-operability and combinatorial comparison. In: Proceedings of the First International Conference on Global Interoperability for Language Resources (ICGL). Hong Kong, 122-129:

37. Kano Y, Dobson P, Nakanishi JM, Tsujii, Ananiadou S (2010) Text mining meets workflow: Linking u-compare with taverna. Bioinformatics.

38. Townsend JA, Downing J, Murray-Rust P (2009) Chic - converting hamburgers into cows. In: Proceedings of the 2009 Fifth IEEE International Conference on e-Science. Washington, DC, USA: IEEE Computer Society, E-SCIENCE '09. pp. 337–343. doi:http://dx.doi.org/10.1109/e-Science http://dx.doi.org/10.1109/e-Science.2009.54.

39. Wilbur WJ, Hazard GF, Divita G, Mork JG, Aronson AR, et al. (1999) Analysis of biomedical text for chemical names: A comparison of three methods. AMIA Symposium. pp. 176–180.

40. Rupp CJ, Copestake A, Teufel S, Waldron B (2006) Flexible interfaces in the application of language technology to an escience corpus.

41. Corbett P, Copestake A (2008) Cascaded classifiers for confidence-based chemical named entity recognition. BMC Bioinformatics 9: S4.

42. Copestake A, Teufel S, Murray-Rust P, Parker A (2010) Extracting the science from scientific publications. URL http://www.cl.cam.ac.uk/research/nl/sciborg/www/. [last accessed on 02-May-2011].

43. NIH (2010) Pubmed. URL http://www.ncbi.nlm.nih.gov/pubmed/. [last accessed on 02-May-2011].

44. EBI (2010) Chemical entities of biological interest (chebi). URLhttp://www.ebi.ac.uk/chebi/. [last accessed on 02-May-2011].

45. .Kano Y, McCrohon L, Ananiadou S, Tsujii J (2009) Integrated nlp evaluation system for pluggable evaluation metrics with extensive interoperable toolkit. In: SETQA-NLP '09: Proceedings of the Workshop on Software Engineering, Testing, and Quality Assurance for Natural Language Processing. Morristown, NJ: Association for Computational Linguistics. pp. 22–30.

46. Medline (2010) Medline. URLhttp://www.nlm.nih.gov/databases/databases_medline.html. [last accessed on 02-May-2011].

47. Nobata C, Sasaki Y, Okazaki N, Rupp CJ, Tsujii J, et al. (2009) Semantic search on digital document repositories based on text mining results.

Chapter 5

DEFINED NANOSCALE CHEMISTRY INFLUENCES DELIVERY OF PEPTIDO-TOXINS FOR CANCER THERAPY

Santosh K. Misra, Mao Ye, Sumin Kim, Dipanjan Pan

Department of Bioengineering, University of Illinois at Urbana-Champaign, Urbana, IL, 61801, United States of America, Beckman Institute of Advanced Science and Technology, University of Illinois at UrbanaChampaign, Urbana, IL, 61801, United States of America, Department of Materials Science and Engineering University of Illinois at Urbana-Champaign, Urbana, IL, 61801, United States of America, Carle Foundation Hospital, Urbana, IL, 61801, United States of America

ABSTRACT

We present an *in-silico-to-in-vitro* approach to develop well-defined, self-assembled, rigid-cored polymeric (Polybee) nano-architecture for controlled delivery of a key component of bee venom, melittin. A competitive formulation with lipid-encapsulated (Lipobee) rigid cored micelle is also synthesized. In a series of sequential experiments, we show how nanoscale chemistry influences the delivery of venom toxins for cancer regression and help evade systemic disintegrity and cellular noxiousness. A relatively weaker association of melittin in the case of lipid-based nanoparticles is compared to the polymeric particles revealed by energy minimization and docking studies, which are supported by biophysical studies. For the first time, the authors' experiment results indicate that melittin can play a significant role in DNA association-dissociation processes, which may be a plausible route for their anticancer activity.

INTRODUCTION

Host defense peptides (HDPs) are a class of evolutionarily conserved substances of the innate immune response that are recognized as chief players in the defense system found among all classes of life. They are usually amphipathic, have a net positive charge (generally +2 to +9) and are short in sequence (10–100 aa); furthermore, HDPs have recently been explored for their anticancer property [1–4]. This class of peptides features many characteristics ideal for anticancer treatment applications, such as i) high water solubility, ii) a broad spectrum of cytotoxicity, and iii) the ability to overcome multidrug resistance, which has developed in cancer cells treated with conventional chemotherapy drugs [5]. Several biophysical studies have shown that small proteins or peptides (20–40 amino acid residues) can penetrate the cell membranes of microorganisms. Melittin, a cationic amphipathic peptide made up of 26 amino acid (aa) residues, has been found to be a potent component of bee venom *Apis mellifera* [6]. It has been proven to have a direct cytotoxic effect on a wide range of cancer cell lines *in vitro*. It has been reported that melittin inhibited cell growth in two ovarian cancer cells via induction of death receptors and down regulation of JAK2/STAT3 [7, 8]. It exerts its toxic activity by disrupting plasma membranes following pore formation. Cationic aa residues of melittin interact directly with anionic cellular membranes via electrostatic interactions and hydrophobic regions; this interaction is responsible for membrane permeation and disruption [6]. A comparably short protein, with an end-to-end distance of ~3.5 nm, the dimension perfectly serves as a single transmembrane-spanning alpha-helix. Numerous computational studies have demonstrated that melittin forms transmembrane pores from its interaction with lecithin PC membranes (2~3 nm in diameter). Its potent activity has attracted researchers to utilize melittin for the next generation anticancer therapeutic agent. However, the therapeutic potential has not been fully achieved in clinic due to their off-target toxicity, rapid degradation and clearance *in vivo*. Melittin has been incorporated into lipid coated perfluorocarbon particles to accumulate in multiple tumor targets, dramatically reducing tumor growth [7].

Although a few of these approaches clearly promise impending success in preclinical studies, their translational potential has not been fully realized. None or very little information can be found in the literature regarding their translational use in human studies. To improve selectivity and reduce toxicity, delivery vehicle implementation in human subjects will require great care. It is, therefore, imperative that we emphasize the fundamental chemical strategy and rationally approach the design of the vehicle suited for translational use. A better understanding of the interaction of venom toxins at the nanoscale is critical, which may dictate its overall stability, systemic integrity and cellular noxiousness. A carefully structured study to comprehend the interactions of melittin with the functional components at the shell and shell-surface will drive the design of next-generation delivery vehicles. Towards this end, we have adopted an in-silico-to-in-vitro approach and developed a well-defined nanoparticulate system for controlled delivery of melittin. The goal of this work was to provide a rational nanoparticle-based design for venom delivery through computational studies and support our theoretical findings with physico-chemical and biological studies. Thus, following the syntheses and physico-chemical characterization, a series of sequential experiments were carried out to study how nanoscale chemistry influences the delivery of venom toxins for cancer regression and help evade systemic disintegrity and cellular noxiousness (Fig 1).

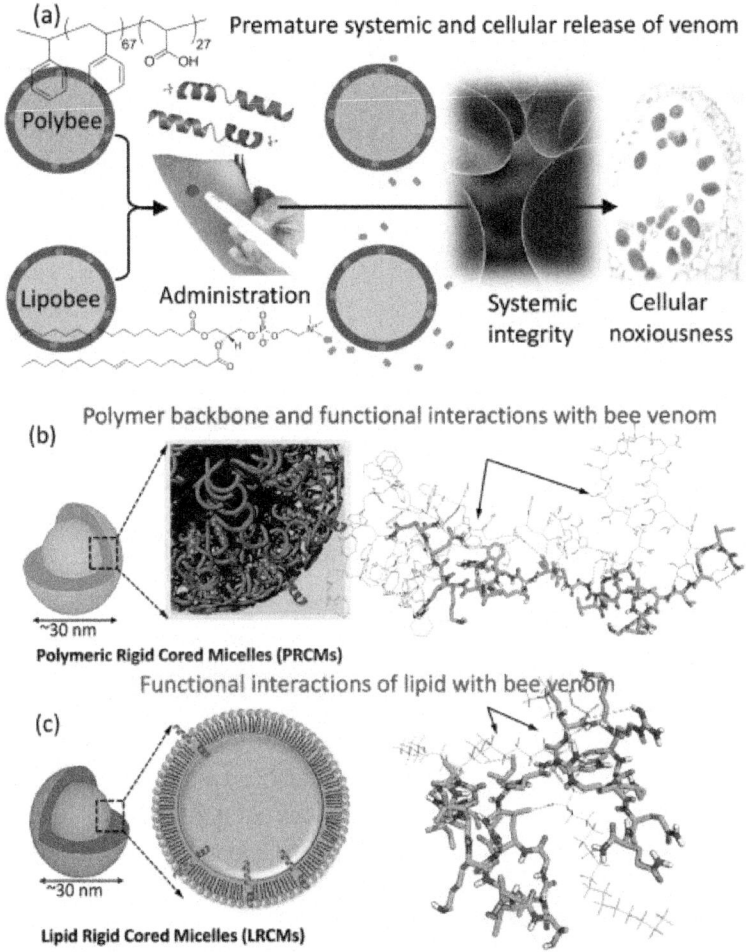

Figure 1: Graphical representation of peptidotoxin delivery.

(a) Schematic of administrative protocol for Lipobee and Polybee; (b) Initial docking images of the PS_{67}-b-PAA_{27} and melittin systems showing how structures of melittin are bound to single amphiphilic polymers and (c) showing the docking structure of melittin and lecithin PC. PS_{67}-b-PAA_{27} and lecithin are depicted by white lines with explicit oxygen atoms depicted in red. The melittin peptide is shown in a green chain link style with oxygen atoms depicted as red and nitrogen depicted as blue.

In silico studies revealed the higher stability response of melittin towards amphiphilic block polymers compared to lipid molecules. Experimental study confirmed the better stability of polymeric system over lipidic assembly. To introduce micellar stability, a concept of rigid core was introduced [9]. Studies exploring change in hydrated size and inertness against serum proteins revealed the higher stability of rigid core particles. Experiments on melittin leaching in the presence of serum concentration revealed the higher stability of a melittin-polymer system (Polybee) compared to a melittin-lipid (Lipobee) system. An in silico study on melittin-DNA interaction was performed and verified by experimental data. It was found that free melittin could bring significant change in inter-helix hydrogen bonding to potentially influence cell growth mechanisms. Melittin in its protected form as Polybee and Lipobee were inactive. Significant changes in the hydrated size of Polybee and Lipobee upon incubation with sodium dodecyl sulfate was observed but not a lower pH. This pointed to the anionic membrane interaction as the responsible factor inside the cytoplasm as a plausible melittin release mechanism. Breast cancer cells of a different estrogen receptor status were used as model *in vitro* cancers for growth inhibitions studies. Irrespective of the cell line, Polybees were found to be better anti-cancer formulations compared to Lipobee and free melittin control.

Results and Discussion

To design a safer as well as efficacious delivery system, we pursued a rigid core nanosystem that can potentially retain their integrity in blood circulation following systemic administration [10a-c]. At the nanoscale level, rigid core micellar (RCM) systems can either be stabilized by amphiphilic PS_{67}-*b*-PAA_{27} (polystyrene-b-polyacrylic acid) (PRCM) or by phospholipids (lecithin PC) (LRCM) encapsulation (Fig 1A). We anticipate that this system will provide model architectures, since the majority of nanomedicine platforms are dominated by lipid and polymeric systems. Furthermore, this strategy can also be extended to a series of peptide-toxins of different natural sources, chemistries and sizes.

To investigate the key interactions in the melittin- PS_{67}-*b*-PAA_{27} polymer and melittin-lecithin PC lipid systems, molecular docking simulations were performed and analyzed (Fig 1B and 1C). All molecules were minimized (Sybyl-X 2.0) [11] before docking with MOE 2013.08. To investigate the key interactions in the melittin- PS_{67}-*b*-PAA_{27} polymer and melittin-lecithin PC lipid system, MOE-Dock [12] was employed to dock the minimized melittin to both PS_{67}-b-PAA_{27} polymer and lecithin PC lipid systems. The five best docking poses with highest S scores (lowest docking energy) were retained and listed in tabular form. Superimpositions of the five best docking poses in both systems are showed in Fig 2. From Fig 2, it can be found that in melittin-PS_{67-b}-PAA_{27} polymer docking structure, the five docking poses are in great diversity, which results in the great difference in the docking score between pose 1 and pose 5 (14.5 kcal/mol). However, in melittin-lecithin PC lipid docking structure, five docking poses can superimpose much better, which resulted in a small docking score difference between pose1 and pose 5 (1 kcal/mol). The conformation differences of docking poses of melittin to PS_{67}-b-PAA_{27} polymer and lecithin PC lipid was caused by the different electric fields and steric fields for these two systems.

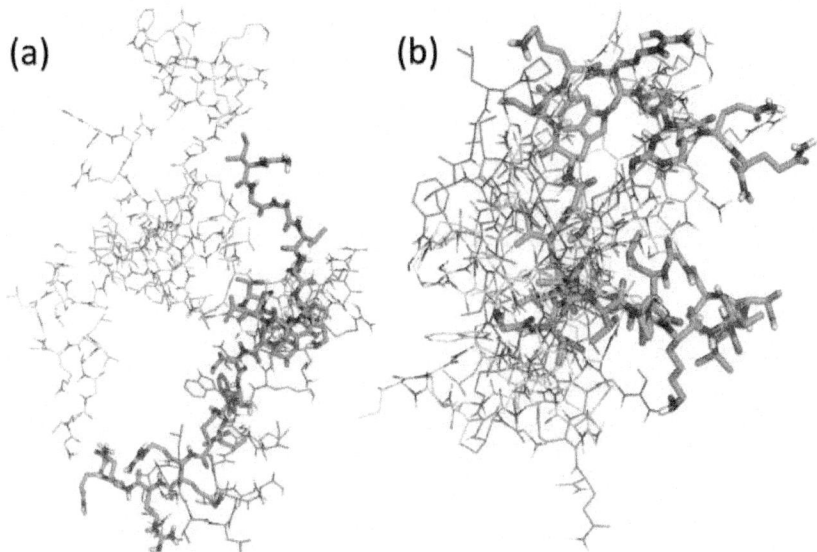

Figure 2: Molecular docking studies.

Super-imposition of the five best docking poses of melittin with lecithin PC and PS_{67}-b-PAA_{27} polymer: (a) Docking poses of melittin to PS_{67}-b-PAA_{27} polymer; (b) docking poses of melittin to lecithin PC. The best scored pose is in the green linked chains with the following smaller attachments listed in order according to their score as indicated by their color: Magenta, 2nd; yellow, 3rd; white, 4th, and cyan, 5th.

Fig 1**A** shows the docking structures of best poses of melittin to PS_{67}-b-PAA_{27} polymer (Fig 1B) and lecithin PC lipid system (Fig 1C). In Fig 1B, it can be seen that melittin peptide stays close and parallels well with the hydrophilic end and a part of hydrophobic section of the polymer. The hydrophilic acrylic acid residues of the polymer formed critical hydrogen bond interactions with amino and hydroxyl groups of aa residues. Hydrophobic phenyl moiety near the hydrophilic terminus of the polymer formed hydrophobic interactions with side chains of aa residues of melittin. Gly1 to Ile17 lie in the hydrophilic end whereas Ser18 to Gln26 lie in the hydrophobic end of the polymer. In detail, the amino groups of the backbone of Gly1, Gly3, Ala4, Val5, Leu6, Gly12, Ala15 and Ile17, the oxygen both in side chain and backbone of Thr11 formed hydrogen bond interactions with oxygen of carboxylic acid of the polymer. The side chains of Leu17, Trp19, Lys21, Arg24 and Gln26 form hydrophobic interactions with phenyl and carbon atoms of the backbone of the polymer. To compare the key interactions of the docking poses with lecithin PC-peptide model, we docked the melittin peptide to the lipid lecithin PC (Fig 1C). Though melittin was found to be well intertwined with the lipid, this peptide-lipid docking structure was much unfastened. The distance between the two molecules was not close enough; therefore, not many hydrogen bond interactions and hydrophobic interactions exist as observed for peptide-polymer complex. The only interactions identified were the amino groups in Arg22 and Arg24 forming hydrogen bond interactions with oxygen in keto and phosphono groups in lecithin PC lipid. The side chains of Ala4, Leu13 and Ile17 formed hydrophobic interactions with the alkyl group in the lipid.

To further explore the sequence and size dependence of peptides for Polybee and Lipobee carriers, we chose several melittin peptidic fragments for computational modeling studies (S1 Fig). We selected (1) right 17-residues peptide, (2) left 9-residues peptide, and (3) middle 16-residues peptide to dock to PS_{67}-b-PAA_{27} polymer and lecithin PC lipid system, respectively. In the docking structure of right 17-residues peptide with PS_{67}-b-PAA_{27} polymer, the amino groups of the backbone of Thr1, the side chain of Lys12, Arg13 and Arg15, the oxygen carbonyl group of Lys12, Gln17 and side chain of hydroxyl group of Thr1, Thr2 and Ser10 formed hydrogen bond interactions with oxygen of carboxylic acid of the polymer. In the lecithin PC lipid docking structure, the amino groups of the backbone of Lys14 and side chain of Arg13, and side chain of hydroxyl group of Thr1 formed hydrogen bond interactions with oxygen in phosphono groups in lecithin PC lipid. In the docking structure of left 9-residues peptide with PS_{67}-b-PAA_{27} polymer, the amino groups of the backbone of Gly1, Ile2, Ala4 and Leu9, the oxygen of carbonyl group of Val8 formed hydrogen bond interactions with oxygen of carboxylic acid of the polymer. In the lecithin PC lipid docking structure, the amino groups of the backbone of Gly1, Ile2 and the oxygen of carbonyl group of Ile2 formed hydrogen bond interactions with oxygen in phosphono groups in lecithin PC lipid. In the docking structure of the middle 16-residues peptide with PS_{67}-b-PAA_{27} polymer, the amino groups of the backbone of Thr5, Gly7, Lys16, the side chain of Trp14, the oxygen of carbonyl group of Gly7 and side chain of the hydroxyl group of Ser13 formed hydrogen bond interactions with oxygen of carboxylic acid of the polymer. In the lecithin PC lipid docking structure, the amino groups of the backbone of Leu1, Lys2, Val3, Thr5 and the side chain of hydroxyl group of Thr5 formed hydrogen bond interactions with oxygen in phosphono groups in lecithin PC lipid.

The analysis of docking poses can clearly explain why the peptide-polymer structure is more stable than the peptide-lipid structure (Table 1). We noticed that on the contrary to the docking poses and interactions analysis, the docking energy in lecithin PC lipid system is lower than PS_{67}-b-PAA_{27} polymer system. This might be caused by differences in average entropy loss/gain due to the conformational flexibility and desolvation energy of each atom rather than that of maximum energy of H-bond between PS_{67}-b-PAA_{27} polymer and lecithin PC lipid whose sizes are greatly different. In the PS_{67}-b-PAA_{27} polymer system, the hydrophilic acrylic acid and hydrophobic styrene unit ratio will affect the hydrogen bond and hydrophobic interactions between melittin and polymer systems, whereas the unit length does not change these interactions greatly. S2 Table shows the scores of docking results of three melittin fragments to polybee and lipobee. As evident, the longer the peptide sequences, the better the docking scores were obtained. Docking scores were better for peptide with PS_{67}-b-PAA_{27} polymer than with lethicin PC lipid because more hydrogen bond interactions are involved in a PS_{67}-b-PAA_{27} polymer system.

Table 1: Scores of different docking poses of melittin superimposed and energy minimized with lipid or amphiphilic polymer system.

Docking Pose	Melittin docked to polymer (Kcal/mol)	Melittin docked to lecithin lipid (Kcal/mol)
Pose 1	-5.07	-8.04
Pose 2	-3.55	-7.51
Pose 3	+2.37	-7.47
Pose 4	+2.81	-7.31
Pose 5	+9.56	-7.14

doi:10.1371/journal.pone.0125908.t001

A post-preparative one-pot insertion method was used for stable entrapment of melittin and a generation of lipidated rigid core micellar melittin (Lipobees) from lipidated rigid core micelles (LRCMs). Similarly, polymerized rigid core micellar melittin (Polybees) were prepared from polymerized rigid core micelles (PRCMs). A typical preparation of LRCMs and PRCMs involved a preparation of 'rigid core' of polyoxyethylene20 cetyl ether (PECE) followed by stable coating with lecithin PC or PS_{67}-b-PAA_{27} [13]. The stability of the micelles was achieved by curing the core at 4°C (PECE mp: 32°C). RCMs were then subjected to post-preparative incubation of the melittin in aqueous suspension for 30 min at ambient temperature with mild vortexing. The hydrodynamic diameter, morphology, layered arrangements; topography, electrophoretic potential, and particle stability were established using various physico-chemical experiments. To find out the loading of melittin in LRCM and PRCM, UV-absorbance spectroscopy was performed. It was seen that signature absorbance for melittin at 290 nm dropped down in the case of Lipobee and Polybee, most likely due to the surface internalization of melittin in LRCM and PRCM with post-interaction methodology.

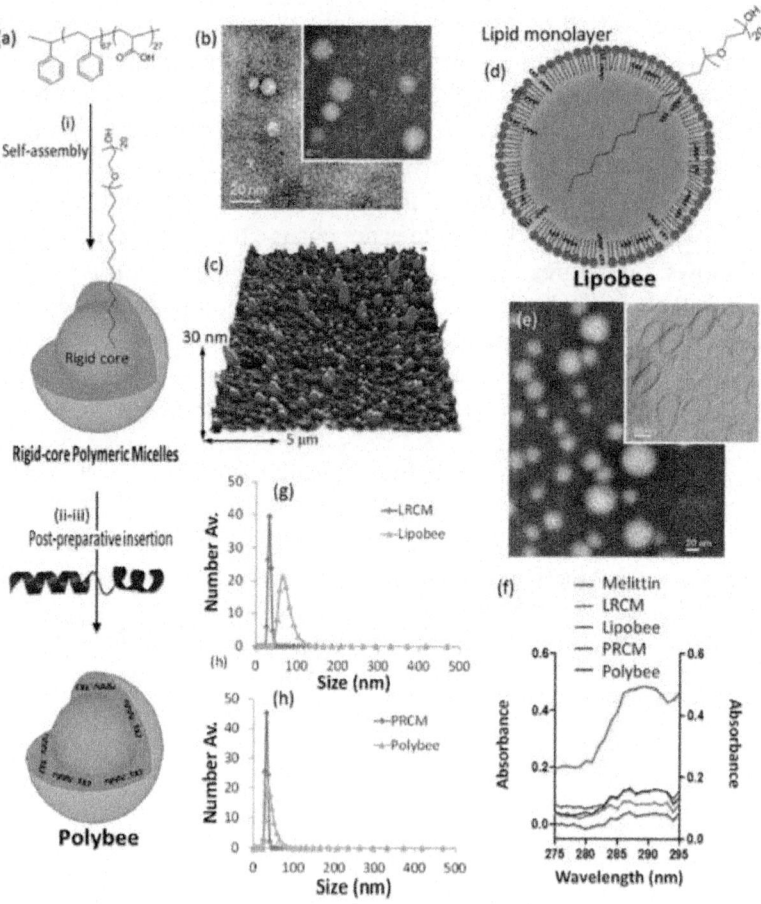

Figure 3: Preparation and physico-chemical characterization studies.

The LRCM had an average hydrodynamic particle size of 23 ± 2 nm, which grew to 83 ± 3 nm in Lipobee primarily due to the surface interaction of melittin with RCM (Fig 3). Similarly, PRCM showed an average hydrodynamic diameter of 25 ± 5 which increased to 40 ± 8 nm in Polybee (Fig 3). Stability of these PRCM particles across various time points at rt and pH 7.4 was measured using DLS measurements, which showed a nominal change in the size of PRCM and Polybee of less than 10%. Similarly, LRCM size did not change to any significant level while Lipobee showed size changes of ~40%, emphasizing the significant instability of Lipobees compare to the high stability of Polybee. Stability of carrier vehicles has always been major concern in the success of nano-delivery protocols. Hence, Polybee promises the probable better melittin delivery response compare to Lipobee particles during *in vitro* and *in vivo* uses. The surface charge density for PRCMs was -12 ± 1 mV, which dropped down to -6 ± 1 mV in Polybee after incubation with the bee toxins.

Synthesis and characterization of rigid core micelles and melittin loaded particles: (a) Synthesis of PRCM and Polybee nanoparticles; (b) representative TEM images of Polybee; (c) representative AFM images of Polybee; (d) Synthesis of LRCM and Lipobee nanoparticles; (e) representative TEM images of Lipobee; (c) representative AFM images of Lipobee; (f) UV-vis spectroscopy of melittin, LRCM, PRCM, Lipobee and Polybee; (g) hydrodynamic diameter distribution (number averaged, nm). TEM samples (20 μL) were prepared on formvar-coated carbon grids and negatively stained with uranyl acetate and vacuum dried before performing the microscopy. Samples (20 μL) were drop casted on freshly cleaved mica sheets and air dried for >24h before performing the tapping mode AFM.

On the other hand, zeta potential of LRCMs showed a nominal change in zeta potential when converted to Lipobee. This signifies the efficiency of making Coulombic interactions of the peptide with the outer corona of block polymers comprised of poly(acrylic acid) residues. Anhydrous state morphology of the Lipobee and Polybee particles was obtained at 25 ± 5 nm size compared to 22 ± 6 nm for LRCM and PRCM as studied by transmission electron microscopy (TEM, Fig 3F). The representative atomic force microscopy (AFM) images were acquired from drop casted samples on mica sheets to study the morphology pattern of these RCM particles. Average height values (H_{av}) of a representative sample were 25 ± 5 nm (Fig 3G). Physico-chemical characterizations of Polybees and Lipobees suggest a potential over-edge for Polybees over Lipobees in formulation stability and other prerequisites to make them better agents for systemic application (Table 2).

Table 2: Hydrodynamic diameter distribution, anhydrous state particle size, particle height and electrophoretic potential distribution of PRCM, Polybee and LRCM and Lipobee in tabular form

Nanoparticle	D_{av}/DLS (nm)	D_{ah}/TEM (nm)	H_{av}/AFM (nm)	ζ/Zeta (mV)
LRCM	23±2	35±5	20±6	-12±1
Lipobee	83±3	25±5	25±7	-10±1
PRCM	25±5	25±7	25±5	-12±1
Polybee	40±8	22±6	26±8	-06±3

doi:10.1371/journal.pone.0125908.t002

To verify the cancer cell regression affinity of these formulations, cytotoxicity assays were performed. As a model system for *in vitro* cancer culture, we chose estrogen positive (MCF-7) and estrogen negative breast cancer cells (MD-MB231) to evaluate the functional therapeutic potential of Polybees and Lipobees. MCF-7 and MD-MB231 cell lines represent early-stage and invasive human breast cancer cell lines, respectively. Irrespective of the cell line, Polybee showed significantly higher efficacy in comparison to Lipobee and melittin as evident from MTT assays (Fig 4). At the 48h incubation point, in MD-MB231 cells, the IC50 value for Polybee has been found at ca. 40 ± 4 nM compared to ca. 70 ± 7 nM in case of Lipobee and ca. 110 ± 10 M for free melittin; moreover, in MCF-7, IC50 value for Polybee was found to be ca. 80 ± 8 nM compared to ca. 100 ± 10 nM in the case of Lipobee and 105 ± 10 nM for free melittin. Meanwhile, LRCM and PRCM showed IC50 >> 1000 nM irrespective of the cell line, (Fig 4G). For the cells treated with free melittin (100 nM), cell growth density and morphological changes were indicative of cell death, whereas LRCM and PRCM did not alter to any significant level (Fig 4).

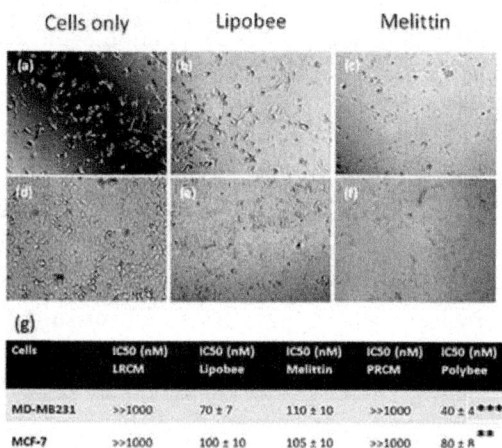

Cells	IC50 (nM) LRCM	IC50 (nM) Lipobee	IC50 (nM) Melittin	IC50 (nM) PRCM	IC50 (nM) Polybee
MD-MB231	>>1000	70 ± 7	110 ± 10	>>1000	40 ± 4***
MCF-7	>>1000	100 ± 10	105 ± 10	>>1000	80 ± 8**

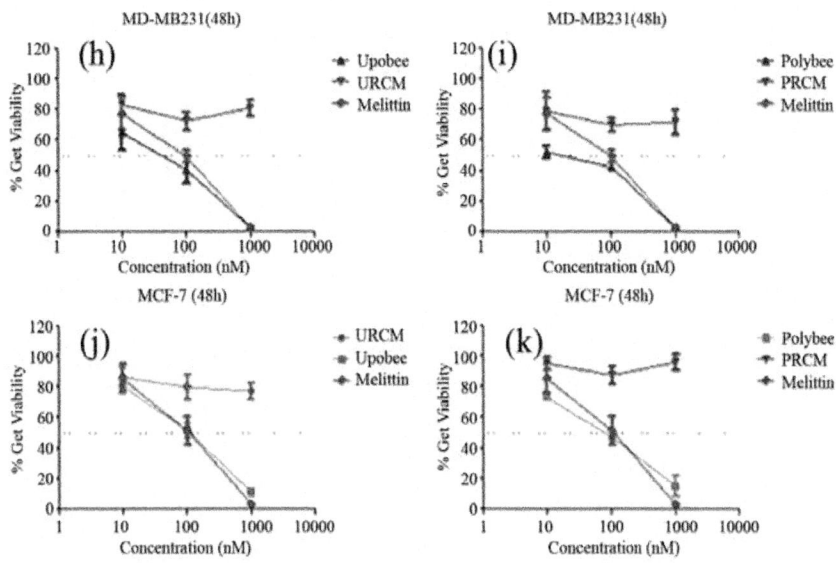

Figure 4: Functional characterization in vitro.

Representative bright field images of cell growth density and cancer cell morphology variation for MCF-7 (a-c) and MD-MB231 (d-f) after 48 h of incubation treated with melittin, LRCM and Lipobee, (g) IC50 values for various formulations in tabular form; biostatistical analysis on IC50 values for Polybee respect to melittin representing *** for p value < 0.001 and ** for p value < 0.005 after ONE way ANOVA with Bonferroni post test and (h-i) % cell viability variations by different formulation in MD-MB231 and (j-k) MCF-7 cells for (h-j) polymeric and (i-k) lipidic formulations.

The CH50 values for all the used formulations were found to be 6 ± 1 for PRCM, LRCM, Polybee, Lipobee and melittin and 8 ± 1 and 3 ± 1 for Reference 1 and Reference 2, respectively (Fig 5A). Although, *in vitro* experiments established Polybee and Lipobee as potent anti-cancer formulations, their presumed behavior for *in vivo* applications still remained unclear.*In vivo* success of such formulations very much depend on two major factors i) neutrality toward blood complement and ii) sustainable passaging of payload through systemic circulation. Our computational studies are indicative of stronger, tighter interaction of melittin with the amphiphilic polymer chains in direct comparison with the lipid, making Polybees; hence, melittin with its rigid core and polymeric shell is a better candidate for *in vivo* application. To confirm this observation experimentally, we explored the complement activation and melittin sustaining ability of these formulations in blood serum.

Complementing this system, a group of proteins will activate to lead target cell lysis and facilitate phagocytosis through opsonisation on exposure to solid foreign materials in circulatory systemic fluid. The CH50 assay, which screens the activation of classical complementary pathways found to be sensitive to the reduction, absence and/or inactivity of any component of the pathway. The complementary CH50 assay is based on lysis of sensibilized sheep erythrocytes in the presence of Ca^{2+} and Mg^{2+}. When sensibilized sheep erythrocytes are incubated with test serum of different treatments, different levels of haemolysis are achieved. CH50 complement activation assay was performed for all the formulations used here and found to be very inert in activation of complementary proteins, showing no significant change in CH50 values compare to normal complementary level plasma (Reference 1, i.e., 8 ± 1).

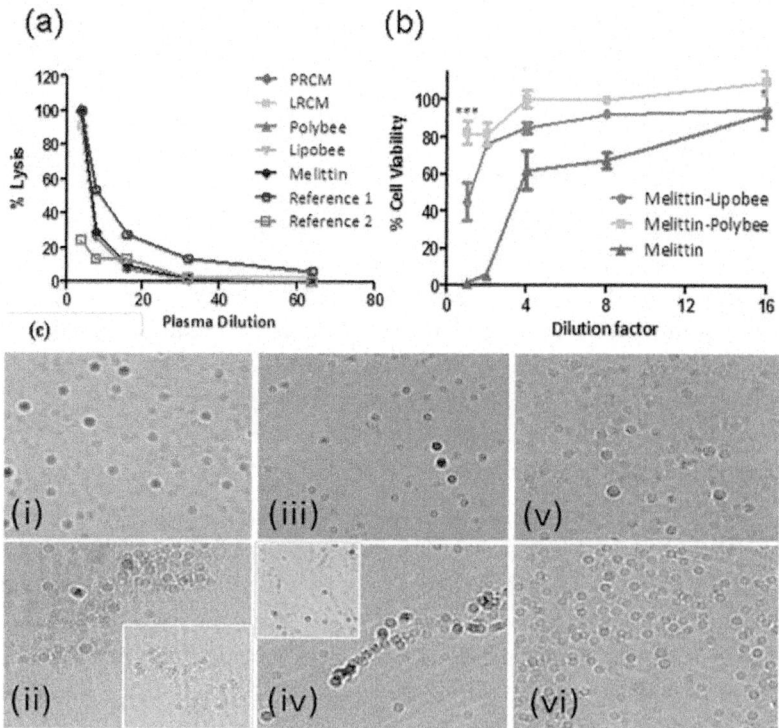

Figure 5: Release mechanism, systemic toxicity and stability studies.

(a) Complementing activation and (b) melittin leaching behavior of Lipobee and Polybees. Free melittin, LRCM and PRCM were used as controls; (c) optical microscopy images of blood smear untreated (i) and treated with melittin (1:10) (ii), LRCM (1:10) (iii), Lipobee (1:10) (iv), PRCM (1:10) (v) and polybee (1:10) (vi), respectively, (with 20x magnification). Melittin- and Lipobee-treated pig blood in the severely clumped, morphologically distorted state are shown in (ii) and (iv). Insets in (ii) and (iv) show red blood cell morphology to emphasize other similar morphological patterns throughout the sample.

The CH50 values for all the used formulations were found to be 6 ± 1 for PRCM, LRCM, Polybee, Lipobee and melittin and 8 ± 1 and 3 ± 1 for Reference 1 and Reference 2, respectively. It indicates that formulation PRCM, LRCM, Polybee, Lipobee and melittin did not induce any complement to any significant level. It supports the feasibility of using these formulations *in vivo* without risk of inducing immune response.

To further assess the benefits of using Polybee over Lipobee for systemic delivery, melittin leaching characteristic of these formulations were evaluated and estimated by performing an MTT assay on MD-MB231 cells. Leached melittin obtained after incubating Polybee and Lipobee with 10% fetal bovine serum (FBS) in the DMEM buffer was used at dilutions 1, 2, 4, 8 and 16 for MTT assays. A known melittin concentration was used as a positive control ranging from 20–1.25 µM. MTT assays exhibited a high amount of melittin leaching from Lipobee causing a high percentage of cell deaths at each dilution compared to cells treated with melittin leaching from Polybee which gives a highly significant bio-statistical significance ($p < 0.001$) at dilution factor 1. On the other hand, at the same dilution, melittin leached out from Polybee, resulting in a significant decline in cell population death with no significant change in cell viability (Fig 5B). These findings indicated that during systemic administration of Lipobee, a high amount of melittin might leach out in the circulatory fluid before reaching the cancer cells thereby causing a significant loss in anti-cancer efficiency.

Polybee nanoparticles have also exhibited noteworthy vigor when admixed with pig blood. A 'blood smear' preparation was made to identify any morphological variations in lymphocytes and blood clumping monitored by a clinical optical microscopy technique (conventional light microscopy) under a high power field (Fig 5C). No superficial plodding or morphological vicissitudes in blood cells were observed in fresh pig blood treated with PRCM, LRCM and Polybee (blood: NP = 10:1). However, pig blood treated with free melittin and Lipobee exhibited significant clumping and morphological alterations (Fig 5C, ii and iv). To understand the plausible mechanism of melittin release from Lipobee and Polybee *in vitro*, further studies were conducted. Release of therapeutic agents from nanoparticles has been reported as pH responsive and/or interactive with the anionic layer of the endosomal compartment in cellular systems. These factors can investigate the responsiveness of used nanoparticles for reaching a probable release mechanism. We used this strategy to narrow down our selection of preferential pathways for melittin release from Lipobee and Polybee nanoformulations. Lipobee and Polybee were prepared as discussed earlier followed by incubation at pH 4.6 and in the presence of sodium dodecyl sulphate (SDS, 1 mM) for 2 h. At the end of the incubation period, a hydrodynamic diameter was acquired for different formulations through the use of dynamic light scattering. Sizes at 0 h were considered as control for the experiments (S2 Fig).

DLS results clearly show the variation in size of Lipobee and Polybee only in presence of SDS (5 mM) without any significant effect at a lower pH (S2 Fig). It supports the plausible occurrence of anionic interaction in an endosomal compartment responsible for melittin release from Polybee and Lipobee. Additionally, a higher change in size of Lipobee occurred upon incubation with SDS signifying a higher lipid-surfactant interaction.

Peptide-polynucleotide interactions have always attracted the interest of medicinal chemists. It is of special interest to decipher melittin DNA interactions and understand their role in dissociation from the DNA secondary structure. Gel electrophoresis was performed to enable observation of the changes in electrophoretic mobility patterns of pBR322 incubated with various formulations in the presence (melittin, Polybee and Lipobee) and in the absence of melittin (LRCM and PRCM) as well as various concentrations of free melittin (50–0.0005 μM).

It was found that only free melittin was able to dissociate the plasmid DNA. In turn, the loss of the electrophoresis band was accomplished either by retarding the DNA migration or by expelling the intercalated EtBr (Ethidium bromide) due to major groove binding of melittin in the DNA duplex. Formulations with protected melittin in either of the cases, Lipobee and Polybee, did not influence the DNA bands to any significant extent (S3A Fig).

A further gel electrophoresis investigation showed that ~0.05 μM free melittin was sufficient to start the dissociation of a DNA secondary structure (S3B Fig). The interactions of melittin with DNA in free form signify that a release from Lipobee and Polybee can target genomic DNA, which might extend to the level of hindering the transcription process. However, no interaction can take place if melittin is stably incorporated. Here no significant effect from LRCM and PRCM on DNA mobility showed that these particles do not have an active role to play in DNA interaction and melittin is the only component in Polybee and Lipobee to participate in interaction with DNA duplex.

Proteins are major constituents of blood serum and counter the load vehicle and pharmacoactive agents while delivering through systemic circulations.

Any obstruction in normal behavior of such blood serum proteins might lead to harmful consequences to the subject. As a model system of study, we chose fetal bovine serum to establish effects of free melittin and its nanoformulations, Lipobee and Polybees on normal spectroscopic properties. UV absorbance study was performed on 50 μM melittin incubated (free or nanoformulation form) 10% FBS solution. No hypso- or bathochromic shift in UV absorbance pattern was noted from FBS solution but absorbance was increased after incubation with free melittin. A decrease in absorbance was seen in case of Lipobee, which reached a level similar to nontreated FBS when compared to Polybee (S3C Fig). This observation could be explained due to the additional absorbance from added formulations, signifying no specific interaction of melittin in free or nanoformulation form with serum proteins.

The fate of released melittin can be rationalized in regards to its interaction with genomic pool cellular components. One among them could have resulted from an interaction of melittin with DNA. To explore the interactions of melittin with DNA in-silico, we performed docking studies of melittin into double stranded DNA (Fig 6). It has been found that melittin intertwines well and remains within a close proximity of DNA structures, forming 17 H-bond interactions and 3 hydrophobic interactions with phosphate groups and base pairings of DNA. It has also been noticed that nitrogen and oxygen in both backbones and side chains formed an H bond only with oxygen of phosphate groups, not with nitrogen and oxygen in paired bases. However, these strong H-bond interactions changed the conformations of paired bases, making the H-bond distance between paired bases mostly enlarged (See S1 Table). The low p-value (0.000415) implied that the conformation of bases of DNA was statistically different under the interactions of melittin. Side chains of Leu and Ala formed hydrophobic interactions with thymine. Docking energy was also found to be quite low (-12.12 mmol/kcal) due to multi-interactions and stable docked structures.

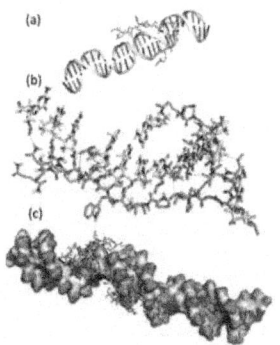

Figure 6: DNA interaction studies.

(a) Illustration of docked structure of melittin with DNA. (b) Key interactions of melittin with DNA. Melittin is shown as green links. (c) Molcad surface picture of docked structure.

doi:10.1371/journal.pone.0125908.g006

We have established that PRCMs are optimized assemblies for melittin, which can improve the inhibition of cancer cell growth based on various hydrogen bonding connections, which originate between the polymer molecule PS_{67}-b-PAA_{27} and melittin. Further studies with various fragments of melittin and its sequences reveal the variation in both docking energies and in the resultant final outcomes of the interactions. These measures could be extrapolated for other peptides too, interacting in totally different ways resulting in different morphology, size, stability and extent of delivering the peptide. These properties can be optimized in many ways whereby final outcomes can show results similar to those discussed here.

Methods

Polyoxyethylene (20) cetyl ether and poly(styrene)-block-poly(acrylic acid) (PS_{67}-b-PAA_{27}) were obtained from Sigma Life Sciences (St. Louis, MO, U.S.A). Tetrahydrofuran (THF) was obtained from Avantor Performance Materials (Center Valley, PA, U.S.A.). Bee venom peptide melittin was obtained from Sigma Aldrich, Inc. (St. Louis, MO, USA). The hydrodynamic diameter was measured on a Malvern Zeta sizer machine equipped with a 633-nm laser. Zeta potential measurement was performed on a Malvern Zeta sizer instrument. Atomic force microscopy was performed on MFP-3D AFM from Asylum Research using Igor Pro software. The TEM images were acquired on JEOL 2100 Cryo TEM machine and imaged by Gatan UltraScan 2kx2k CCD.

Computational Modeling

Mellitin, lethicin PC and and PS_{67}-b-PAA_{27} polymer were built using a SKETCH module in Sybyl-X 2.0. Structural energy minimization was performed with the Tripos force field until a gradient convergence of 0.05 kcal/mol was achieved. The NB Cutoff was set to 8.00. Distance was set to the dielectric function, and the diaelectric constant was set to 1.00. MOE-Dock was used for the docking process with a MOE 2013.08 program. InducedFit was chosen as the docking protocol. The selected active sites were whole target molecules, which are PS_{67}-b-PAA_{27} polymer and Lethicin PC, respectively. Triangle Matcher, which generated poses by aligning ligand triplets of atoms on triplets of alpha spheres, was used as the placement method with default settings.

London dG was chosen as the scoring function. The free energy change upon binding of ligand to a receptor was calculated as Eq.1.

$$\Delta G = c + E_{flex} + \sum_{h-bonds} c_{HB} f_{HB} + \sum_{m-lig} c_M f_M + \sum_{atoms\ i} \Delta D_i \quad (1)$$

where c is average entropy loss/gain due to rotational/translational motion; Eflex is entropy loss due to conformational flexibility; cHB is H-bond maximum energy, and H-bond fHB measures geometric imperfections. cM is metal ligation maximum energy. Metal ligation fM measures geometric imperfections; ΔDi estimates the desolvation energy of each atom where i.5 poses were retained for the docking results analysis. The 1st pose with best score, i.e., lowest S value, was chosen as the docking pose. H-bond distance changes between bases of DNA after docking with melittin. (Fig 7)

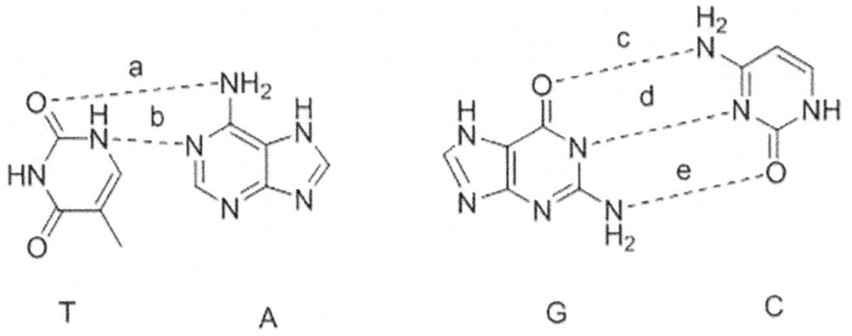

Figure 7:

MATERIALS PREPARATION

Preparation of PRCM

Polyethylene glycol cetyl ether (2 mg) was melted at 65°C for 5 min followed by the dropwise addition of 2 ml of water (approximately 1 drop/sec). The solution was allowed to stir for 20 min at 1150 rpm. Simultaneously, a solution of poly(styrene) 67-block-poly (acrylic acid) 27 (PS$_{67}$-b-PAA$_{27}$, Mn 1,600–1,950 (poly(acrylic acid)), Mn 6,500–7,000 (polystyrene), Mn 8,100–9,100, average Mn 8,700, mp: 192–197°C; Mw/Mn = 1.2) was prepared by adding 2 mg of the amphiphilic polymer and 1 ml of tetrahydrofuran (THF) to a glass vial. After the polyethylene glycol cetyl ether miceller suspension was stirred for 20 min, 250 µl of PS67 PS$_{67}$-b-PAA$_{27}$/THF solution was added drop-wise (approximately 1 drop/10 sec) to the solution. The solution was left for stirring overnight to allow THF evaporation. At the end of the procedure, volume was increased up to 2 ml with autoclaved nanopure water (0.2 µM). The suspension was further allowed to stir for 10 min at room temperature. The hydrodynamic diameter of prepared nanoparticles was measured using a nano series Zetasizer. Finally, the suspension was stored at 4°C overnight for curing the core of the particle and the particle size measurement was repeated. The nano particles were purified by dialysis against nanopure (0.2 μM) water using a 20,000 Da MWCO cellulose membrane for a prolonged period of time and then passed through a 0.45 μm Acrodisc Syringe filter. The nanoparticles were stored under argon atmosphere typically at 4°C in order to prevent any bacterial growth.

$DLS\,(D_{av})/nm\ =\ 25\,\pm\,5\,nm;\ TEM\,(D_{ah})/nm\ =\ 25\,\pm\,7\,nm\ AFM\,(H_{av})/nm$
$=\ 25\,\pm\,5\,nm;\ Zeta\,(\zeta)/mV\ =\ -\,12\,\pm\,1\,mV.$

Preparation of LRCM.

Polyethylene glycol cetyl ether (1 mg) was melted at 65°C for 5 min and 1.33 mg of lecithin PC was added. The mixture was allowed to stir for 20 min at 1150 rpm. Simultaneously, 1 ml of water was warmed up at 60°C and added to the mixture dropwise (approximately 1 drop/sec). The solution was left for stirring for 30 min and subsequently cooled down at room temperature. The hydrodynamic diameter of as-synthesized nanoparticles was measured using a nano series Zetasizer. Finally, the suspension was stored at 4°C overnight to allow the core of the particle to be cured and the particle size measurement was repeated for 5 days. The nanoparticles were purified by dialysis against nanopure (0.2μM) water using a 20,000 Da MWCO cellulose membrane for a prolonged period of time and then passed through a 0.45 μm Acrodisc Syringe filter. The nanoparticles were stored under argon atmosphere typically at 4°C in order to prevent any bacterial growth and characterized by various physico-chemical techniques [14, 15].

$DLS\,(D_{av})/nm\ =\ 23\,\pm\,2\,nm;\ TEM\,(D_{ah})/nm\ =\ 35\,\pm\,5\,nm\ AFM\,(H_{av})/nm$
$=\ 20\,\pm\,6\,nm;\ Zeta\,(\zeta)/mV\ =\ -\,12\,\pm\,1\,mV.$

Preparation of Polybee and Lipobee.

5 mM of Melittin solution was made from a dilution of 20 mM of melittin aqueous solution in 100 mM KCL in nanopure water. 40 μl of 250 μM melittin and 750 μl of PRCMs and LRCMs (as synthesized above) were mixed to make the 5 μM melittin containing Polybee and Lipobee, respectively. The mixed suspension was subsequently vortexed and incubated at room temperature for 30 min before storing at 4°C. The nanoparticles were purified by dialysis against nanopure (0.2 μM) water using a 20,000 Da MWCO cellulose membrane for a

prolonged period of time. Melittin loaded formulations were characterized by various physic-chemical techniques [14, 15].

Polybee : $DLS(D_{av})/nm = 40 \pm 8\,nm$; $TEM\,(D_{ah})/nm = 22 \pm 6\,nm\,AFM\,(H_{av})/nm$
$= 26 \pm 8\,nm$; $Zeta\,(\zeta)/mV = -6 \pm 1\,mV$.

Lipobee : $DLS(D_{av})/nm = 83 \pm 3\,nm$; $TEM\,(D_{ah})/nm = 25 \pm 5\,nm\,AFM\,(H_{av})/nm$
$= 25 \pm 7\,nm$; $Zeta\,(\zeta)/mV = -10 \pm 1\,mV$.

Dynamic light scattering measurements.

Hydrodynamic diameter distribution and distribution averages for Polybee, Lipobee, PRCM and LRCMs in aqueous solutions were determined using a Malvern Zetasizer nano series–Nano ZS90. Scattered light was collected at a fixed angle of 90°. A photomultiplier aperture of 400 mm was used, and the incident laser power was adjusted to obtain a photon counting rate between 200 and 300 kcps. Only measurements for which the measured and calculated baselines of the intensity autocorrelation function agreed to within +0.1% were used to calculate nanoparticle hydrodynamic diameter values. The measurements for the particles were made at 0 h, 24 h, 48 h, 72 h, 96 h, and 120 hr after synthesis to evaluate the stability. All determinations were made in multiples of five consecutive measurements.

Zeta potential measurements.

Zeta potential (ζ) values for Polybee, Lipobee, PRCMs and LRCMs were determined with a nano-series Malvern Zetasizer zeta potential analyzer. Measurements were made following dialysis (MWCO 20 kDa dialysis tubing, Spectrum Laboratories, Rancho Dominguez, CA) of nanoparticle suspensions into water. Data were acquired in the phase analysis light scattering (PALS) mode following solution equilibration at 25°C. Calculation of ζ from the measured nanoparticle electrophoretic mobility (μ) employed the Smoluchowski equation: $\mu = \varepsilon\zeta/\eta$, where ε and η are the dielectric constant and the absolute viscosity of the medium, respectively. Measurements of ζ were reproducible to within ±5 mV of the mean value given by 20 determinations of 10 data accumulations

Atomic force microscopy measurements (AFM)

Atomic force microscopy (AFM) was performed to observe the morphological topography in PRCM and LRCM structures. The samples were drop cast onto freshly cleaved mica sheets and air-dried for 24 h. Topographic imaging of all the formulations was obtained by operating the AFM in a tapping mode with an Asylum Cypher AFM instrument. The average particle height (H_{av}) values and standard deviations were generated from the analyses of a minimum of 50 particles from three micrographs. Analysis of the AFM images was processed using ImageJ.

Transmission electron microscopy measurements (TEM)

The Transmission electron microscopy (TEM) was performed on Polybee, Lipobee, PRCMs and LRCMs to evaluate their morphologies. Imaging was performed on samples prepared on copper grids coated with a formvar plastic and then coated with carbon for stability followed by negative staining with 7% Uranyl acetate.

Human transformed cancer cell culture

MD-MB231 cells (ER (-) breast cancer cells) and MCF-7 cells (ER (+) breast cancer cells) were cultured in Dulbecco's Modified Eagle's Medium (DMEM; Sigma) supplemented with 10% fetal bovine serum (FBS) in T25 culture flasks (Cellstar; Germany) and were incubated at 37°C in a 99% humidified atmosphere containing 5% CO_2. Cells were regularly passaged by trypsinization with 0.1% trypsin (EDTA 0.02%, dextrose 0.05%, and trypsin 0.1%) in DPBS (pH 7.4). Non-synchronized cells were used for all the experiments.

MTT Assay

The cell viability of Polybee, Lipobee, PRCM, LRCM and melittin formulations in used MD-MB231 and MCF-7 cells were investigated by using 3-(4,5-dimethylthiazole-2-yl)-2,5-diphenyltetrazolium bromide (MTT) [16] reduction assay in presence of 10% FBS in antibiotic free media. Experiments were performed in 96 well plates (Cellstar; Germany) growing 8,000 cells per well 24 h before treatments. Experiments were performed for various concentrations of melittin ranging from 10 to 1000 nM present in free or Polybee and Lipobee forms while the same volume of LRCM and PRCM was used as negative controls. Cells were incubated for 48 h before performing the MTT assay. After incubation period, cells were imaged for investigating growth density and morphology variations. Cells were further treated with MTT as 20 µl (5 mg/mL) per well and further incubated for 5 h. At the end of the incubation, the entire medium was removed from wells and 200 µL of DMSO was added to dissolve blue colored formazan crystals. The percentage cell viability was obtained from a plate reader and was calculated using the formula % Viability = {[A630(treated cells)- (background)]/[A630(untreated cells)-background]}x100.

Plausible mechanism of melittin release from Lipobee and Polybee in vitro

To establish the possible pathway of melittin release from Lipobee and Polybee formulations, experiments were performed by incubating the formulations with SDS (5 mM) at pH 4.6 for 2 h before acquiring their hydrodynamic diameter [17, 18]. A significant change in hydrous diameter was achieved by interaction with SDS and by not lowering pH to 4.6, which revealed the repackaging of structural components only in presence of anionic assemblies, during which melittin might get released.

Blood-smear experiment

A single smear was made per slide by putting a drop of fresh pig blood on the slide (near the end). The drop was spread by using another slide ("spreader"), placing the spreader at a 45° angle and backing into the drop of blood [19, 20]. The spreader catches the drop and it spreads by capillary action along its edge. Smear is allowed air dry for 10 min and cover-slip before placed directly on the microscope and observed under 40 x magnifications. A ratio of 1:9 Polybee, Lipobee, PRCM, LRCMs and melittin (10 µl; melittin conc. 50 nM) and pig blood was used for preparation of the smear.

Melittin-DNA interaction and dissociation of primary and secondary structure

The most favorable interactions between amino acids and nucleobases are toward arginine and lysine with guanine and also for lysine with thymine. These preferences could explain partial charge interactions between amino acid side chains and base functionality in the major groove [21]. It is interesting that electrostatic interactions are of critical importance but an additional contribution might come from the flexible side chain interactions viz. hydrogen bonding and hydrophobic interaction, which might also play significant role. Interactions between amino acid residues and DNA bases are introduced by H-bonding or hydrophobic packing interactions in the major groove. Amino acid residues Glu and Asp are known to have H-bond acceptors but not donors, and they only interact with C or A bases. Ser, Cys, and Thr play dual roles as both donor and acceptor, and Arg and Lys play H-bond donor but not acceptor. Hydrophobic interaction in the major groove of DNA is the single methyl group of the T base. Hydrophobic interactions can occur between Ala, Val, Ile, Leu, Met, Phe, Tyr, Trp and Thr with three methyl groups on T base. Similarly two ring CH groups of the C bases can also interact with hydrophobic residues; however, interaction with the hydrogen atoms of C will not be as strong as with the methyl group of the thymine. On the other side, Ala appears to be insufficiently hydrophobic to contact the C base.

Keeping these possible interactions in mind, we want to explore interaction patterns and possible dissociation in the secondary structure of plasmid DNA, as a model duplex system with melittin [22]. An experiment was performed by incubating plasmid DNA pBR322 with melittin in free or Lipobee and Polybee forms while LRCM and PRCM were used as negative controls. Another experiment was performed on 1% agarose gel in TAE buffer. The pBR322 plasmid DNA (0.3 µg/cocktail) was used to mix with melittin, Polybee and Lipobee (with a melittin concentration of 50 µM). LRCM and PRCM were used with equal volume as in case of Polybee and Lipobee formulations. To know the effect of melittin concentration, another set of cocktails were also prepared using the same amount of DNA per cocktail but with melttin concentration varying as 50, 5, 0.5, 0.05, 0.005 and 0.0005 µM. At the end of 1 h of incubation at room temperature, cocktails were mixed with 6x DNA loading dye and loaded on gel for running electrophoretically. Untreated DNA was run as control for the experiment. After running the gel, it was stained in ethidium bromide (20 mg/100 ml) for 10 min followed by washing for 5 min to remove excess EB staining. Gels were imaged under UV light.

Interaction of melittin with serum proteins

To investigate the interaction pattern of melittin and melittin-based Lipobee and Polybees with serum proteins, FBS (fetal bovine serum) was used for the model system [23]. A 10% FBS solution was incubated with 50 µM Melittin in free or Lipobee and Polybee forms. Samples were incubated for 4 h before measuring the UV absorbance.

Biostatistical Analysis

To evaluate the extent of improvements in activity of melittin against cancer cell growth when formulated as Polybee, ONE way ANOVA was performed. It was performed on IC50 values with a Bonferroni post test to calculate p values ay 0.001 and 0.005 and represented as *** and ** values next to the values.

CONCLUSIONS

We have established an *in-silico-to-in-vitro* approach to synthesize a well-defined, self-assembled, rigid-cored polymeric (polybee) nano-architecture for controlled delivery of a key component of bee-venom, melittin. A competitive formulation with lipid-encapsulated (Lipobee) rigid cored micelle was synthesized. In a series of sequential experiments, we studied how nanoscale chemistry influences the delivery of venom toxins for cancer regression and helps evade systemic disintegrity and cellular noxiousness. Our experimental and computational results indicated that Polybees were better cancer cell growth inhibitors than Lipobees in two breast cancer cell lines, presumably due to their stable and tighter association with melittin. Our results indicated insignificant to no complementary activation from these particles and free melittin. Studies with sodium dodecyl sulfate and lower pH revealed anionic membrane interactions as probable mechanisms for melittin release from Lipobee and Polybee nanoparticles. UV absorption studies for fetal bovine serum revealed their inactivity against serum proteins while gel electrophoretic assays described very strong interaction of free melittin with plasmid DNA. Extensive molecular docking studies revealed significant changes in base pair H-bonding distances after interaction with melittin. The main reason behind the differential behavior of peptide incorporation in lipid and amphiphilic polymer-based nanoassemblies can be found in the different types of interactions they face. Model lipidic membranes are known to be disrupted after interaction with polypeptides, [24] which are supposed to expand the lipid assembly in case of Lipobee and in turn increase the hydrodynamic diameter by a significant extent. Here none of the components of PRCM and LRCM are designed to respond to lower pH although PS-b-PAA could have improved electrostatic repulsion between the deprotonated PAA chains resulting in the thinning of the vesicle membrane at higher pH only.[25] On the other hand, melittin is known to have different fusion extent to model vesicles of phospahtidyl choline with maximum at 5.1 which could play role in case of cellular interactions of Lipobee and Polybee formulations, [26] but as LRCM and PRCM making major fractions of Lipobee and Polybee, anionic assemblies/membranes are supposed to play major role. On the other hand, we have demonstrated through melittin leaching studies that Polybees are very stable as such and do not change size as much as Lipobee. After release from Polybee particles, melittin could play a significant role in DNA association-dissociation processes, too, which might assist in growth inhibition of the desired cell population. Thus, we conclude that the use of amphiphilic polymer in preparation may provide a better strategy to produce stable formulation of melittin for systemic application compared to lipid-based amphiphiles.

ACKNOWLEDGMENTS

AFM, TEM and Zeta potential measurements were carried out in part in the Frederick Seitz Materials Research Laboratory Central Research Facilities, University of Illinois. We thank University of Illinois at Urbana-Champaign for study support.

AUTHOR CONTRIBUTIONS

Conceived and designed the experiments: DP. Performed the experiments: SM MY SK. Analyzed the data: SM MY. Contributed reagents/materials/analysis tools: SM MY. Wrote the paper: DP SM MY.

REFERENCES

1. Keramidas A, Moorhouse AJ, Schofield PR, Barry PH. Ligand-gated ion channels: mechanisms underlying ion selectivity. Prog Biophys Mol Biol. 2004; 86: 161–204. pmid:15288758 doi: 10.1016/j.pbiomolbio.2003.09.002

2. King G. Venoms to Drugs: Translating Venom Peptides into Therapeutics. Aust. Biochem. 2013; 44: 13–16.

3. Saez NJ, Senff S, Jensen JE, Yan S, Herzig V, Rash LD, et al. Spider-Venom Peptides as Therapeutics. Toxins 2010; 2: 2851–2871. doi: 10.3390/toxins2122851. pmid:22069579

4. Hmed BN, Serria HT, Mounir ZK. Scorpion Peptides: Potential Use for New Drug Development. J Toxicol. 2013: 1–15.

5. Garcia ML, Gao Y, McManus OB, Kaczorowski GJ. Potassium channels: from scorpion venoms to high-resolution structure. Toxicon. 2001; 39: 739–748. (b) Han YY, Liu HY, Han DJ, Zong XC, Zhang SQ, Chen YQ. Role of glycosylation in the anticancer activity of antibacterial peptides against breast cancer cells. Biochem Pharmacol. 2013; 86: 1254–1262. (c) Kondo E, Saito K, Tashiro Y, Kamide K, Uno S, Furuya T, et al. Tumour lineage-homing cell-penetrating peptides as anticancer molecular delivery systems. Nature Commun. 2012; 3: 1–13.

6. Possani LD, Merino E, Corona M, Bolovar F, Becerril B. Peptides and genes coding for scorpion toxins that affect ion-channels. Biochimie. 2000; 82: 861–868. pmid:11086216 doi: 10.1016/s0300-9084(00)01167-6

7. Zuo XP, Ji YH. Molecular mechanism of scorpion neurotoxins acting on sodium channels. Mol Neurobiol. 2004; 30: 265–278. pmid:15655252 doi: 10.1385/mn:30:3:265

8. Elgar D, Plessis JD, Plessis LD. Cysteine-free peptides in scorpion venom: geographical distribution, structure-function relationship and mode of action. Afr J Biotechnol. 2006; 5: 2495–2502.

9. Misra SK, Ye M, Kim S, Pan D. Highly efficient anti-cancer therapy using scorpion 'NanoVenin'. Chem Commun. 2014; 50:13220–13223. doi: 10.1039/c4cc04748f. pmid:25061638

10. Guo X, Ma C, Du Q, Wei R, Wang L, Zhou M, et al. Two peptides, TsAP-1 and TsAP-2, from the venom of the Brazilian yellow scorpion, Tityus serrulatus: Evaluation of their antimicrobial and anticancer activities. Biochimie. 2013; 95: 1784–1794. (b) Isa L, Amstad E, Schwenke K, Del Gado E, Ilg P, Kroeger M, et al. Adsorption of core-shell nanoparticles at liquid–liquid interfaces. Soft Matter. 2011; 7: 7663–7675; (c) Padovan-Merhar O, Lara FV, Starr FW. Stability of DNA-linked nanoparticle crystals: Effect of number of strands, core size, and rigidity of strand attachment. J Chem Phy. 2011; 134: 244701/1–7.

11. SYBYL-X 2.0, Tripos International, 1699 South Hanley Rd., St. Louis, Missouri, 63144, USA.

12. Molecular Operating Environment (MOE), 2013.08; Chemical Computing Group Inc., 1010 Sherbooke St. West, Suite #910, Montreal, QC, Canada, H3A 2R7, 2013.

13. Wu J, Eisenberg A. Proton Diffusion across Membranes of Vesicles of Poly(styrene-b-acrylic Acid) Diblock Copolymers. J Am Chem Soc. 2006; 128: 2880–2884. pmid:16506766 doi: 10.1021/ja056064x

14. Elsabahy M, Wooley KL. Design of polymeric nanoparticles for biomedical delivery applications. Chem Soc Rev. 2012; 41: 2545–2561. doi: 10.1039/c2cs15327k. pmid:22334259

15. Kim B, Schmieder AH, Stacy AJ, Williams TA, Pan D. Sensitive Biological Detection with a Soluble and Stable Polymeric Paramagnetic Nanocluster. J Am Chem Soc. 2012; 134: 10377–10380. doi: 10.1021/ja3040366. pmid:22693958

16. Fischer D, Bieber T, Li Y, Elsasser HP, Kissel T. A Novel Non-Viral Vector for DNA Delivery Based on Low Molecular Weight, Branched Polyethylenimine: Effect of Molecular Weight on Transfection Efficiency and Cytotoxicity. Pharm Res. 1999; 16: 1273–1279. pmid:10468031

17. Chen Y, Dong CM. pH-Sensitive Supramolecular Polypeptide-Based Micelles and Reverse Micelles Mediated by Hydrogen-Bonding Interactions or Host-Guest Chemistry: Characterization and In Vitro Controlled Drug Release, J Phys Chem B. 2010; 114: 7461–7468. doi: 10.1021/jp100399d. pmid:20469900

18. Bhat PA, Rather GM, Dar AA. Effect of Surfactant Mixing on Partitioning of Model Hydrophobic Drug, Naproxen, between Aqueous and Micellar Phases. J Phys Chem B. 2009; 113: 997–1006. doi: 10.1021/jp807229c. pmid:19123827

19. Pan D, Williams TA, Senpan A, Stacy AJ, Scott MJ, Gaffneya PJ, et al. Detecting Vascular Biosignatures with a Colloidal, Radio-Opaque

Polymeric Nanoparticle. J Am Chem Soc. 2009; 131: 15522–15527. doi: 10.1021/ja906797z. pmid:19795893

20. Wheater PR, Burkitt HG, Daniels VG. Functional Histology: A text and colour. Longman Group; UK: 1987; p. 407–408.

21. Suzuki M. A framework for the DNA–protein recognition code of the probe helix in transcription factors: the chemical and stereochemical rules. Structure 1994; 2: 317–326. pmid:8087558 doi: 10.1016/s0969-2126(00)00033-2

22. Yadav S, Mahato M, Pathak R, Jha D, Kumar B, Deka SR, et al. Multifunctional self-assembled cationic peptide nanostructures efficiently carry plasmid DNA in vitro and exhibit antimicrobial activity with minimal toxicity. J Mater Chem B. 2014; 2: 4848–4861. doi: 10.1039/c4tb00657g

23. Yang Y, Yang Y, Xie X, Cai X, Zhang H, Gong W, et al. PEGylated liposomes with NGR ligand and heat-activable cell-penetrating peptide–doxorubicin conjugate for tumor-specific therapy. Biomaterials 2014; 35: 4368–4381. doi: 10.1016/j.biomaterials.2014.01.076. pmid:24565519

24. Cao P, Abedini A, Wang H, Tu L-H, Zhang X, Schmidt AM, et al. Islet amyloid polypeptide toxicity and membrane interactions. PNAS 2013; 110: 19279–19284. doi: 10.1073/pnas.1305517110. pmid:24218607

25. Chen Q, Vancso GJ. pH dependent elasticity of polystyrene-block-poly(acrylic acid) vesicle shell membranes by atomic force microscopy. Macromol Rapid Commun. 2011; 32:1704–1709. doi: 10.1002/marc.201100332. pmid:21994204

26. Murata M, Nagayama K, Ohnishi S. Membrane fusion activity of succinylated melittin is triggered by protonation of its carboxyl groups. Biochemistry 1987; 26:4056–4062. pmid:2820482 doi: 10.1021/bi00387a047

Chapter 6

COAL CHEMISTRY AND MORPHOLOGY OF THAR RESERVES, PAKISTAN

Anila Sarwar[1], M. Nasiruddin Khan[2], Kaniz Fizza Azhar[3]

[1] Fuel Research Centre, Pakistan Council of Scientific & Industrial Research, Karachi, Pakistan

[2] Department of Chemistry, University of Karachi, Karachi, Pakistan

[3] Scientific Information Centre, Pakistan Council of Scientific & Industrial Research, Karachi, Pakistan

ABSTRACT

The surface of Thar coal has been characterized by spectroscopic, microscopic and chemical methods using atomic absorption spectroscopy, fourier transform infrared analysis, X-ray diffraction, scanned electron microscopy and pH titration. The samples contained high moisture, low volatile and low to moderate sulfur content and ranked as lignite (heating value 2541 - 4289 kcal/kg on moist, mineral-matter-free basis). Scanned electron micrographs show porous matrix with calcium, potassium or sodium minerals. Fourier transform infrared analysis also confirmed the presence of aluminum, silica and hydrate mineral constituents. The spectra showed C=C aromatic groups at 1604 - 1609 cm−1. Phenolic ester and carboxylic acid are identified by C=O stretching vibration peaks at 1702 cm−1. The peaks of quartz and kaolinite were observed at 900 - 1100 cm−1. Point of zero charge of Thar coal has been estimated as 6.00 to 6.27 through adsorption of H+ and OH− ions by suspending coal particles in aqueous electrolyte solution. Oxygen containing functional groups, mineral matter, and metal oxides are found to have a remarkable impact on point of zero charge. The surface characterization study will be helpful in the separation of hydrophilic impurities during coal preparation processes considering pzc as the controlling factor.

INTRODUCTION

Run-of-mine coal consists of coal, minerals and contaminants of large particle size. It needs a series of coal preparation steps after pulverization. During mining, crushing and other mechanical operations a significant amount of coal fines (particle size < 0.5 - 0.6 mm) is produced. The handling of fine particles is difficult, expensive and needs special attention [1]. There are several physical methods for the separation of minerals and other impurities from coal [2,3]. Physical cleaning through gravity separation has been recommended for lumps of coal but it is considered inefficient for fine particles [4].

Of the existing fine coal cleaning techniques, froth flotation method is the most common and effective. The process consists of bubbling air through coal/water slurry. The separation occurs by preferential physical attachment of air bubbles to the coal. The coal particles floated at the surface are removed, while the unwanted particles completely wetted and stay in the water phase. The ability of air bubbles to selectively adhere to coal is concerned with the surface chemistry.

Therefore, surface characterization of coal plays a vital role in coal preparation processes (such as floatation, dispersion, wettability, and coal-water slurry) prior to its utilization. Generally fourier transform infrared (FT-IR) analysis, X-ray diffraction (XRD), and scanned electron microscopy (SEM) techniques are used for surface characterization of coal and coal-derived products [5,6]. Various industrial operations depend on the pH where electrical charge density on coal surface becomes zero (pzc); but limited data is reported in which pzc of coal has been considered as an important surface parameter [7,8]. When coal is immersed in a liquid environment, a charge is developed on its surface by dissociation of functional groups or by the adsorption of H+ or OH– ions [9]. The surface charge together with counter ions constitutes the electrical double layer and responsible to control the stability of coal suspensions (Figure 1). During adsorption phenomenon, if the pH is equal to pzc, the surface of coal acts as neutral specie. Below pzc, the surface of adsorbent becomes positively charged and attracts anions.

Conversely, above pzc the surface becomes negatively charged and attracts cations. Therefore, the charge on the surface of coal plays a significant role in the removal of unwanted species. Thar coalfield has been recently discovered in Tharparker District of Sindh Province. It is the largest coalfield of the country (covering an area of 9100 km2) with estimated reserves of about 175.5 billion tones. Unfortunately the huge reserves are not exploited yet. However, now serious efforts are going on at government level to utilize these reserves. Our present understanding about the nature of Thar coal is surprisingly limited, which restricted from its efficient utilization [10].

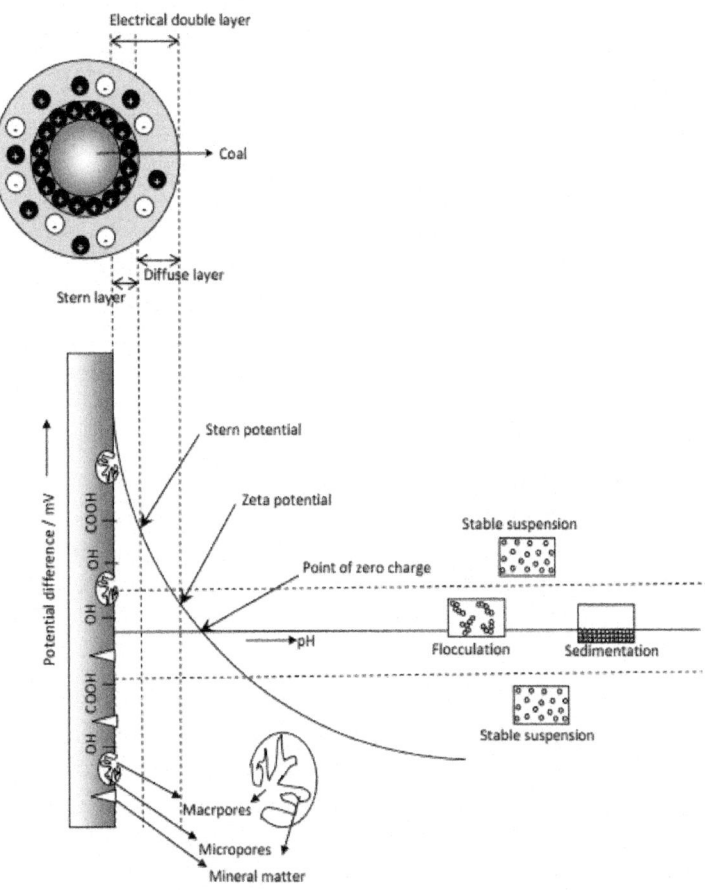

Figure 1: Schematic representation showing role of pH on surface dependent processes of coal.

The main objective of this study is surface characterization of Thar coal. Direct examination of surface of Thar coal by SEM, FT-IR, and XRD has been performed to obtain some useful qualitative information related to morphology, functional groups and mineral matters identification. These information are useful in the utilization of the reserve for various applications. SEM describes the suitability of coal surface for gasification process because diffusion of gases is easily permitted from porous surface. FT-IR analysis is helpful for the determination of functional groups which are the most reactive components in conversion processes. XRD tells about the mineral constituents of coal which are converted into ash during combustion. Mineral composition determines the mode of ash removal. An important outcome of the present study is surface quantification by measuring pzc. With the help of this important parameter one can identify the pH where the coal surface behaves as the least stable state. It is a major controlling factor in the separation of oxides, silicates, etc. Proper choice of pH is a requirement for selective flotation of one mineral from another.

EXPERIMENTAL

Eight samples of coal were obtained from Thar coalfield of Pakistan. The samples were air dried (ASTM D-3302- 00a) and ground into fine powder of particle size 60 mesh (ASTM D2013-00a). Thermogravimetric analyzer (TGA- 2000 A, Las Navas Instruments) was used to measure proximate analysis of coal according to ASTM D7582. Isoperibol bomb calorimeter (Parr 6300, USA) was used to measure the heating value in accordance with ASTM D5865-00. The elemental composition of the samples (C, H and N) was calculated using empirical formula derived by Carpenter and Diederichs [11]. Total sulfur was estimated using SC-32 LECO sulfur determinator (ASTM D 4239-00). The percentage of oxygen was taken by difference. Each analysis was performed in three replicates to check the reproducibility. Experimental data was converted into different basis using ASTM D388-99.

Fourier Transform Infrared (FT-IR) spectra of coal samples were recorded using NicoletTM 380, Fourier Transform Infrared Spectrophotometer (Thermo Electron Corporation, USA), with spectral resolution of 4 cm–1. Major and minor oxides in the combustion residue of coal were analyzed by Perkin Elmer-2380 model atomic absorption spectrophotometer (ASTM D-3286). National Bureau Standards 1633a and 1635, Washington DC were used as the standard reference materials. Mineralogical characterization was performed by X-ray diffraction using monochromatic Cu Kμ radiation at 40 kV and 30 mA (1 = 1.5406 Å).

Microscopic observations were made using a JEOL JSM-6380A type scanning electron microscope at the accelerating voltage of 10 - 15 kV. The surface charge at coal/water interface was determined by pH titration [12]. Two experiments were conducted for the purpose, each with 10 g of coal suspended in 35 ml of 0.1 M $NaNO_3$. After waiting 15 min to attain the equilibrium, one suspension was titrated with 0.1 M HNO_3 and the other with 0.1 M NaOH. The surface charge qH and qOH for HCl and NaOH additions respectively was estimated using Equations (1) and (2).

Table 1: Classification of Thar coal samples (basis are shown in superscript)

Parameters	Sample No:							
	1	2	3	4	5	6	7	8
Proximate Composition^a (%)								
Moisture	41.19	37.62	42.09	48.10	45.80	48.80	46.38	45.98
Volatile matter	16.47	14.38	13.45	13.69	18.03	11.95	12.37	17.64
Ash yield	5.06	19.17	0.63	2.49	1.90	3.98	6.78	3.40
Fixed carbon	37.28	28.83	43.83	35.72	34.27	35.27	34.47	32.98
Ultimate Composition (%)								
Carbon^b	27.49	21.28	17.88	29.16	32.03	26.85	25.41	29.88
Hydrogen^b	6.06	7.49	5.01	5.45	5.65	5.52	5.73	5.76
Nitrogen^b	2.22	2.95	1.91	1.99	1.88	2.13	2.21	1.93
Sulfur^b	1.42	0.19	0.22	0.40	0.21	0.23	1.20	1.24
Oxygen^c	58.17	48.27	74.39	60.26	58.58	60.81	56.94	56.96
Others								
GCV^d (kcal/kg)	4289.46	3194.16	2541.36	3279.75	3850.16	2729.25	2980.58	2978.76
ASTM Rank	Lignite							

^{a}ar basis = as-received basis; ^{b}daf basis = dry, ash-free basis; ^{c}db basis = dry basis; ^{l}m, mmf basis = moist, mineral-matter-free basis.

$$q_H = \frac{C_a - \left[H^+\right]}{m} \tag{1}$$

$$q_{OH} = \frac{\left[OH^-\right] - C_b}{m} \tag{2}$$

where C_a and C_b are the concentration (mol·L^{-1}) of added acid and base, respectively, H$^+$ and OH$^-$ are the molar proton and hydroxide ion concentration, and m is the mass of solid in g. The surface charge as a function of pH is calculated and plotted using Equations (1) and (2). The pzc was estimated as the pH where the surface charge crosses the x-axis (q = 0).

RESULTS AND DISCUSSION

Physico-chemical properties of coal samples are shown in Table 1. The quality of coal was assessed on as-received basis [13].

The samples show high moisture, low volatile matter and low ash content (except sample 2 of high ash). High moisture content is attributed to the water aquifers which are present at Thar at an average depth of 50 m, 120 m and more than 200 m. The sulfur content in the samples is low to moderate. Ultimate analysis shows total carbon 17.88% - 32.03%, hydrogen 5.01% - 7.49%, nitrogen 1.88% - 2.95% and oxygen 48.27% - 74.39% on dry-ashfree (daf) basis. The samples are classified as lignite on the basis of gross calorific value (GCV) on moist, mineralmatter-free basis (ASTM D388-99).

SEM Micrographs

SEM Images of Thar samples are shown in Figure 2 (a to h). The samples show interconnected and open micropores (300 Å) [14]. The SEM micrograph shows that coal matrix is covered with bright and dark luminous materials indicating the presence of minerals. The bright glow is due to the presence of calcium, aluminum, potassium or sodium. The dark luminosity is mainly due to the presence of chalcophiles [15]. The minerals are in the form of irregular shaped aggregates. Non luminous portion on the surface is mainly made up of carbon content. Randomly distributed fissures, cracks and etched pits could also be seen on the micrograph. These might be produced from the calcinations of dolomite and calcites as a result of thermal shock during metamorphism [16]. It is evident from the images that Thar coal contains large proportions of silica, calcium carbonates and dolomite, as well as some proportions of elements such as aluminum, potassium and sulfur.

FT-IR Analysis

Figure 3 shows FT-IR spectra of the samples. It is evident that Thar coal contains aliphatic CH, CH_2, and CH_3 groups, as well as aromatic ring systems. It also contains C-O-, C-O-C and associated -OH or NH bonds, and few C=O bonds. The band at 3350 - 3385 cm^{-1} is assigned to associated OH and NH groups (hydrogen bonded). All samples exhibit band at 2916 cm-1 with a shoulder peak at 2850 cm^{-1} showing the presence of aliphatic C-H stretching vibration. The strong band at 1604 - 1609 cm^{-1} is attributed to aromatic ring vibrations, enhanced by oxygen groups [14]. The shoulder peak at 1702 cm^{-1} (C=O stretching vibration) is represented phenolic ester and carboxylic acid [15,16]. All samples of Thar show lower intensity of the aliphatic methylene band at 1460 cm^{-1} as compared to the methyl band at 1370 cm^{-1} indicates that coal is composed of fewer aliphatic methylene groups than methyl groups [14]. Weak bands at 1560 cm^{-1} was identified showing condensed aromatic ring C=C at their surfaces. The prominent peak at 1030 – 1035 cm^{-1} with a shoulder peak at 1005 - 1010 cm^{-1} is represented to Si-O bending vibration. Quartz and kaolinite were identified by the presence of bands in the region of 912 - 917 cm–1 [5]. The aromatic character of coal due to C=C stretching vibration is found to be more pronounced than the aliphatic character as the band intensity at 1620 cm^{-1} is higher than the bands at 2960 cm^{-1} (stretching vibration of methyl's group). The same finding has been observed in FT-IR spectrum of Pittsburgh No. 8 coal [5].

Point of Zero Charge (pzc) Determination

The surface of coal is generally considered as negatively charged due to the presence of polar functionalities, dominantly carboxylic and phenolic groups. pzc of coal depends on the relative affinity of the coal surface for H$^+$ and OH$^-$. Figure 4 describes that Thar coal behave at the least stable state at pH 6.00 - 6.27. At this pH range noncharged coal particles are supposed to be unable to bond with water. The maximum hydrophobicity of coal at this pH range supports the maximum floatability [17, 18]. pzc of Thar coal is attributed to the presence of weakly acidic oxygen groups, and ash forming minerals such as kaolinite and quartz etc. pzc of pure carbon is 6.5 - 7.0 while pzc of silicate minerals is approximately 2.0. The presence of silicate minerals has been confirmed by X-ray diffraction analysis (Figure 5). pzc of Thar coal also depends on the minor oxides present in significant amount in the inorganic part of coal such as Fe_2O_3 (pzc 8.5) and CaO (pzc 8.1) [19]. The pzc of Thar coal will be of great importance for process engineers to control wet washing processes which depends on stability and coagulation of colloidal dispersion.

XRD Analysis

A typicalXRD spectrum of Thar coal has been shown in Figure 5. The samples were composed of hexagonal shaped quartz low, dauphine-twinned SiO2, monoclinic carbon oxide hydrate $(COOH)_2·2H_2O$ and triclinic Kaolinite $_1$ A $Al_2(Si_2O_5)(OH)_4$. The mineralogical analysis was supplemented with the metal oxide quantification by atomic absorption spectrophotometer (Table 2). Higher amounts of SiO_2 and Al_2O_3 verified XRD observations.

Comparison of Thar Coal with Other Lignite Coalfields

In addition to characterization of the samples, chemical composition of Thar coal was compared with other lignite coals of the world. The samples of Thar are seems to closely resembles with the lignite coal mines of North Dakota with ultimate composition (C 30.46% - 34.25%, H 6.49% - 7.12% and N 0.50% - 0.56%) and oxide composition of ash (23.16% - 39.13% SiO2, 7.66% - 13.14% Al_2O_3, 1.94% - 14.87% Fe_2O_3, 13.14% - 18.39% CaO, 4.77% - 6.44% MgO, 4.53% - 6.61% Na2O, and 0.40% - 0.52% K_2O) [20]. Chemical composition of Thar coal is also similar to the coal mine of Velenje, Slovenia with total moisture 38%, ash content 20%, heating value 3956 btu/lb, S 1.4%, C 27.1%, H 2.1%, O 11.0% and N 0.4% [21]. The heating value of Thar coal is higher to a large extent than the lignite coal of Greece mined from the basin of Megalopolis (1620 - 1980 btu/lb) and Ptolemais Basin (2070 - 2520 btu/lb) [22]. Thar coal has an advantage of low to moderate sulfur compare to the world's average sulfur (2.42%) in lignite coal [6]. Oxides of sulfur are the most concerned emission pollutants from the regulatory stand point. Low sulfur content in Thar coal makes it acceptable for power generation without exceeding the emission standard for sulfur oxides ($\leq$1%).

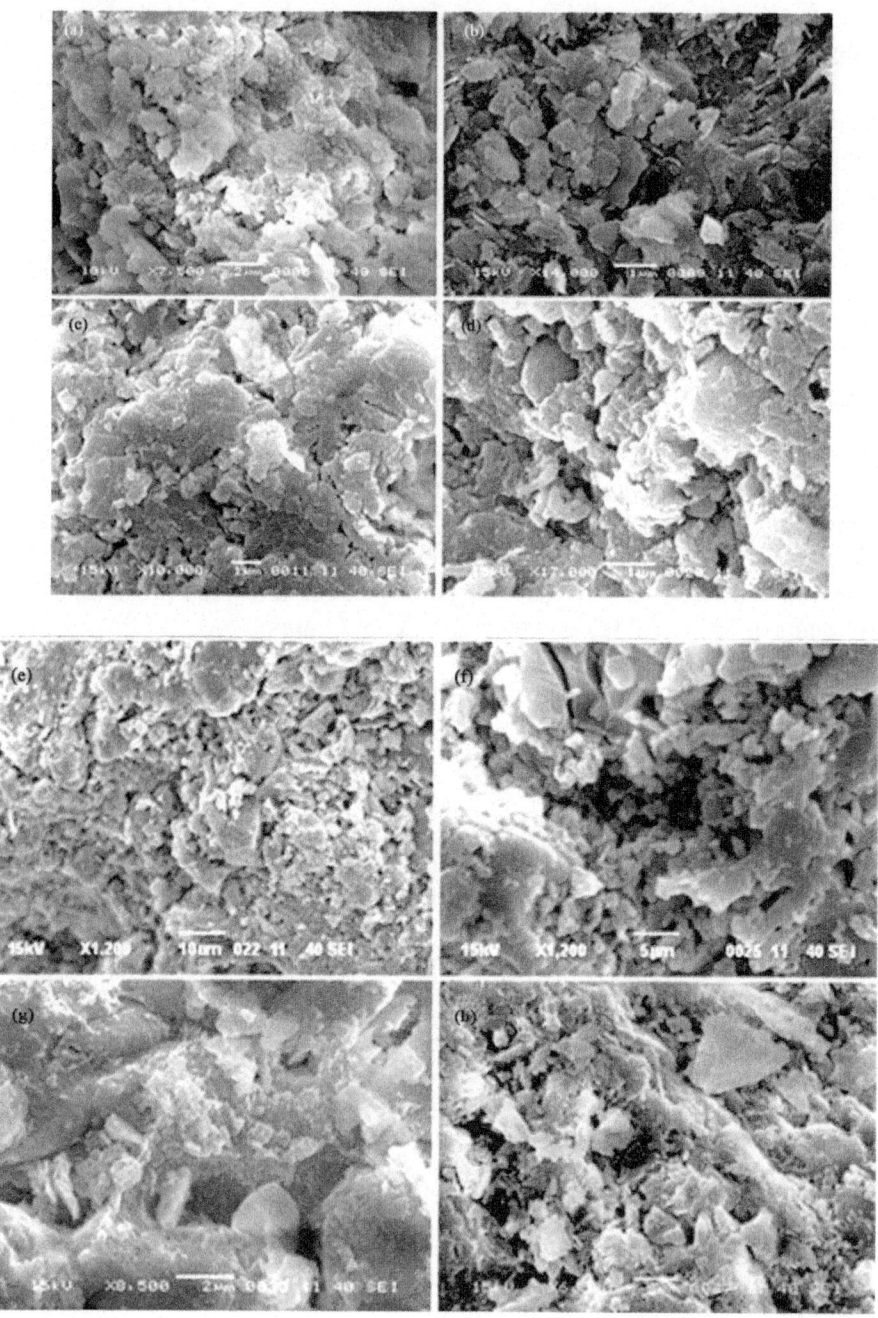

Figure 2: SEM images of Thar coal: (a) to (h) represents sample 1 to 8 respectively.

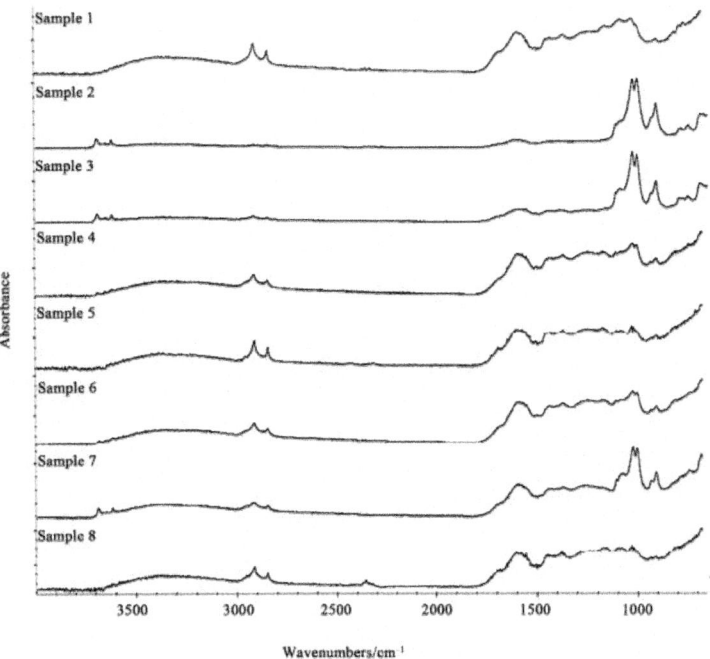

Figure 3: FT-IR spectra of Thar coal.

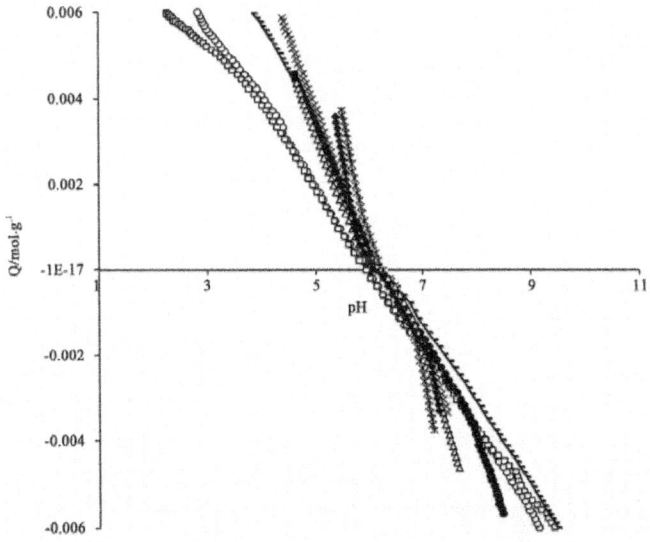

Figure 4: Point of zero charge of Thar samples, Sample No: 1(○), 2 (□), 3 (-), 4(×), 5 (●), 6(♦), 7 (△), 8 (♦).

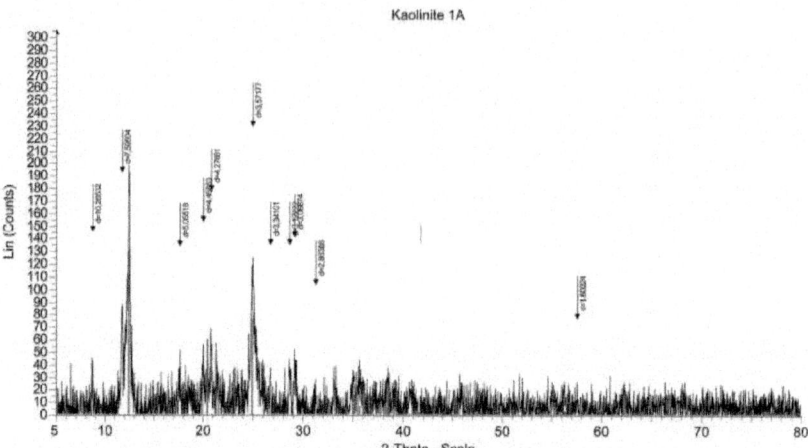

File sample: 11 12-02-08.raw -Type: 2Th/Th locked -Start: 5.000° -End: 80.000° -Step: 0.020° -Step time: 1.s -Temp.: 25°C(Room) -Time Sharted: 10 s -2-Theta: 5.000° -Theta: 2.500° -Chi: 0.00° -Phi: 0.00° -X: Operations: Background 1.000,1.000 | Import

01-080-0885(C)-Kaolinite 1a-A12(Si2O5)(OH)4 -Y: 65.50% -d x by: 1. -WL: 1.5406 -Triclinic -a 5.15550 - b 8.94380 -c 7.40510 -alpha 91.700 -beta 104.840 -gamma 89.830 -Base-centered -C1(0) -2-329.909- | / |

Figure 5: A typical XRD spectrum of Thar coal.

Table 2: Major and minor elements in Thar coal.

Oxides (%)	Sample No:							
	1	2	3	4	5	6	7	8
Major								
SiO_2	44.45	36.2	27.45	32.34	29.95	31.48	31.59	25.35
Al_2O_3	10.52	18.49	19.94	24.44	24.45	19.32	20.36	17.77
Minor								
Fe_2O_3	21.22	17.38	20.25	15.44	17.85	20.34	18.34	5.53
Na_2O	1.48	1.67	1.04	1.54	1.27	1.11	1.05	3.49
K_2O	0.2	0.19	0.14	0.02	0.24	0.31	0.34	1.04
MgO	4.31	4.27	5.55	4.3	4.99	2.56	5.36	8.73
CaO	14.81	18.24	22	17.45	18.44	23.26	20.34	31.85
MnO_2	0.01	0.01	0.03	0.01	0.01	0.02	0.02	0.03

CONCLUSION

Thar coal has been ranked as lignite. SiO2 and Al2O3 were identified as the major and Fe2O3 and CaO as the minor metal oxides in the ash residue. The surface of coal has been characterized by macro, meso and micropores with the irregular aggregates of minerals of calcium, sodium, potassium and aluminum. pzc of Thar coal ranges from pH 6.00 - 6.27. At this pH fines of coal float relatively easily, so the coal washing of low-rank coal of hydrophilic nature may work fast.

ACKNOWLEDGEMENTS

The authors are grateful to Dean of Science, University of Karachi, for the financial support.

REFERENCES

1. A. Das, B. Sarkar and S. P. Mehrotra, "Prediction of Separation Performance of Floatex Density Separator for Processing of Fine Coal Particles," International Journal of Mineral Processing, Vol. 91, No. 1-2, 2009, pp. 41-49

2. A. U. Kurniawan, O. Ozdemir, A. V. Nguyen, P. Ofori and B. Firth, "Flotation of Coal Particles in MgCl2, NaCl, and NaClO3 Solutions in the Absence and Presence of Dowfroth 250," International Journal of Mineral Processing, Vol. 98, No. 3-4, 2011, pp. 137-144. doi:10.1016/j.minpro.2010.11.003

3. X. Li, R. Shaw and P. Stevenson, "Effect of Humidity on Dynamic Foam Stability," International Journal of Mineral Processing, Vol. 94, No. 1-2, 2010, pp. 14-19. doi:10.1016/j.minpro.2009.10.002

4. K. S. Birdi, "Handbook of Surface and Colloid Chemistry," 3rd Edition, CRC Press, Taylor & Francis Group, Boca Raton, London, New York, 2009, pp. 655-679.

5. Y. D. Abreu, P. Patil, A. I. Marquez and G. G. Botte, "Characterization of Electroxidized Pittsburgh No. 8 Coal," Fuel, Vol. 86, No. 4, 2007, pp. 573-584. doi:10.1016/j.fuel.2006.08.021

6. V. Bouska and J. Pesek, "Quality Parameters of Lignite of the North Bohemian Basin in the Czech Republic in Comparison with the World Average Lignite," International Journal of Coal Geology, Vol. 40, No. 2-3, 1999, pp. 211-235. doi:10.1016/S0166-5162(98)00070-6

7. S. E. Kuh and D. S. Kim, "Effects of Surface Chemical and Electrochemical Factors on the Dewatering Characteristics of Fine Particle Slurry," Journal of Environmental Science and Health Part A: Toxic/Hazardous Substances & Environmental Engineering, Vol. 39, No. 8, 2004, pp. 2157-2182. doi:10.1081/ESE-120039382

8. M. Kosmulski, "pH-Dependent Surface Charging and Points of Zero Charge IV Update and New Approach," Journal of Colloid and Interface Science, Vol. 337, No. 2, 2009, pp. 439-448. doi:10.1016/j.jcis.2009.04.072

9.   M. N. Khan and A. Sarwar, "Determination of Points of Zero Charge of Natural and Treated Adsorbents," Surface Review and Letters, Vol. 14, No. 3, 2007, pp. 461-469. doi:10.1142/S0218625X07009517

10. A. Sarwar, M. N. Khan and K. F. Azhar, "Kinetic Studies of Pyrolysis and Combustion of Thar Coal by Thermo-gravimetry and Chemometric Data Analysis," Journal of Thermal Analysis and Calorimetry, Vol. 109, No. 1, 2012, pp. 97-103. doi:10.1007/s10973-011-1725-0

11. R. C. Carpenter and H. Diederichs, "Experimental Engi-neering," 8th Edition, Wiley, New York, 1913.

12. M. Davranche, S. Lacour, F. Bordas and J. C. Bollinger, "An Easy Determination of the Surface Chemical Properties of Simple and Natural Solids," Journal of Chemical Education, Vol. 80, No. 1, 2003, pp. 76-78. doi:10.1021/ed080p76

13. The International Coal Encyclopedia, "Coal Services In-ternational," Vol. 1, Time off set Pte Ltd., 1990.

14. S. C. Tsai, "Coal Science and Technology Series 2: Fundamentals of Coal Beneficiation and Utilization," Elsevier Scientific Publishing Company, Amsterdam, 1982.

15. 15.   M. Shakirullah, I. Ahmad, M. A. Khan, M. Ishaq, H. Rehman and U. Khan, "Leaching of Minerals in Degari Coal," Journal of Minerals & Material Characterization and Engineering, Vol. 5, No. 2, 2006, pp. 131-142.

16. B. Manoj, A. G. Kunjomana and K. A. Chandrasekharan, "Chemical Leaching of Low Rank Coal and Its Characterization Using SEM/ EDAX and FTIR," Journal of Min-erals & Materials Characterization & Engineering, Vol. 8, No. 10, 2009, pp. 821-832.

17. K. H. Nimerick and B. E. Scolt, "New Method of Oxidised Coal Flotation," Mining Congress Journal, Vol. 66, 1980, pp. 21-22.

18. A. J. Rubin and R. J. Kramer, "Recovery of Fine-Particle Coal by Colloid Flotation," Separation Science and Technology, Vol. 17, No. 4, 1982, pp. 535-560. doi:10.1080/01496398208060257

19. K. Y. Zhang, H. P. Hu, L. J. Zhang and Q. Y. Chen, "Surface Charge Properties of Red Mud Particles Generated from Chinese Diaspore Bauxite," Transactions of Nonferrous Metals Society of China, Vol. 18, No. 5, 2008, pp. 1285-1289. doi:10.1016/S1003-6326(08)60218-6

20. B. C. Folkedahl and C. J. Zygarlicke, "Sulfur Retention in North Dakota Lignite Coal Ash," Preprints Papers— American Chemistry Society, Division of Fuel Chemistry, Vol. 49, No. 1, 2004, pp. 167-168.

21. J. Oman, A. Senegacnik and B. Dejanovic, "Influence of Lignite Composition on Thermal Power Plant Performance: Part 2: Results of Tests," Energy Conversion and Management, Vol. 42, No. 3, 2001, pp. 265-277. doi:10.1016/S0196-8904(00)00062-5

22. M. J. Galetakis and F. F. Pavloudakis, "The Effect of Lignite Quality Variation on the Efficiency of On-Line Ash Analyzers," International Journal of Coal Geology, Vol. 80, No. 3-4, 2009, pp. 145-156. doi:10.1016/j.coal.2009.09.002

Chapter 7

A REVIEW OF THE AQUEOUS AEROSOL SURFACE CHEMISTRY IN THE ATMOSPHERIC CONTEXT

Kalliat T. Valsaraj

Department of Chemical Engineering, Louisiana State University, Baton Rouge, USA

ABSTRACT

In this review the surface chemistry and properties of aqueous atmospheric aerosols are explored. Water plays a major role in scavenging pollutants. Reactions occur on thin water films in atmospheric aerosols. The study of the aerosol water surface is important to properly account for chemical transformations in the troposphere. The thermodynamics of adsorption of organic molecules and oxidant species on the aqueous surface and, the techniques employed to quantify the adsorption isotherms are summarized. Experimental techniques for elucidating the reactions on the water surface are described. Field and laboratory data for oxidation reactions of compounds at the air-water interface are summarized. The Langmuir-Hinshelwood reaction mechanism is useful in quantifying the reaction rate on the aqueous aerosol surface. A hypothesis for the large heterogeneous reaction rate on the water surface over the homogeneous bulk aqueous phase reaction is presented.

INTRODUCTION

Atmospheric aerosols are colloidal systems with one phase (solid or liquid) dispersed in another phase (gas). Solid aerosols are generally organic or inorganic particles dispersed in air ranging from sub-micron to a few microns in diameter; they may either be dry or wet. Liquid aerosols comprise of rain, fog, mist, cloud droplets, snow and ice. Table 1 provides the essential characteristic dimensions and properties of atmospheric aerosols. It is noteworthy that they have lifetimes ranging from a few minutes to days. The presence of surfaces such as cloud, fog and fine particulates can influence atmospheric processes with characteristic times of the order of hours.

Table 1: Typical properties of atmospheric aerosols [1,2]

Type of aerosol	Size range/ μm	Surface area/ $m^2 \cdot m^{-3}$	Average life time
Aerosol particles	10^{-2} - 10	~1×10^{-3}	4 - 7 d
Fog droplets	1 - 10	~8×10^{-4}	3 - 6 h
Cloud droplets	10 - 10^2	~2×10^{-1}	7 - 10 h
Rain	10^2 - 10^3	~5×10^{-4}	3 - 15 min
Snow	10^3 - 10^5	~0.3	15 - 50 min

Aerosols play a large role in tropospheric chemistry and climate. Aerosols take part in numerous heterogeneous reactions with gases (organic and inorganic) and atmospheric oxidants (ozone, hydroxyl radical, singlet oxygen, nitrate radical, chlorine). Aerosols absorb and reflect sunlight and contribute to direct and indirect effects on climate forcing. The 2007 IPCC assessment [3] on global climate change stated that the largest uncertainty is the effect of aerosols on climate forcing. Over the last few decades an impressive array of data has been accumulated on the role of atmospheric aerosols.

Atmospheric aerosols consist of a core of inorganic or organic nucleus (e.g., soluble salts, insoluble black carbon, soil or soot) that forms the condensation site for water as well as organic and inorganic molecules. Water effects the aerosol properties in various ways: 1) it determines the nature of aerosol (solid, liquid, or glassy/ amorphous), 2) the hygroscopic growth and size of the aerosol, 3) uptake of gases and oxidants, and 4) multiphase reactions in aerosol droplets. The water content of the aerosol is determined by environmental factors such as the type of condensation nucleus (CN), the relative humidity, temperature, and oxidation state of molecules on the surface of the aerosol. The water either plays the role of a competitive adsorbate, or acts as a substrate to provide the surface area for reaction. If the water content is substantially less than that required for a monomolecular layer, then the individual water molecules compete for adsorption sites on the surfaces of CN. If, on the other hand, the water forms a thin layer (1 nm to 50 μm) of high specific surface area it provides the site for adsorption and reaction on the aerosol. For water films of ~1 - 2 nm on silica, the surface-induced modifications to the adsorbed film propagate to ~1 - 4 nm from the solid-water interface [4]. When the water layer thickness exceeds few monolayers (>100 μm), the bulk water properties predominate.

Thin layers of water are important from the atmospheric chemistry perspective. The thin water layer in aerosols may have properties different from those of bulk water. Questions that are relevant in this regard include the following: 1) Is there any difference between "bound" and "unbound" water? 2) Is the H-bonding characteristic of water near surfaces different from bulk water? 3) Are the activity coefficients of solutes in thin layers of water (γ_i) different from their infinite dilute activity coefficients ($\gamma_{i\infty}$)?

Figure 1 is the schematic of an aqueous aerosol. The aqueous phase consists of dissolved inorganic ions, small particles and dissolved organic species, some of which are surface active and of large molecular weight that form a distinct phase on the water surface. The adsorption and reaction of trace gases and oxidants in the atmospheric aerosol will be influenced by the nature and composition of the aqueous phase. There have been several reports that describe the relevance of the aqueous surface in atmospheric droplets. For example, one of the earliest reports involved the conversion of sulfur dioxide in fog and cloud droplets to sulfates in the context of the so-called "acid rain" issue [2]. More recently, aqueous aerosol surface reaction was invoked for the depletion of ozone near sea surface aerosols [5]. There is also evidence for the real-time production of secondary organic aerosols in fog droplets [6]. We have provided evidence for the formation of both oxygenated PAHs and other compounds in fog sampled in the United States [7]. A variety of questions remain on the mechanisms of transport, adsorption and reaction on aqueous aerosol droplets [8].

In the following review I will summarize the various physico-chemical aspects pertaining to the adsorption and reaction of organic molecules on aqueous aerosol droplets and water thin films.

THE AIR-WATER INTERFACE

The surface of water is unique in its energy characteristics. The surface tension of water (72 mN·m^{-1}) is one of the highest for any liquid. Hence, the air-water interface is considered "the most hydrophobic, non-polar surface known" [9]. The "hyper hydrophobicity of the air-water interface" attracts hydrophobic moieties to the surface

[10]. The hydrophobic interactions are the basis for a number of well-known phenomena such as [11]: 1) the low aqueous solubility of hydrophobic molecules in water, 2) the ability of proteins to fold in water, 3) the formation of micelles by surfactants in water, 4) the inability of water to spread on hydrophobic surfaces, 5) the ability to separate hydrophobic particles and compounds by attachment to air bubbles (foam flotation) and, 6) dimerization of hydrophobic chains in water.

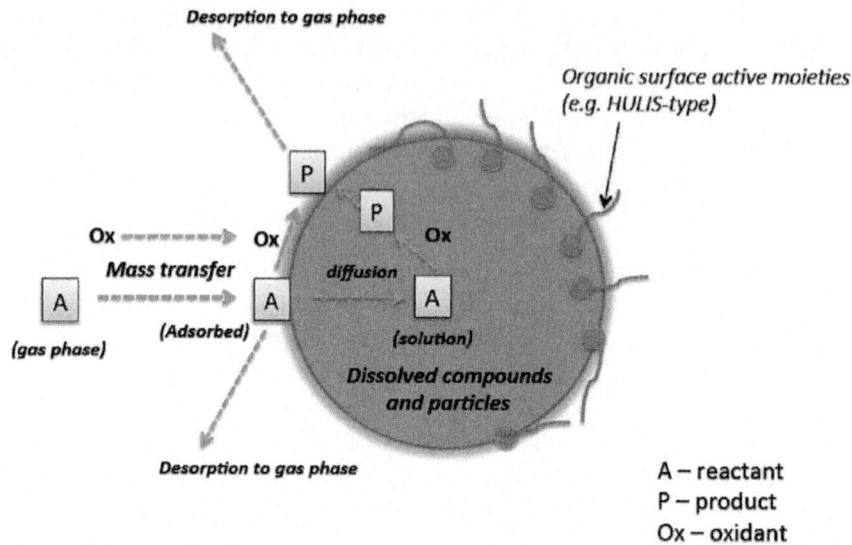

Figure 1: A pictorial representation of an atmospheric aqueous aerosol and its interactions with the gas-phase molecule and oxidant species.

The separation of hydrophobic organic compounds (HOC) from water by adsorption on air bubbles, viz. solvent sublation, was developed in our laboratory for wastewater treatment [12]. It was observed that the inverse process, viz., water droplets in air (as in a spray column), was able to remove high concentrations of HOC by adsorption [13]. The latter process was also invoked to explain the high concentrations of HOC observed in fog and dew waters collected from urban areas [14]. Further there has been several observations regarding the unique reactions of gas-phase HOC in fog and cloud droplets leading to the formation of products that are precursors to secondary organic aerosols (SOA) in the aqueous phase [15]. Field data on real-time SOA formation in a London fog event was recently reported [6].

Two aspects concerning trace gas interactions with atmospheric water surfaces need study [16]. Firstly, we need data on the thermodynamics of adsorption at the air-water interface. Secondly, we need information on the reactivity of adsorbed molecules towards gas-phase and liquid-phase oxidant species. In the following discussion, we focus on both aspects.

Thermodynamics of Adsorption at the Air-Water Interface

Utilizing the thermodynamic criteria of equal fugacities between the air and the air-water interface, we can derive the following equation for the free energy of partitioning for an HOC [17]

$$\Delta_{ads} G^0 = -RT \ \ell n \left(K_{\sigma a} / \delta_0 \right) \tag{1}$$

where $K_{\sigma a}$ is the Kemball-Rideal standard state surface thickness (=6 × 10⁻¹⁰ m). Experimental values of $K_{\sigma a}$ can be obtained from a variety of techniques. These include direct and indirect methods and are summarized in Table 2. The first two methods in Table 2 are direct methods that are suitable for compounds that have high vapor pressure. The last two indirect methods are suitable for compounds with low vapor pressure. They are especially useful for obtaining thermodynamic adsorption parameters (free energy, enthalpy and entropy) on thin water films of a few microns in thickness.

The experimental methods described above are complimentary to correlations with compound properties such as sub-cooled liquid vapor pressure, octanol-air partition constant, and molecular properties (e.g., molecular connectivity index, molecular surface area, molar volume) [23,24]. Further, the correlations were extended to include polar and non-polar compounds via the use of H-bond acceptor and donor indices [25,26]. Ab-initio and quantum chemistry based calculations of the partition constants using a Universal Surface Area and Solvation Model [27] and the use of the Conductor-Like Screening Model (COSMOTherm) [28] were also reported to provide good comparisons to experimental data.

Molecular dynamics (MD) and Monte Carlo (MC) simulations have been performed for the transfer of HOCs from air to water to determine the free energy of adsorption and solvation of molecules [29-33]. MD simulations provide good estimates of the basic thermodynamic properties, which can be compared to the experimental values. These calculations have been made not only for the HOCs but also for various gas-phase oxidant species (OH, O_3, H_2O_2) at the interface [34]. MD simulations for the air-to-water transfer of a series of PAHs showed that a free energy minimum exists at the air-water interface which is conducive to surface adsorption. The free energy difference was larger between the bulk air and the interface than between the bulk air and bulk water phases, indicating a large energy for complete solvation of the molecule. The free energy values obtained from the calculations agree with the experimental values obtained (Table 3). The presence of surfactants at the interface increased the free energy minimum [33]. The MD and MC simulations also allowed an understanding of the differences in surface orientation of the HOC in the presence of the surfactant [33]. It is clear that the adsorption at the air-water interface is a favorable process, but that the complete solvation (solution) of the molecule involves considerable energy penalty for HOCs. However, for the small oxidant species the two free energy values are similar, although there is a slight preference for the interfacial adsorption.

Table 2: Methods suitable for determining the partition constants of HOC from the gas phase to the aqueous surface

Method	Description	Reference
Surface tension-static	Wilhelmy plate method	[18]
Surface tension-dynamic	Axisymmetric drop shape analysis method	[19]
Gas chromatography	Inverse gas chromatography method	[20,21]
Flow reactor	Adsorption on a thin water film	[22]

Table 3: Experimental and MD simulated values of free energy of adsorption and solution for HOC and Oxidant species [29,35,36]

Compound	$\Delta_{ads}G^0$/kJ·mol^{-1}		$\Delta_{soln}G^0$/kJ·mol^{-1}	
	MD simulation	Experimental	MD simulation	Experimental
Benzene	−15	−16.3 ± 0.4	−3	−4 to −3
Naphthalene	−24	−26.5 ± 0.1	−11	−11 to −7
Phenanthrene	−32	−44.2	−15	−17 to −11
Anthracene	−33		−15	−17 to −10
Ozone	−5		+3	+2 to +3
•OH	−24		−18	−17 to −16
HO$_2$•	−31		−28	−31 to −25
H$_2$O$_2$	−44		−40	−37 to −36

Reactions of HOCs at the Air-Water Interface

As described above, both gas-phase oxidants and gaseous HOCs show a surface free energy minimum. We conclude that surface interactions between them should be feasible. The question is whether "surface" water is different from "bulk" water in supporting surface interactions. For example, theoretical work show that there is an increase in energy of surface water molecules of approximately 10 - 15 kcal·mol^{-1} over bulk water molecules [37] and, hence surface contains, on average, far more reactive sites than the bulk.

A number of techniques have been employed to study the reactions of gaseous HOC molecules with typical gas-phase oxidants at the air-water interface, which are summarized in **Table 4**. The air-water interface is difficult to probe compared to a gas-solid interface due to features peculiar to the former. For example, techniques suitable to study reactions of molecules on solids under vacuum are not useful since water has a high partial pressure. Unlike a solid surface, a water surface is constantly renewed and highly fluxional. The mobility of molecules adsorbed on water surfaces is larger than on solid surfaces with sub-nanosecond time scales for surface to bulk exchange of molecules. The delineation of a true surface phase for water is problematic since composition and density are not necessarily sharp and extend over a significant depth. Thus, the probing of water surfaces had to wait till suitable non-linear techniques were available to selectively amplify the molecular signals. One such technique is the sum frequency generation (SFG) method. This technique has been utilized to study the orientation of molecules at the water surface as well as the potential for configurational changes upon adsorption [46,47]. SFG has also been used to explore the dynamics of adsorption of molecules on the water surface [48].

A useful method to simultaneously study the adsorption and reaction involves the tracking of a molecular signal from the adsorbed species. For example, the adsorption and reaction of PAHs can be explored by following their characteristic fluorescence signals. Glancing angle laser induced fluorescence (LIF) of PAHs adsorbed at the air-water interface and its changes as a result of exposure and reaction with gas-phase oxidants (e.g., ozone) can be followed to obtain the progress of the reaction at the interface [40]. LIF can be also used to directly sample other reactions at the surface such as singlet oxygen with PAH molecules [48]. A recent innovation is the use of droplet electrospray mass spectroscopy [40], which involves the introduction of droplets of water containing the probe molecule into an electron spray mass spectrometer. The droplets are then exposed for a very short burst of an oxidant and the surface concentration directly ascertained. The principle of exposing water droplets in a falling droplet reactor to gas phase PAH and ozone and liquid analysis by HPLC/MS was used in our laboratory to study the surface reactions as a function of droplet size [43]. The reaction of aqueous surface adsorbed halide species to gas phase chlorine was reported recently using glancing angle Raman spectroscopy [44]. The pseudo-first order reaction rate constants for various molecules with gas phase oxidants have been explored using the above methods. Two examples are shown in Figures 2 and 3. An interesting observation is the saturation-type rate behavior. This can be justified by invoking a Langmuir-Hinshelwood (L-H) type surface reaction mechanism. Figure 3 shows the schematic of the same. The mechanism involves the reaction of adsorbed organic molecules with the oxidant on the surface of the water. The overall reaction rate is given by

Table 4: A sampling of various techniques used for studying the adsorption and reaction at the water surface

Technique used	Reference
Vibrational sum frequency generation (VSFG)	[38]
Glancing angle laser induced fluorescence	[39]
Droplet Electrospray ionization mass spectroscopy (ESI/MS)	[40]
Reflection absorption infrared spectroscopy (RAIRS)	[41]
Glancing angle Raman spectroscopy	[42]
Falling droplet reactor with HPLC/MS	[43]
Heterodyne-detected electronic sum frequency generation (HD-ESFG)	[44]
Reactive scattering technique	[45]

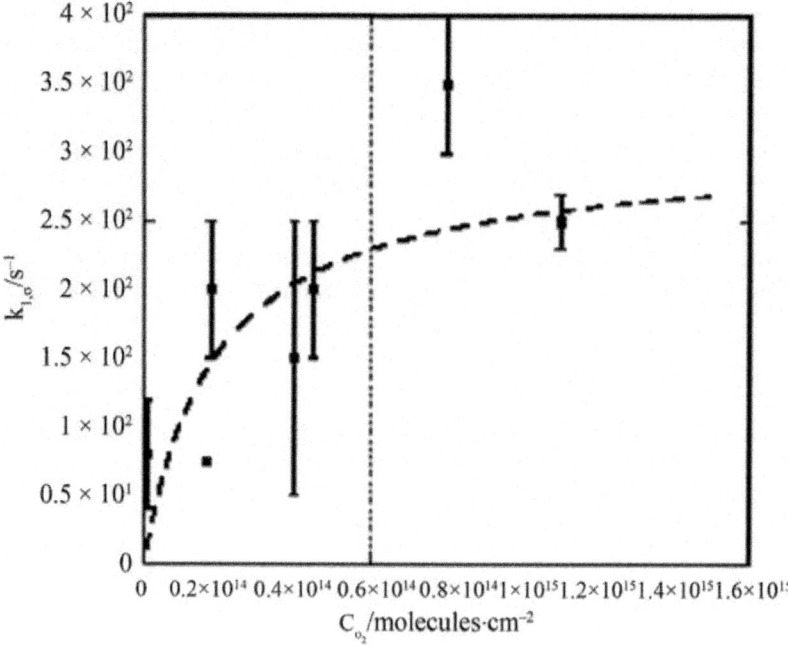

Figure 2: The pseudo first order reaction rate of naphthalene with gaseous ozone on aqueous droplets of 91 mm diameter in a falling droplet reactor. Data obtained from Raja et al [43]. The data fit gave k_{max} = 303 s^{-1} and $C_{Ox,1/2}$ = 1.92 × 10^{14} molecules.cm^{-3}. Note the asymptotic approach to a constant rate constant at high ozone concentration in the gas phase.

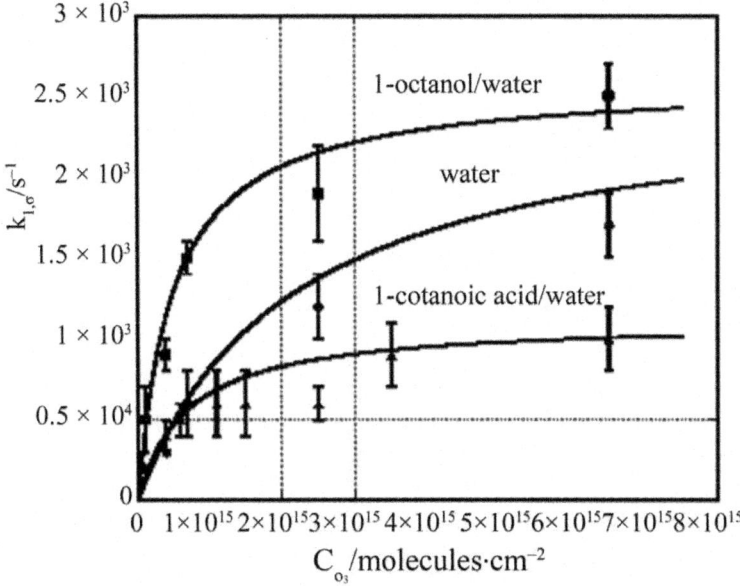

Figure 3: Pseudo first order reaction rate constants for gaseous anthracene reaction with gaseous ozone on different types of aqueous surfaces, viz., 2.5 mM aqueous solution of octanol, pure uncoated water, and 3.8 mM aqueous solution of 1-octanoic acid. Data obtained from Mmereki et al [39]. Note that in each case the rate constant reaches an asymptotic value at high concentrations of gaseous ozone. Values of kmax and $C_{ox,1/2}$ are given in Table 5.

$$r_i^s = k_{2,\sigma} \cdot \left(\Gamma_{ox}^{max} C_{ox}(g) \big/ \{ C_{ox,1/2} + C_{ox}(g) \} \right) \cdot \Gamma_i^\sigma \qquad (2)$$

Thus, a pseudo first-order reaction rate constant can be defined

$$k_{1,\sigma} = k_{max} \cdot \left(C_{ox}(g) \big/ \{ C_{ox,1/2} + C_{ox}(g) \} \right). \qquad (3)$$

The above equation explains the saturation-type behavior for the reaction rate constant with oxidant concentration as shown in Figures 2 and 3. The values of the two constants are given below in Table 5 for different water surfaces. Note that $C_{ox,1/2}(g)$ relates to the binding energy of ozone with the water surface and k_{max} depends on the different bimolecular reaction rate constants on specific surfaces. Thus, these values change depending on the nature of the water surface [49]. A problem with the above treatment is that molecular O_3 has only a small residence time ($\sim 10^{-9}$ s) on surfaces. The molecular mechanism leading to the L-H surface reaction of ozone with PAH can be reconciled only by invoking the presence of long-lived (>100 s) reactive oxygen intermediates (ROI) on the surface [50].

The overall pseudo-first order reaction rate constant should depend on the surface thickness since there are two parallel channels in which reactions occur in a water film once the organic molecules adsorb (and solvate) on the surface. This is shown in Figure 4. The overall reaction rate is given by the following equation [22]

$$k_1 = k_{\text{bulk}} + \left(k_{1,\sigma}/\delta\right)$$

(4)

where k_{bulk} and $k_{1,\sigma}$ are respectively the homogenous bulk aqueous phase and heterogeneous surface reaction rate constants. It is evident that as the film thickness becomes smaller, the heterogeneous reaction rate will make a larger contribution towards the overall reaction rate. This fact was demonstrated in our laboratory with regard to the UV photooxidation of two gaseous PAHs (naphthalene and phenanthrene). In both cases the major product concentrations were followed in the aqueous films and the first order rate constants for the degradation of PAH into respective products were obtained [36,51]. For example, the reaction rate constants for phenanthrene conversion to major products are given in **Figure 5**. It is clear that for all the three major products of oxidation the reaction rate increases as the film thickness decreases. Similar behavior was also observed for naphthalene degradation.

Table 5: Pseudo first order rate constant parameters for the reaction of anthracene with gas phase ozone on various water surfaces

Surface	k_{max}/s^{-1}	$C_{\text{ox},1/2}/\text{molecule·m}^{-3}$
Water	$(2.55 \pm 0.17) \times 10^{-3}$	$(21.4 \pm 0.4) \times 10^{-20}$
1-octanol on water	$(2.54 \pm 0.14) \times 10^{-3}$	$(5.0 \pm 0.9) \times 10^{-20}$
Octanoic acid on water	$(1.11 \pm 0.14) \times 10^{-3}$	$(6.8 \pm 2.9) \times 10^{-20}$
Hexanoic acid on water	$(0.48 \pm 0.07) \times 10^{-3}$	$(11.8 \pm 3.6) \times 10^{-20}$

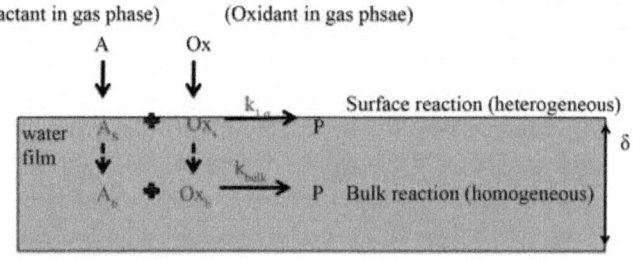

Figure 4: Schematic of the two reaction channels (heterogeneous and homogeneous) for gaseous organic species in water films.

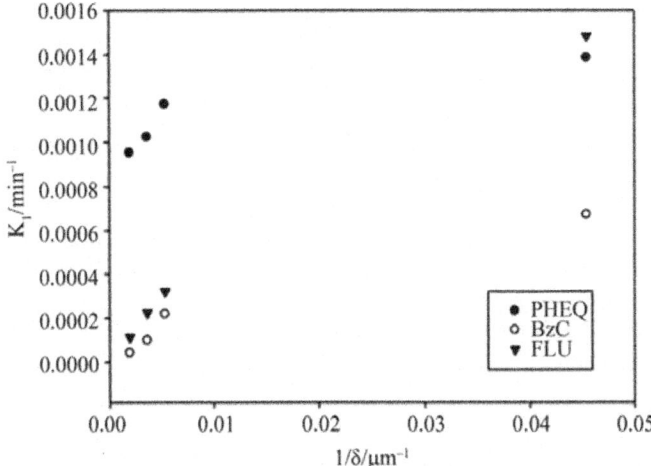

Figure 5: The photooxidation rates of gaseous phenanthrene on water films as a function of film thickness. The formation rate constants for three different compounds (PHEQPhenanthraquinone, BzC-Benzocoumarin, FLU-Fluoranthone) from phenanthrene are given. Data obtained from Chen et al. [36].

Others have also confirmed the surface reactions on water [52]. The reaction of $Cl_2^{-\bullet}$ with ethanol in water films was investigated using UV diffuse reflectance laser flash photolysis technique. It was demonstrated that the surface reaction rate is different from the bulk reaction rate. The surface reaction rate varied as the square root of the ethanol concentration and reached an asymptote at high concentrations. However, the bulk reaction rate was much slower and varied linearly with ethanol concentration.

Experimental field observations also lend support to the importance of surface reactions. For example, in analyzing the loss of ozone in coastal regions, it was observed that the surface reaction of ozone on the air-water interface of sea salt particles had to be invoked in models to correctly explain the observations [5]. This observation was confirmed with laboratory work and from molecular dynamics and ab-initio calculations. Fog and cloud droplets can process organic trace gases and particulates via aqueous phase photochemistry. Field observations have shown that fog droplets provide a microreactor environment for the formation of secondary organic aerosols (SOA) by photochemical transformations of organic compounds [6]. Real-time formation of secondary aerosol was detected using on-line aerosol mass spectrometry immediately upon the detection of fog. It was observed that the post-fog atmospheric environment contained particles and gases that were produced by evaporation of water from the fog droplet.

In the organic chemistry literature there is the term "on-water" heterogeneous chemistry that refers to orders of magnitude increase in the rates of catalysis of organic reactions at oil-water interfaces compared to homogeneous reactions in organic solvents [53]. This has been attributed to the breaking of an existing H-bonding network in homogeneous aqueous solution in order to permit catalysis, but not for the "on-water" reactions [54].

In order to explain the preference of surface reactions over bulk reactions for gaseous species on aerosols in the atmosphere, a hypothesis has been proposed [55,56]. For bulk aqueous phase reactions to occur the molecular species that are fully solvated (in a H-bonded water network) have to break their solvation shells. For surface species that are only partly solvated [57] and such an energy penalty does not exist. Diffusion also limits the encounter of the fully solvated species in the case of the bulk solution. There is evidence that the adsorbed molecules at the interface display faster orientational dynamics than in the bulk [58]. Moreover, even when reactions occur within the bulk solution, the diffusion of molecules out of the solvent cage is in competition with recombination reactions. At the surface recombination of reactants is inhibited as shown in photolysis reactions at the surface compared to the bulk [59].

SUMMARY

It is evident from principal observations that "surface" water is different from "bulk" water in reactivity. Trace gases (HOC) and atmospheric oxidant species adsorb and react on aqueous surfaces. These processes are of significance for aqueous atmospheric aerosols. Experimental work and molecular simulations support the above conclusion. It is evident that heterogeneous reactions deserve special attention in tropospheric chemistry and is important for thin films of water existing on aqueous aerosols.

ACKNOWLEDGEMENTS

The US National Science Foundation is gratefully acknowledged for support of this work through research grants (AGS 0907261 and AGS 1106569). This material was also presented in part as a Plenary Lecture at the "Recent Advances in Chemical Engineering and Technology" meeting organized by the Indian Institute of Chemical Engineers, Kochi Regional Centre, in Kochi, Kerala, India on 10-12 March 2011.

REFERENCES

1. J. H. Seinfeld and S. N. Pandis, "Atmospheric Chemistry and Physics," John Wiley and Sons, Inc., New York, 2006.

2. K. T. Valsaraj, "Elements of Environmental Engineering," 3rd Edition, Taylor and Francis Publishers, New York, 2009.

3. S. Solomon, S. Qin, D. Chen, Z. Marquis, K. B. Averyt, M. Tignor and H. L. Miller (Eds.), "Climate Change 2007: The Physical Basis, Contributions of Working Group I to the Fourth Assessment Report of the Intergovernmental Panel on Climate Change," IPCC, Cambridge University Press, Cambridge, 2007.

4. S. R.-V. Castrillon, N. Giovambattista, I. A. Aksay and P. Debenedetti, "Structure and Energetics of Thin Water Films," Journal of Physical Chemistry C, Vol. 115, 2011, pp. 4624-4635. doi:10.1021/jp1083967

5. E. M. Knipping, M. J. Lakin, K. L. Foster, P. Jungwirth, D. J. Tobias, R. B. Gerber, D. Dabdub and B. J. Finlayson-Pitts, "Experiments and Simulations of Ion-Enhanced Interfacial Chemistry on Aqueous NaCl Aerosols," Science, Vol. 288, No. 5464, 2000, pp. 301-306. doi:10.1126/science.288.5464.301

6. M. Dall'Osto, R. M. Harrison, H. Coe and P. Williams, "Real-Time Secondary Aerosol Formation during a Fog Event in London," Atmospheric Chemistry and Physics, Vol. 9, No. 7, 2009, pp. 2459-2469. doi:10.5194/acp-9-2459-2009

7. S. Raja, R. Ravikrishna, X.-Y. Yu, T. Lee, J. Chen, R. R. Kommalapati, K. Murugesan, X. Shen, K. T. Valsaraj and J. L. Collett, "Fog Chemistry in the Texas-Louisiana Gulf Coast Corridor," Atmospheric Environment, Vol. 42, No. 9, 2008, pp. 2048-2061.doi:10.1016/j.atmosenv.2007.12.004

8. J. George and J. P. D. Abbatt, "Heterogeneous Oxidation of Atmospheric Aerosol Particles by Gas-Phase Radicals," Nature Chemistry, Vol. 2, No. 9, 2010, pp. 713-722.doi:10.1038/nchem.806

9. C. J. van Oss, "Interfacial Forces in Aqueous Media," Taylor and Francis, New York, 2006.

10. E. E. Meyer, K. J. Rosenberg and J. N. Israelachvili, Proceedings of the National Academy of Sciences, Vol. 103, 2006, pp. 15739-15746.

11. J. N. Israelachvili, "Intermolecular and Surface Forces," 3rd Edition, Academic Press, New York, 2010.

12. J. S. Smith and K. T. Valsaraj, "Solvent Sublation for Industrial Wastewater Treatment," Chemical Engineering Progress, Vol. 94, No. 5, 1998, pp. 69-76.

13. H. F. Rafson (Ed.), "Odor and VOC Control Handbook," McGraw Hill Pub Co., New York, 1998.

14. K. T. Valsaraj, G. J. Thoma, D. D. Reible and L. J. Thibodeaux, "On the Enrichment of Hydrophobic Organic Compounds in Fog Droplets," Atmospheric Environment, Vol. 27A, No. 2, 1993, pp. 203-210. doi:10.1016/0960-1686(93)90351-X

15. J. D. Blando and B. J. Turpin, "Secondary Organic Aerosol Formation in Cloud and Fog Droplets: A Literature Evaluation of Plausibility," Atmospheric Environment, Vol. 34, No. 10, 2000, pp. 1623-1632. doi:10.1016/S1352-2310(99)00392-1

16. D. J. Donaldson and K. T. Valsaraj, "Adsorption and Reaction of Trace Gas-Phase Organic Compounds on Atmospheric Water Film Surfaces: A Critical Review," Environmental Science & Technology, Vol. 44, No. 3, 2010, pp. 865-873. doi:10.1021/es902720s

17. K. T. Valsaraj, "Trace Gas Adsorption Thermodynamics at the Air-Water Interface: Implications in Atmospheric Chemistry," Pure and Applied Chemistry, Vol. 81, No. 10, 2009, pp. 1889-1901. doi:10.1351/PAC-CON-08-07-06

18. D. J. Donaldson and D. Anderson, "Adsorption of Atmospheric Gases at the Air-Water Interface. 2. C_1-C_4 Alcohols, Acids, and Acetone," Journal of Physical Chemistry A, Vol. 103, No. 1, 1999, pp. 871-876. doi:10.1021/jp983963h

19. R. Braunt and M. J. Conklin, "Dynamic Determination of Vapor/Water Interface Adsorption for Volatile Hydrophobic Organic Compounds (VHOCs) Using Axisymmetric Drop Shape Analysis: Procedure and Analysis of Benzene Adsorption," Journal of Physical Chemistry B, Vol. 104, No. 47, 2004, pp. 11146-11152. doi:10.1021/jp001140y

20. Hartkopf and B. L. Karger, "Study of the Interfacial Properties of Water by Gas Chromatography," Accounts of Chemical Research, Vol. 6, 1973, pp. 209-221. doi:10.1021/ar50066a006

21. S. Raja, F. S. Yaccone, R. Ravikrishna and K. T. Valsaraj, "Thermodynamic Parameters for the Adsorption of Aromatic Hydrocarbon Vapors at the Gas-Water Interface," Journal of Chemical & Engineering Data, Vol. 47, No. 5, 2002, pp. 1213-1219. doi:10.1021/je025520j

22. J. Chen, F. S. Ehrenhauser, K. T. Valsaraj and M. J. Wornat, "Uptake and UV-Photooxidation of Gas-Phase PAHs on the Surface of Atmospheric Water Films. 1. Naphthalene," Journal of Physical Chemistry A, Vol. 110, No. 29, 2006, pp. 161-916. doi:10.1021/jp062560b8

23. K. T. Valsaraj, "On the Physico-Chemical Aspects of Partitioning of Non-Polar Hydrophobic Organics at the AirWater Interface," Chemosphere, Vol. 17, No. 5, 1988, pp. 875-887.doi:10.1016/0045-6535(88)90060-4

24. K. T. Valsaraj, "Binding Constants for Non-Polar Hydrophobic Organics at the Air-Water Interface: Comparison of Experimental and Predicted Values," Chemosphere, Vol. 17, No. 10, 1988, pp. 2049-2061. doi:10.1016/0045-6535(88)90015-X

25. K.-U. Goss and R. P. Schwarzenbach, "Linear Free Energy Relationships Used to Evaluate Equilibrium Partitioning of Organic Compounds," Environmental Science & Technology, Vol. 35, No. 1, 2001, pp. 1-9. doi:10.1021/es000996d

26. K.-U. Goss, "Conceptual Model for the Adsorption of Organic Compounds from the Gas Phase to Liquid and Solid Surfaces," Environmental Science & Technology, Vol. 31, No. 12, 1997, pp. 3600-3605. doi:10.1021/es970361n

27. C. P. Kelly, C. J. Cramer and D. G. Truhlar, "Predicting Adsorption Coefficients at Air-Water Interfaces Using Universal Solvation and Surface Area Models," Journal of Physical Chemistry B, Vol. 108, No. 34, 2004, pp. 12882-12897. doi:10.1021/jp037210t

28. K.-U. Goss, "Predicting Adsorption of Organic Chemicals at the Air-Water Interface," Journal of Physical Chemistry A, Vol. 113, No. 44, 2009, 12256-12259.doi:10.1021/jp907347p

29. R. Vacha, P. Jungwirth, J. Chen and K. T. Valsaraj, "Adsorption of Polycyclic Aromatic Hydrocarbons at the Air-Water Interface: Molecular Dynamics Simulations and Experimental Atmospheric Observations," Physical Chemistry Chemical Physics, Vol. 8, No. 38, 2006, pp. 4461-4467. doi:10.1039/b610253k

30. R. Vacha, K. Cwiklik, J. Rezac, P. Hobza, P. Jungwirth, K. Valsaraj, S. Bahr and V. Kempter, "Adsorption of Aromatic Hydrocarbons and Ozone at Environmental Aqueous Surfaces," Journal of Physical Chemistry A, Vol. 112, No. 22, 2008, pp. 4942-4950.doi:10.1021/jp711813p

31. T. Somasundaram, R. M. Lyndon-Bell and C. H. Patterson, "The Passage of Gases through the Liquid Water/ Vapour Interface: A Simulation Study," Physical Chemistry Chemical Physics, Vol. 1, No. 1, 1999, pp. 143-148. doi:10.1039/a805067h

32. C. D. Wick, B. Chen and K. T. Valsaraj, "Computational Investigation of the Influence of Surfactants on the AirWater Interfacial Behavior of Polycylic Aromatic Hydrocarbons," Journal of Physical Chemistry C, Vol. 114, No. 34, 2010, pp. 14520-14527.doi:10.1021/jp1039578

33. J. Chen, F. Ehrenhauser, T. Arachi, F. Hung, K. Valsaraj and M. Wornat, "Adsorption of Gas-Phase Phenanthrene on Atmospheric Water and Ice Films," Polycyclicarom Atichydrocarbons, Vol. 31, 2011, pp. 1-26. doi:10.1080/10406638.2011.585370

34. R. Vacha, P. Slavaicek, M. Mucha, B. Finlayson-Pitts and P. Jungwirth, "Adsorption of Atmospherically Relevant Gases at the Air/Water Interface: Free Energy Profiles of Aqueous Solvation of N_2, O_2, O_3, OH, H_2O, HO_2, and H_2O_2," Journal of Physical Chemistry A, Vol. 108, 2004, pp. 11573-11579.

35. R. Vacha, P. Slavicek, M. Mucha, B. J. Finlayson-Pitts and P. Jungwirth, "Adsorption of Atmospherically Relevant Gases at the Air/Water Interface: Free Energy Profiles of Aqueous Solvation of N_2, O_2, O_3, OH, H_2O, HO_2, and H_2O_2," Journal of Physical Chemistry A, Vol. 108, No. 52, 2004, pp. 11573-11579. doi:10.1021/jp046268k

36. J. Chen, F. Ehrenhauser, K. T. Valsaraj and M. J. Wornat, "Adsorption and UV Photooxidation of Gas-Phase Phenanthrene on Atmospheric Films," ACS Symposium Series, Vol. 1005, 2009, pp. 127-146. doi:10.1021/bk-2009-1005.ch009

37. I.-F. W. Kuo and C. J. Mundy, "An Ab initio Molecular Dynamics Study of the Aqueous Liquid-Vapor Interface," Science, Vol. 303, No. 5658, 2004, pp. 658-662.doi:10.1126/science.1092787

38. X. D. Zhu, H. Suhr and Y. R. Shen, "Surface Vibrational Spectroscopy by Infrared-Visible Sum Frequency Generation," Physical Review B, Vol. 35, No. 6, 1987, pp. 3047-3050.doi:10.1103/PhysRevB.35.3047

39. B. T. Mmereki, D. J. Donaldson, J. B. Gilman, T. L. Eliason and V. Vaida, "Kinetics and Products of the Reaction of Gas-Phase Ozone with Anthracene Adsorbed at the Air-Aqueous Interface," Atmospheric Environment, Vol. 38, 2004, pp. 6091-6103.doi:10.1016/j.atmosenv.2004.08.014

40. E. S. Enami, M. R. Hoffmann and A. J. Colussi, "Ozonolysis of Uric Acid at the Air/Water Interface," Journal of Physical Chemistry B, Vol. 112, No. 14, 2008, pp. 4153-4157.doi:10.1021/jp712010k

41. R. Vacha, L. Cwiklik, J. Rezac, P. Hobza, P. Jungwirth, K. Valsaraj, S. Bahr and V. Kempter, "Adsorption of Aromatic Hydrocarbons and Ozone at Environmental Aqueous Surfaces," Journal of Physical Chemistry A, Vol. 112, No. 22, 2008, pp. 4942-4947.doi:10.1021/jp711813p

42. S. N. Wren and D. J. Donaldson, "Glancing-Angle Raman Spectroscopic Probe for Reaction Kinetics at Water Surfaces," Physical Chemistry

Chemical Physics, Vol. 12, No. 11, 2010, pp. 2648-2654. doi:10.1039/b922254e

43. S. Raja and K. T. Valsaraj, "Heterogeneous Oxidation by Ozone of Naphthalene Adsorbed at the Air-Water Interface of Micron-Size Water Droplets," Journal of the Air & Waste Management Association, Vol. 55, No. 9, 2005, pp. 1345-1355.

44. V. Stiopkin, H. D. Jayathilake, A. N. Bordenyuk and A. V. Benderskii, "Heterodyne-Detected Vibrational Sum Frequency Generation Spectroscopy," Journal of the American Chemical Society, Vol. 130, No. 7, 2008, pp. 2271-2275. doi:10.1021/ja076708w

45. C. Waring, P. A. J. Bagot, M. L. Costen and K. G. McKendrick, "Reactive Scattering as a Chemically Specific Analytical Probe of Liquid Surfaces," The Journal of Physical Chemistry Letters, Vol. 2, No. 1, 2011, pp. 12-18. doi:10.1021/jz1013032

46. Y. R. Shen and V. Ostroverkhov, "Sum-Frequency Vibrational Spectroscopy on Water Interfaces: Polar Orientation of Water Molecules at Interfaces," Chemical Reviews, Vol. 106, No. 4, 2006, pp. 140-154. doi:10.1021/cr040377d

47. K. Harper, B. Minofar, M. R. Sierra-Hernandez, N. N. Casillas-Ituarte, M. Roeselova and H. C. Allen, "Surface Residence and Uptake of Methyl Chloride and Methyl Alcohol at the Air/Water Interface Studied by Vibrational Sum Frequency Spectroscopy and Molecular Dynamics," Journal of Physical Chemistry A, Vol. 113, No. 10, 2009, pp. 2015-2024. doi:10.1021/jp808630v

48. K. B. Eisenthal, "Equilibrium and Dynamic Processes at Interfaces by Second Harmonic and Sum Frequency Generation," Annual Review of Physical Chemistry, Vol. 43, 1992, pp. 627-661. doi:10.1146/annurev.pc.43.100192.003211

49. N.-O. A. Kwamena, M. G. Staikova, D. J. Donaldson, I. J. Goerge and J. P. D. Abbatt, "Role of the Aerosol Substrate in the Heterogeneous Ozonation Reactions of Surface-Bound PAHs," Journal of Physical Chemistry A, Vol. 111, No. 43, 2007, pp. 11050-11058.doi:10.1021/jp075300i

50. M. Shiraiwa, Y. Sosedova, A. Ronviere, H. Yang, Y. Zhang, J. P. D. Abbatt, M. Ammann and U. Poschl, "The Role of Long-Lived Reactive Oxygen Intermediates in the Reaction of Ozone with Aerosol Particles," Nature Chemistry, Vol. 3, 2011, pp. 291-295.doi:10.1038/NCHEM.988

51. J. Chen and K. Valsaraj., "Uptake and UV-Photooxidation of Gas-Phase PAHs on the Surface of Atmospheric Water Films. 1. Naphthalene," Journal

of Physical Chemistry A, Vol. 110, 2006, pp. 9161-9166. doi:10.1021/jp062560b

52. R. S. Strekowski, R. Remorov and Ch. George, "Direct Kinetic Study of the Reaction of Cl_2Radical Anions with Ethanol at the Air-Water Interface," Journal of Physical Chemistry A, Vol. 107, 2003, pp. 2497-2504. doi:10.1021/jp026174f

53. S. Narayan, J. Muldoon, M. G. Finn, V. V. Fokin, H. C. Kolb and K. B. Sharpless, "On Water: Unique Reactivity of Organic Compounds in Aqueous Suspension," Angewandte Chemie International Edition, Vol. 44, No. 21, 2005, pp. 3275-3279.doi:10.1002/anie.200462883

54. Y. Jung and R. A. Marcus, "On the Theory of Organic Catalysis 'on Water'," Journal of the American Chemical Society, Vol. 129, No. 17, 2007, pp. 5492-5502.doi:10.1021/ja068120f

55. K. T. Valsaraj, "Trace Gas Adsorption Thermodynamics at the Air-Water Interface: Implications in Atmospheric Chemistry," Pure and Applied Chemistry, Vol. 81, No. 10, 2009, pp. 1889-1901. doi:10.1351/PAC-CON-08-07-06

56. P. Nissensson, C. J. X. Knox, B. J. Finlayson-Pitts, L. F. Phillips and D. Dabdub, "Enhanced Photolysis in Aerosols: Evidence for Important Surface Effects," Physical Chemistry Chemical Physics, Vol. 8, No. 40, 2006, pp. 4700-4710. doi:10.1039/b609219e

57. H. Watanabe, S. Yamaguchi, S. Sen, A. Morita and T. Tahara, "'Half-Hydration' at the Air/Water Interface Revealed by Heterodyne-Detected Electronic Sum Frequency Generation Spectroscopy, Polarization Second Harmonic Generation, and Molecular Dynamics Simulation," Journal of Chemical Physics, Vol. 132, No. 14, 2010, Article ID: 144701. doi:10.1063/1.3372620

58. M. L. Johnson, C. Rodriguez and I. Benjamin, "Rotational Dynamics of Strongly Adsorbed Solute at the Water Surface," Journal of Physical Chemistry A, Vol. 113, 2009, pp. 2086-2091. doi:10.1021/jp808842k

59. P. Nissenson, D. Dabdub, R. Das, V. Maurino, C. Minero, D. Vione, "Evidence of the Water-Cage Effect on the Photolysis of and $FeOH^{2+}$. Implications of This Effect and of H_2O_2Surface Accumulation on Photochemistry at the Air-Water Interface of Atmospheric Droplets," tmospheric Environment, Vol. 44, 2010, pp. 4859-4866.

Chapter 8

BIOMIMETIC MODELING OF COPPER COMPLEXES: A STUDY OF ENANTIOSELECTIVE CATALYTIC OXIDATION ON D-(+)-CATECHIN AND L-(-)-EPICATECHIN WITH COPPER COMPLEXES

Francesco G. Mutti, Roberta Pievo, Maila Sgobba, Michele Gullotti, and Laura Santagostini

Dipartimento di Chimica Inorganica, Metallorganica e Analitica "Lamberto Malatesta", Università di Milano, Istituto ISTM-CNR, Via Venezian 21, 20133 Milano, Italy

ABSTRACT

The biomimetic catalytic oxidations of the dinuclear and trinuclear copper(II) complexes versus two catechols, namely, D-(+)- catechin and L-(−)-epicatechin to give the corresponding quinones are reported. The unstable quinones were trapped by the nucleophilic reagent, 3-methyl-2-benzothiazolinone hydrazone (MBTH), and have been calculated the molar absorptivities of the different quinones. The catalytic efficiency is moderate, as inferred by kinetic constants, but the complexes exhibit significant enantio-differentiating ability towards the catechols, albeit for the dinuclear complexes, this enantio-differentiating ability is lower. In all cases, the preferred enantiomeric substrate is D-(+)-catechin to respect the other catechol, because of the spatial disposition of this substrate.

INTRODUCTION

Reproducing complex biological reactivity within a simple synthetic molecule is a challenging endeavor with both intellectual and aesthetic goals. The sequence of examining biological reactivity, creating similar chemical architectures, and determining functional reaction conditions for model systems is a process that allows the biological code of reactivity to be deciphered. In the past years, the report on the crystal structures of type 3 copper enzymes (e.g., catechol oxidase, hemocyanins, and tyrosinase) [1–4], as too type 2- type 3 copper enzymes (e.g., ascorbate oxidase, laccase, ceruloplasmin) [5–8] has taken a new turn. The greater availability of such structural information now allows a shift in the role of synthetic modeling from structural and spectroscopic endeavors to the development of functional and catalytic models. Functional models can provide an opportunity to examine a biological reactivity at a small-molecule level of detail through systematic and comparative studies. Although one goal of modeling is reproduction of reactivity, extension of this reactivity beyond the scope of the inspiring system is perhaps an even more important objective. Adequate synthetic models that have similar structural, spectroscopic, and functional properties of active sites of copper proteins are done [9–13]. These models provide many elegant examples of selective and environmentally benign oxidants capable of performing interesting organic transformations, and many of these are copper complexes that use dioxygen as the ultimate oxidant above all in the catecholase activity [14–20]. The interest of our group has mainly focused on dinuclear, and trinuclear [21–28] copper complexes derived from octadentate nitrogen ligands which show catecholase activities. Some of these compounds contain chiral centers [24–28], and we have demonstrated the possibility to induce stereoselectivity in the catalytic oxidation of chiral catechols, such as L- and D-Dopa and their methyl esters. In the present paper we have extended this investigation on our chiral complexes using as substrates other potential catechols, namely, D-(+)-catechin and L-(–)-epicatechin, and we have found that the stereoselective catalytic oxidation of these substrates depends on the chirality of the dinuclear or trinuclear copper compounds, and on the spatial disposition of the catechols.

EXPERIMENTAL

General Remarks

The dinuclear and trinuclear copper(II) complexes $[Cu_2\text{-RDABN-4Bz}_4]$ $[ClO_4]_4 2H_2O$ (1); and $[Cu_3\text{-R-DABN-4Bz}_4]$ $[ClO_4]_6 2H_2O$ (2); $[Cu_2\text{-L-Lys-4Bz}_4]$ $[ClO_4]_4 6H_2O$ (3); $[Cu_3\text{-L-Lys-4Bz}_4][ClO_4]_6 6H_2O$ (4); $[Cu_2\text{-R-DABN-}_3Im_4]$ $[ClO_4]_4 6H_2O$ (5); $[Cu_3\text{-R-DABN-3Im}_4][ClO_4]_6 2H_2O$ (6) were prepared as described previously [27]. The trichiral complexes $[Cu_2\text{-R-DABN-L-Ala-Bz}_4][ClO_4]_4$ (7) and $[Cu_3\text{- R-DABN-L-Ala-Bz}_4][ClO_4]6$ (8) were prepared with standard procedures [29]. The dinuclear complex $[Cu_2L\text{-}66]$ $[ClO_4]_4 6H_2O$ (9) was prepared as described previously [30]. All the compounds are shown in Figure 1.

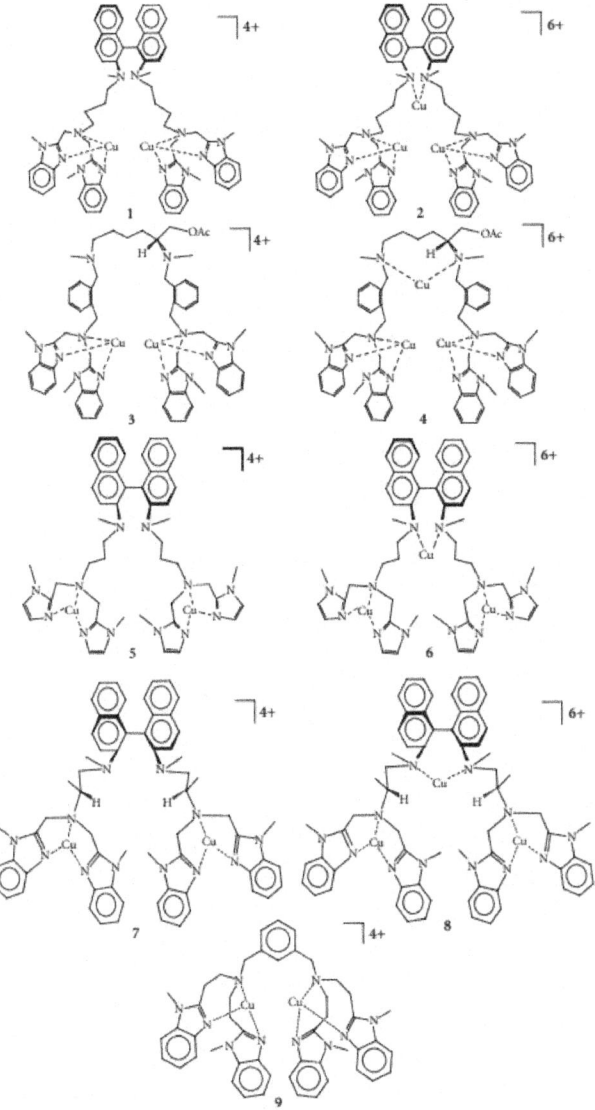

Figure 1: Dinuclear and trinuclear copper complexes used in the biomimetic catalytic oxidations of catechins.

Caution

Although no problems were encountered during the preparation of perchlorate salts, suitable care should be taken when handling such potentially hazardous compounds.

Materials and Physical Methods

Commercial starting materials were used without purifi- cation and the solvents used for the reactions were all spectrophotometric grade. Acetonitrile was distilled from potassium permanganate and sodium carbonate; it was then stored over calcium hydride and distilled before use under nitrogen. The pH of the solutions was measured with an Amel instrument 338. Optical spectra were obtained with HP 8453 diode array spectrophotometer equipped with a thermostated cell holder at the temperature of $20 \pm 0.1 \circ C$. The data were treated with the commercial program FigSys (BioSoft, Cambridge, UK). Formation kinetics was carried out under pseudo-first-order conditions at $20 \pm 0.1 \circ C$.

Determination of Molar Absorptivities of the Quinones

It is well known that dinuclear and trinuclear model complexes, like tyrosinase, oxidizes o-diphenols, triphenols, and flavonoids to quinones, but, in all cases, the resulting quinones may undergo nonenzymatic autopolymerization to produce colored compounds. To prevent further reactions of the quinones initially formed, a nucleophilic reagent, 3- methyl-2-benzothiazolinone hydrazone (MBTH), that traps the quinones and generates chromophoric adducts, was used. Unfortunately, no molar absorptivities of these adducts for the substrates were available, so a spectrophotometric method to determine the λ_{max} and the molar absorptivities of the adducts was performed. In general, the method is based on the oxidation of the substrate by an excess of sodium periodate, condition under which the reaction was very fast [31, 32]. The unstable quinones were trapped by the nucleophilic reagent (MBTH), and related λ_{max} was detected. In all the experiments, only one band developed in the range 300–900 nm, which corresponds to the adducts with MBTH. Based on the recording of λ_{max}, an experimental design can be carried out to determine the molar absorptivities of the different quinones, for example, performing spectra with different substrate concentrations and fitting the data so obtained to a Lambert-Beer equation by linear regression. Plots of the absorbance values obtained versus the different substrate concentrations allow calculating molar absorptivities. For D-(+)-catechin (CQ), the experimental conditions were λ_{max} = 459 nm; 50 mM phosphate buffer (pH 7.0)/MeOH (9:1, v:v) at $20 \pm 0.1 \circ C$; 2 mM NaIO4; 1 mM MBTH; substrate concentrations (CQ) from 5 µM to 40 µM; quartz cell 1 cm path length; final volume in cell 2 mL. The coefficient of determination (r^2) was 0.998 and the molar absorptivity was 17230 M^{-1} cm^{-1}. For L-(−)- epicatechin (EQ), the experimental conditions were λmax = 463 nm; 50 mM phosphate buffer (pH 7.0)/MeOH (9:1, v:v) at $20 \pm 0.1 \circ C$; 2 mM $NaIO_4$; 1 mM MBTH; substrate concentrations (EQ) from 5 µM to 45 μM; quartz cell 1 cm path length; final volume in cell 2 mL. The coefficient of determination (r2) was 0.994 and the molar absorptivity was 18950 M^{-1} cm^{-1}.

Catecholase Activities

The kinetics of catalytic oxidation of D-(+)-catechin and L-(−)-epicatechin were studied by UV-Vis spectroscopy using a magnetically stirred and thermostated 1-cm path length cell. The temperature during the measurements was kept constant at 20 ± 0.1°C. A mixture of aqueous phosphate buffer (50 mM, pH 7.0)-methanol 9:1 (v:v) saturated with atmospheric oxygen was used as solvent. All the kinetic experiments were carried out in duplicate. The experiments performed over a substrate concentration range were initiated by adding a few microlitres of the complexes (final concentrations $0.2-1.4 \times 10^{-5}$ M) to the solution of the substrates; MBTH was maintained 1.0×10^{-3} M; the concentration of the substrate was varied between 4.0×10^{-6} and 8.0×10^{-4} M (final volume 2 mL). The formation of the stable D-(+)-catechin-o-quinone-MBTH and L-(−)-epicatechino-quinone-MBTH adducts was followed through the development of the strong absorption band at 459 nm ($\varepsilon = 17230$ M−1 cm^{-1}) for D-(+)-catechin-o-quinone-MBTH, and an absorption band at 463 nm ($\varepsilon = 18950$ M−1 cm^{-1}) for L-(−)-epicatechin-o-quinone-MBTH, respectively. In all the experiments, the noise was reduced by reading the absorbance difference between λ_{max} and 1100 nm, where the absorption remains negligible during the assay. The initial rates of oxidations were obtained by fitting the absorbance versus time curves in the first seconds of the reactions.

RESULTS AND DISCUSSION

Stereoselective Catalytic Oxidations

The catalytic oxidations of catechols are the most widely employed test reaction to investigate the behavior of tyrosinase and catechol oxidase model complexes. Previous studies [24–27] have shown that chiral dinuclear and trinuclear copper complexes were able to display stereo-discriminating ability towards optically active catechols to give the corresponding o-quinones. To confirm this behavior, new chiral catechols were employed in these catalytic stereoselective oxidations. D-(+)-catechin and L-(−)-epicatechin (flavan-3-ols) (Figure 2) constitute a class of phenolic compounds ubiquitous in plants and widely found in fruits, vegetables, and beverages [33–35]. In particular, they are one of the major quality factors in grapes and then in the resulting wine [36, 37].

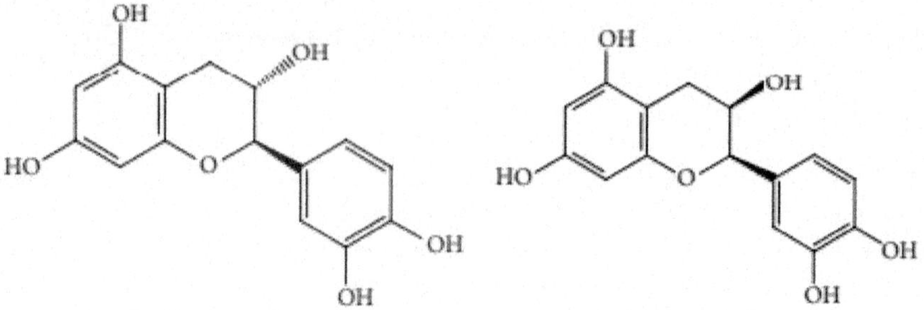

D-(+)-catechin

(a)

L-(−)-epicatechin

(b)

Figure 2: Absolute stereochemistry configuration of the catechols.

The catalytic oxidation of polyphenolic substrates, including catechins, was well studied by many authors [38–41]. These reactions take place in the presence of atmospheric oxygen when polyphenol oxidase (PPO) and the corresponding substrates are mixed at the same time. The fundamental first step is the transformation of o-diphenols to the corresponding o-quinones. The fate and stability of o-quinones vary widely, depending both on the phenolic precursor and on environmental factors. In particular, the o-quinones of D-(+)-catechin and L-(−)-epicatechin were seen to be much less stable than those of other o-quinones.

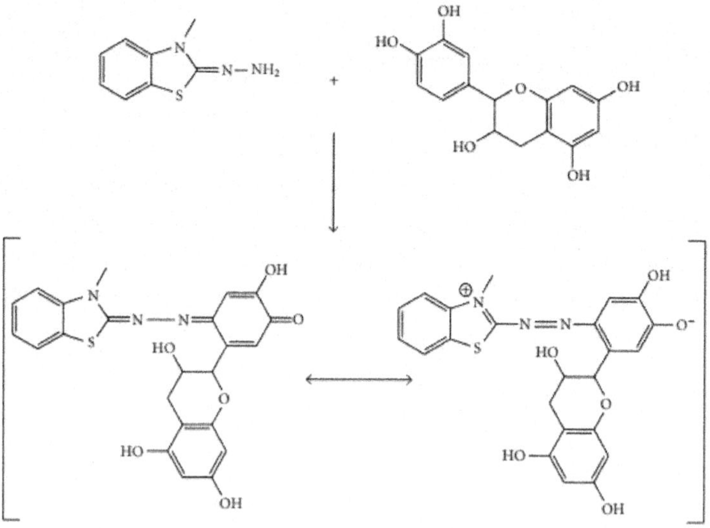

Scheme 1: Nucleophilic attach of the reagent MBTH to the catechols.

The prolonged autoxidation, either chemical or enzymic, led to the formation of polymers resulting from repeated condensation reactions between an aromatic ring of one molecule with an aromatic ring of another ("head to tail" polymerization mechanism). Depending on how phenolic compounds are oxidized, the condensation products formed from catechins may differ. In fact, the pH of the solution influences considerably the obtained products [42, 43], because at low pH values is favored the formation of colorless condensation products, whereas yellow compounds tended to be formed at higher pH values. To avoid any effect due to pH-dependence of oxidation products and to stop the reaction at quinones formation, a nucleophilic reagent MBTH, that traps the quinones and generates chromophoric adducts, was used (Scheme 1).

We have then studied the pH dependence of the reaction rates and have found that the better pH for the catalytic oxidation of catechins in the presence of the chiral copper complexes is pH 7.0. In order to make a comparison of the catalytic activity among the various chiral complexes reported here, we also studied the catalytic oxidation of the catechins in the presence of the achiral dinuclear complex $[Cu2(L-66)]^{4+}$ and MBTH.Assuming that, for the present biomimetic catalytic reactions, a two-step mechanism of catechol oxidation holds as in the case of our previous studies with dinuclear [22], and trinuclear [23] copper(II) complexes, the following simplified catalytic scheme can be hypothesized, where two molecules of (CatH$_2$) per cycle are oxidized to quinone (Q):

$$Cu_2^{II} + CatH_2 \leftrightarrow [Cu_2^{II}/CatH_2] \rightarrow Cu_2^{I} + Q + 2H^+,$$

$$Cu_2^{I} + CatH_2 + O_2 \leftrightarrow [Cu_2/O_2/CatH_2] \rightarrow Cu_2^{II} + Q + 2H_2O. \tag{1}$$

Since, the kinetic experiments showed monophasic behavior and it was impossible to separate the two steps. Thus, either the two steps have a similar rate or the first one is slower. The dependence of the rates of the catalytic reactions as a function of the substrate concentration exhibited a hyperbolic behavior in all cases. However all the complexes exhibited substrate inhibition at high-substrate concentrations, and therefore the kinetic parameters reported in Table 1 were estimated with the equation here reported:

$$v = \frac{V_{max}[S]}{K_M(1 + K_c[S]^2) + [S]}. \tag{2}$$

Table 1: Kinetic parameters for the stereoselective oxidations of -(+)-catechin and -(−)-epicatechin in methanol-aqueous phosphate buffer, pH 7.0 at C.

Complexes Substrate	K_M (M)	k_{cat} (s^{-1})	k_{cat}/K_M (M^{-1} s^{-1})	R%
[Cu$_2$(R-DABN-4Bz$_4$)]$^{4+}$				
D-(+)-catechin	$(2.00 \pm 0.33) \times 10^{-5}$	$(1.21 \pm 0.75) \times 10^{-2}$	604	−32.0
L-(−)-epicatechin	$(1.20 \pm 0.31) \times 10^{-5}$	$(1.41 \pm 0.96) \times 10^{-2}$	1169	
[Cu$_3$(R-DABN-4Bz$_4$)]$^{6+}$				
D-(+)-catechin	$(1.33 \pm 0.32) \times 10^{-5}$	$(1.85 \pm 0.12) \times 10^{-2}$	1387	32.8
L-(−)-epicatechin	$(3.93 \pm 0.94) \times 10^{-5}$	$(2.75 \pm 0.31) \times 10^{-2}$	701	
[Cu$_2$(L-Lys-4Bz$_4$)]$^{4+}$				
D-(+)-catechin	$(1.14 \pm 0.38) \times 10^{-5}$	$(1.86 \pm 0.18) \times 10^{-2}$	1632	46.0
L-(−)-epicatechin	$(2.95 \pm 0.75) \times 10^{-5}$	$(1.78 \pm 0.17) \times 10^{-2}$	604	
[Cu$_3$(L-Lys-4Bz$_4$)]$^{6+}$				
D-(+)-catechin	$(1.37 \pm 0.33) \times 10^{-5}$	$(2.01 \pm 0.02) \times 10^{-2}$	1470	60.4
L-(−)-epicatechin	$(1.02 \pm 0.26) \times 10^{-4}$	$(3.71 \pm 0.58) \times 10^{-2}$	363	
[Cu$_2$(R-DABN-3Im$_4$)]$^{4+}$				
D-(+)-catechin	$(2.19 \pm 0.70) \times 10^{-5}$	$(1.09 \pm 0.15) \times 10^{-2}$	498	7.2
L-(−)-epicatechin	$(1.96 \pm 0.58) \times 10^{-5}$	$(8.44 \pm 0.69) \times 10^{-3}$	431	
[Cu$_3$(R-DABN-3Im$_4$)]$^{6+}$				
D-(+)-catechin	$(1.01 \pm 0.32) \times 10^{-3}$	$(5.10 \pm 0.39) \times 10^{-3}$	507	42.6
L-(−)-epicatechin	$(3.51 \pm 1.05) \times 10^{-5}$	$(7.15 \pm 0.80) \times 10^{-3}$	204	
[Cu$_2$(R-DABN-L-Ala-Bz$_4$)]$^{4+}$				
D-(+)-catechin	$(3.60 \pm 0.36) \times 10^{-5}$	$(2.77 \pm 0.11) \times 10^{-2}$	769	8.8
L-(−)-epicatechin	$(4.81 \pm 0.37) \times 10^{-5}$	$(3.10 \pm 0.11) \times 10^{-2}$	644	
[Cu$_3$(R-DABN-L-Ala-Bz$_4$)]$^{6+}$				
D-(+)-catechin	$(5.05 \pm 0.48) \times 10^{-5}$	$(6.46 \pm 0.15) \times 10^{-2}$	1280	5.3
L-(−)-epicatechin	$(5.08 \pm 0.37) \times 10^{-5}$	$(5.84 \pm 0.10) \times 10^{-2}$	1150	
[Cu$_2$(L66)]$^{4+}$				
D-(+)-catechin	$(6.69 \pm 1.06) \times 10^{-5}$	$(7.57 \pm 0.60) \times 10^{-2}$	1131	1.9
L-(−)-epicatechin	$(4.08 \pm 0.81) \times 10^{-5}$	$(4.44 \pm 0.13) \times 10^{-2}$	1088	

The copper complexes with chiral centers exhibited variable degree of stereoselectivity (Table 1) which has been evaluated using kinetic parameters with the following equation:

$$R\% = \frac{[(k_{cat}/K_M)_D - (k_{cat}/K_M)_L]}{[(k_{cat}/K_M)_D + (k_{cat}/K_M)_L]} \times 100,$$

(3)

where D and L are D-(+)-catechin and L-(−)-epicatechin.

The catalytic activity of all the complexes, except [Cu$_2$(R-DABN-4Bz$_4$)]$^{4+}$, shows that the preferred coordination of the catechols is for D-(+)-catechin. This preference is probably dictated by the chirality of the binaphthyl or lysine residues, as shown by our studies on related complexes [24–26], and especially by the spatial disposition of the catechol substrates. In fact, by simple calculation of molecular energy minimization, D-(+)-catechin shows a disposition almost planar with only the hydroxyl group out of plane (Figure 3(a)).

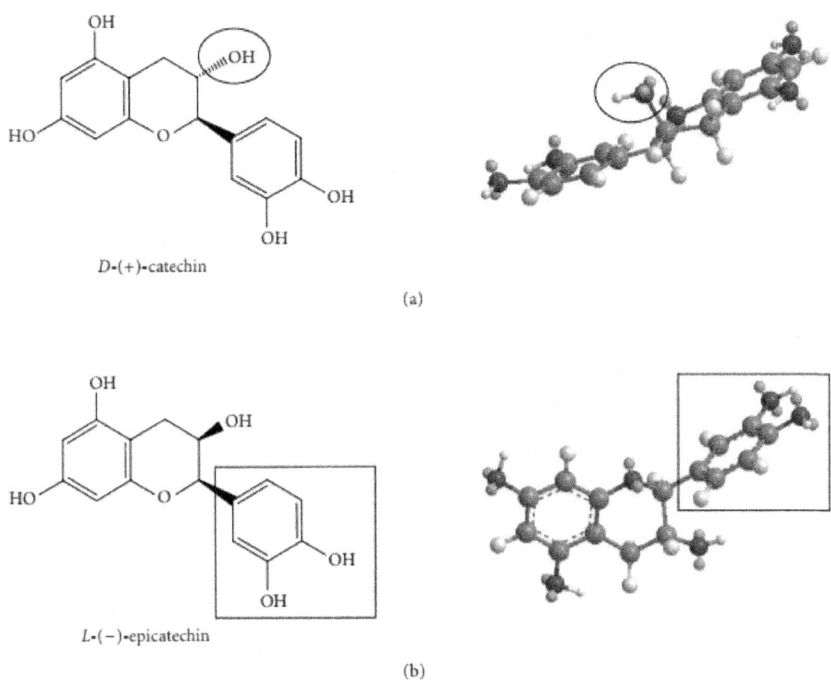

D-(+)-catechin

(a)

L-(−)-epicatechin

(b)

Figure 3: Three-dimensional structure of (a) -(+)-catechin and (b) -(−)-epicatechin with MM2 method.

On the contrary, the L-(−)-epicatechin shows a more bulky spatial structure, because the two aromatic rings are positioned on orthogonal planes, so the hindrance is very greater than in its isomeric form (Figure 3(b)).

Previous studies on the catalytic oxidations of catechol derivatives demonstrated that the reaction needs the cooperation of two close copper centers [22] to enable the binding of the catechol as a bridging ligand and allow a fast two-electron transfer process. In the dinuclear copper complexes, the catechin substrate can only form a productive complex by binding the catechol residue to the two copper ions in the A sites and that forces the resting part of the molecule to approach the optically active residue so that chiral recognition is possible (Scheme 2).

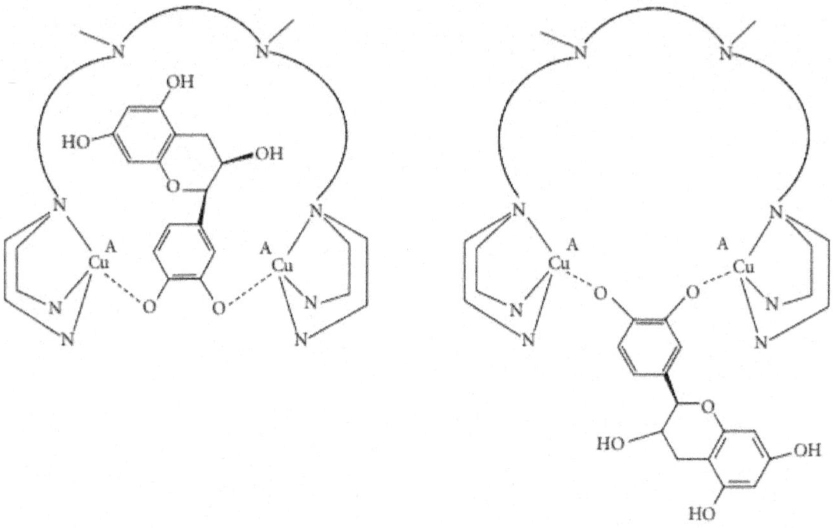

Structure I   Structure II

Scheme 2: Proposed structures for the putative intermediate adducts formed by the dinuclear copper(II) complexes in the catalytic oxidations of the catechins.

However, the dinuclear complexes that contain as central core the 1,1-binaphthyl residue show a rigid and bulky structure that reduces the possibility of effective chiral recognition for the catechins. In this case, the coordination of the catechols could occur on the outside of the complexes (Scheme 2, Structure II). The complex $[Cu_2(L\text{-}Lys\text{-}4Bz_4]^{4+}$ has a different design. It contains a chiral L-Lysine residue as a central unit, which is much more flexible than the 1,1-binaphthyl moiety and this is connected with two arms carrying four benzimidazole donors through a pair of orthoxylyl spacers. The high flexibility of the spacer and the length of the two arms allow a better chiral recognition of the substrates, as inferred by the kinetic constants and by the degree of stereoselectivity (Scheme 2, Structure I).

As reported in the previous papers [26, 27], the trinuclear complexes display a structure in which the Cu(II) center at B site and one of the two centers at A site are mediated by a double hydroxide bridge (Scheme 3, Structure I). In this case, the chiral recognition could depend not only on steric interactions but also by coordination of the free aliphatic hydroxide to the other Cu(II): a site (Scheme 3, Structure II) that allows a significant enantio-differentiating behavior towards optically active substrates. In fact, considering the three-dimensional structures of the two catechins reported before in Figure 3, one notices that the aliphatic hydroxide, in the D-(+)-catechin, is opposite to the two catecholic groups, and therefore able to coordinate at the Cu(II) center at A site. The L-(–)-epicatechin shows the aliphatic hydroxide too far from the Cu(II) center at A site and, in this case, the interaction needs a modification of the structure of the complexes with a strong tension of the ligands.

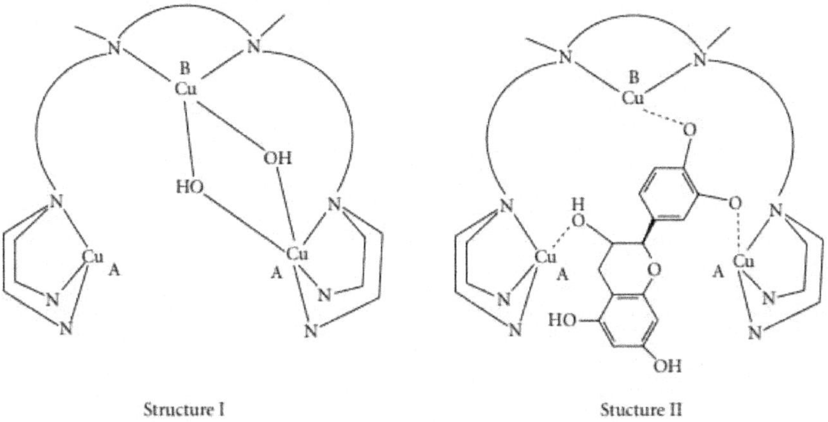

Structure I   Stucture II

Scheme 3: Proposed structures for the trinuclear copper(II) complexes (I) and for the putative intermediate adducts (II) with the catechins.

For $[Cu_2(L-66)]^{4+}$, experimental data evidences a very low enantio-differentiation toward D-(+)-catechin (see Table 1). This behavior could be due to the stacking interaction between the aromatic ring of m-xylene residue and the aromatic ring far from the catecholic one in the substrate; this interaction should be generated by a parallel disposition of the xylene and the plane of the substrate molecule.

ACKNOWLEDGMENT

This work was supported by the Italian Ministero dell'Università e della Ricerca (MIUR), through a Finanziamenti per l'Innovazione, la Ricerca e lo Sviluppo Tecnologico (FIRST) project.

REFERENCES

1.   T. Klabunde, C. Eicken, J. C. Sacchettini, and B. Krebs, "Crystal structure of a plant catechol oxidase containing a dicopper center," Nature Structural Biology, vol. 5, no. 12, pp. 1084–1090, 1998.

2.   M. E. Cuff, K. I. Miller, K. E. van Holde, and W. A. Hendrickson, "Crystal structure of a functional unit from Octopus hemocyanin," Journal of Molecular Biology, vol. 278, no. 4, pp. 855–870, 1998.

3.   K. A. Magnus, B. Hazes, H. Ton-That, C. Bonaventura, J. Bonaventura, and W. G. J. Hol, "Crystallographic analysis of oxygenated and deoxygenated states of arthropod hemocyanin shows unusual differences," Proteins: Structure, Function and Genetics, vol. 19, no. 4, pp. 302–309, 1994.

4.   Y. Matoba, T. Kumagai, A. Yamamoto, H. Yoshitsu, and M. Sugiyama, "Crystallographic evidence that the dinuclear copper center of tyrosinase is flexible during catalysis," Journal of Biological Chemistry, vol. 281, no. 13, pp. 8981–8990, 2006.

5.   Messerschmidt, R. Ladenstein, R. Huber, et al., "Refined crystal structure of ascorbate oxidase at 1.9 Å resolution," Journal of Molecular Biology, vol. 224, no. 1, pp. 179–205, 1992.

6.   K. Piontek, M. Antorini, and T. Choinowski, "Crystal structure of a laccase from the fungus Trametes versicolor at 1.90-Å resolution containing a full complement of coppers," Journal of Biological Chemistry, vol. 277, no. 40, pp. 37663–37669, 2002.

7.   N. Hakulinen, L.-L. Kiiskinen, K. Kruus, et al., "Crystal structure of a laccase from Melanocarpus albomyces with an intact trinuclear copper site," Nature Structural Biology, vol. 9, no. 8, pp. 601–605, 2002.

8.   Zaitseva, V. Zaitsev, G. Card, et al., "The X-ray structure of human serum ceruloplasmin at 3.1 Å: nature of the copper centres," Journal of Biological Inorganic Chemistry, vol. 1, no. 1, pp. 15–23, 1996.

9.   L. Casella and M. Gullotti, "Dioxygen activation by biomimetic dinuclear complexes," in Bioinorganic Chemistry of Copper, K. D. Karlin and Z. Tyeklar, Eds., pp. 292–305, Chapman & Hall, New York, NY, USA, 1993.

10.   N. Kitajima and Y. Moro-oka, "Copper-dioxygen complexes. Inorganic and bioinorganic perspectives,"Chemical Reviews, vol. 94, no. 3, pp. 737–757, 1994.

11.   E. I. Solomon, U. M. Sundaram, and T. E. Machonkin, "Multicopper oxidases and oxygenases,"Chemical Reviews, vol. 96, no. 7, pp. 2563–2605, 1996.

12.   H.-C. Liang, M. Dahan, and K. D. Karlin, "Dioxygen-activating bio-inorganic model complexes,"Current Opinion in Chemical Biology, vol. 3, no. 2, pp. 168–175, 1999.

13.   V. Mahadevan, R. J. M. K. Gebbink, and T. D. P. Stack, "Biomimetic modeling of copper oxidase reactivity," Current Opinion in Chemical Biology, vol. 4, no. 2, pp. 228–234, 2000.

14.   Manzur, A. M. Garcia, V. Rivas, A. M. Atria, J. Valenzuela, and E. Spodine, "Oxidation of 3,5-ditert-butylcatechol catalyzed by copper(II)

complexes. A kinetic study," Polyhedron, vol. 16, no. 13, pp. 2299–2305, 1997.

15. M. R. Malachowski, H. B. Huynh, L. J. Tomlinson, R. S. Kelly, and J. W. Furbee J., "Comparative study of the catalytic oxidation of catechols by copper(II) complexes of tripodal ligands," Journal of the Chemical Society, Dalton Transactions, no. 1, pp. 31–36, 1995.

16. C.-H. Kao, H.-H. Wei, Y.-H. Liu, G.-H. Lee, Y. Wang, and C.-J. Lee, "Structural correlation of catecholase-like activities of oxy-bridged dinuclear copper(II) complexes," Journal of Inorganic Biochemistry, vol. 84, no. 3-4, pp. 171–178, 2001.

17. M. Gupta, P. Mathur, and R. J. Butcher, "Synthesis, crystal structure, spectral studies, and catechol oxidase activity of trigonal bipyramidal Cu(II) complexes derived from a tetradentate diamide bisbenzimidazole ligand," Inorganic Chemistry, vol. 40, no. 5, pp. 878–885, 2001.

18. P. Gentschev, N. Möller, and B. Krebs, "New functional models for catechol oxidases," Inorganica Chimica Acta, vol. 300–302, pp. 442–452, 2000.

19. C. Fernandes, A. Neves, A. J. Bortoluzzi, et al., "A new dinuclear unsymmetric copper(II) complex as model for the active site of catechol oxidase," Inorganica Chimica Acta, vol. 320, no. 1-2, pp. 12–21, 2001.

20. A. Koval, K. Selmeczi, C. Belle, et al., "Catecholase activity of a copper(II) complex with a macrocyclic ligand: unraveling catalytic mechanisms," Chemistry—A European Journal, vol. 12, no. 23, pp. 6138–6150, 2006.

21. E. Monzani, L. Quinti, A. Perotti, et al., "Tyrosinase models. Synthesis, structure, catechol oxidase activity and phenol monooxygenase activity of a dinuclear copper complex derived from a triamino-pentabenzimidazole ligand," Inorganic Chemistry, vol. 37, no. 3, pp. 553–562, 1998.

22. E. Monzani, G. Battaini, A. Perotti, et al., "Mechanistic, structural, and spectroscopic studies on the catecholase activity of a dinuclear copper complex by dioxygen," Inorganic Chemistry, vol. 38, no. 23, pp. 5359–5369, 1999.

23. E. Monzani, L. Casella, G. Zoppellaro, et al., "Synthetic models for biological trinuclear copper clusters. Trinuclear and binuclear complexes derived from an octadentate tetraamine-tetrabenzimidazole ligand," Inorganica Chimica Acta, vol. 282, no. 2, pp. 180–192, 1998.

24. Santagostini, M. Gullotti, R. Pagliarin, E. Monzani, and L. Casella, "Enantio-differentiating catalytic oxidation by a biomimetic

trinuclear copper complex containing L-histidine residues," Chemical Communications, vol. 9, no. 17, pp. 2186–2187, 2003.

25. C. Mimmi, M. Gullotti, L. Santagostini, et al., "Stereoselective catalytic oxidations of biomimetic copper complexes with a chiral trinucleating ligand derived from 1,1-binaphthalene," Journal of Molecular Catalysis A, vol. 204-205, pp. 381–389, 2003.

26. C. Mimmi, M. Gullotti, L. Santagostini, et al., "Models for biological trinuclear copper clusters. Characterization and enantioselective catalytic oxidation of catechols by the copper(II) complexes of a chiral ligand derived from (S)-(−)-1,1'-binaphthyl-2,2'-diamine," Dalton Transactions, no. 14, pp. 2192–2201, 2004.

27. Gullotti, L. Santagostini, R. Pagliarin, A. Granata, and L. Casella, "Synthesis and characterization of new chiral octadentate nitrogen ligands and related copper(II) complexes as catalysts for stereoselective oxidation of catechols," Journal of Molecular Catalysis A, vol. 235, no. 1-2, pp. 271–284, 2005.

28. G. Battaini, A. Granata, E. Monzani, M. Gullotti, and L. Casella, "Biomimetic oxidations by dinuclear and trinuclear copper complexes," in Advances in Inorganic Chemistry, Volume 58, R. van Eldik and J. Reedijk, Eds., pp. 185–233, Academic Press, New York, NY, USA, 2006.

29. F. G. Mutti, M. Gullotti, L. Santagostini, et al., "Biomimetic Modelling of Copper Enzymes: Synthesis, Characterization, EPR Analysis and Enantioselective Catalytic Oxidations by a New Chiral Trinuclear Copper(II) Complex," submitted to European Journal of Inorganic Chemistry.

30. L.Casella,O.Carugo,M.Gullotti,S.Garofani,andP.Zanello,"Hemocyanin and tyrosinase models. Synthesis, azide binding, and electrochemistry of dinuclear copper(II) complexes with poly(benzimidazole) ligands modeling the met forms of the proteins," Inorganic Chemistry, vol. 32, no. 10, pp. 2056–2067, 1993.

31. S. W. Weidman and E. T. Kaiser, "The mechanism of the periodate oxidation of aromatic systems. III. A kinetic study of the periodate oxidation of catechol," Journal of the American Chemical Society, vol. 88, no. 24, pp. 5820–5827, 1966.

32. L. Muñoz, F. García-Molina, R. Varón, J. N. Rodriguez-Lopez, F. García-Cánovas, and J. Tudela, "Calculating molar absorptivities for quinones: application to the measurement of tyrosinase activity," Analytical Biochemistry, vol. 351, no. 1, pp. 128–138, 2006.

33. S. M. MacManus, "The handbook of natural flavonoids, volumes 1 and 2 J. B. Harborne and H. Baxter (editors), Wiley, Chichester, 1999, volume 1: xii+889 pp., volume 2: xvi+879 pp., ISBN 0-471-95893-X,£950.00 (the pair)," Talanta, vol. 54, no. 1, p. 207, 2001.

34. F. C. Richard-Forget, M.-A. Rouet-Mayer, P. M. Goupy, J. Philippon, and J. J. Nicolas, "Oxidation of chlorogenic acid, catechins, and 4-methylcatechol in model solutions by apple polyphenol oxidase,"Journal of Agricultural and Food Chemistry, vol. 40, no. 11, pp. 2114–2122, 1992.

35. C. W. Arts, B. van de Putte, and P. C. H. Hollman, "Catechin contents of foods commonly consumed in The Netherlands. 2. Tea, wine, fruit juices, and chocolate milk," Journal of Agricultural and Food Chemistry, vol. 48, no. 5, pp. 1752–1757, 2000.

36. M. Monagas, C. Gómez-Cordovés, B. Bartolomé, O. Laureano, and J. M. Ricardo da Silva, "Monomeric, oligomeric, and polymeric flavan-3-ol composition of wines and grapes from Vitis vinifera L. cv. Graciano, Tempranillo, and Cabernet Sauvignon," Journal of Agricultural and Food Chemistry, vol. 51, no. 22, pp. 6475–6481, 2003.

37. S. Pérez-Magariño and M. L. González-San José, "Evolution of flavanols, anthocyanins, and their derivatives during the aging of red wines elaborated from grapes harvested at different stages of ripening," Journal of Agricultural and Food Chemistry, vol. 52, no. 5, pp. 1181–1189, 2004.

38. Oszmianski and C. Y. Lee, "Enzymatic oxidative reaction of catechin and chlorogenic acid in a model system," Journal of Agricultural and Food Chemistry, vol. 38, no. 5, pp. 1202–1204, 1990.

39. S. Guyot, J. Vercauteren, and V. Cheynier, "Structural determination of colourless and yellow dimers resulting from (+)-catechin coupling catalysed by grape polyphenoloxidase," Phytochemistry, vol. 42, no. 5, pp. 1279–1288, 1996.

40. F. C. Richard-Forget and F. A. Gauillard, "Oxidation of chlorogenic acid, catechins, and 4-methylcatechol in model solutions by combinations of pear (Pyrus communis cv. Williams) polyphenol oxidase and peroxidase: a possible involvement of peroxidase in enzymatic browning," Journal of Agricultural and Food Chemistry, vol. 45, no. 7, pp. 2472–2476, 1997.

41. M. López-Serrano and A. R. Barceló, "Comparative study of the products of the peroxidase-catalyzed and the polyphenoloxidase-catalyzed (+)-catechin oxidation. Their possible implications in strawberry (Fragaria×ananassa) browning reactions," Journal of Agricultural and Food Chemistry, vol. 50, no. 5, pp. 1218–1224, 2002.

42. S. Guyot, V. Cheynier, J.-M. Souquet, and M. Moutounet, "Influence of pH on the enzymatic oxidation of (+)-catechin in model systems," Journal of Agricultural and Food Chemistry, vol. 43, no. 9, pp. 2458–2462, 1995.

43. M. Jiménez-Atiénzar, J. Cabanes, F. Gandía-Herrero, and F. García-Carmona, "Kinetic analysis of catechin oxidation by polyphenol oxidase at neutral pH," Biochemical and Biophysical Research Communications, vol. 319, no. 3, pp. 902–910, 2004.

Chapter 9

DISTINCT OPTICAL CHEMISTRY OF DISSOLVED ORGANIC MATTER IN URBAN POND ECOSYSTEMS

Nicola A. McEnroe[1], Clayton J. Williams[1], Marguerite A. Xenopoulos[1], Petr Porcal[2], Paul C. Frost[1]

[1] Department of Biology, Trent University, Peterborough, Ontario, Canada

[2] Biology Centre of the Academy of Science of the Czech Republic, v.v.i., Institute of Hydrobiology, České Budějovice, Czech Republic

ABSTRACT

Urbanization has the potential to dramatically alter the biogeochemistry of receiving freshwater ecosystems. We examined the optical chemistry of dissolved organic matter (DOM) in forty-five urban ponds across southern Ontario, Canada to examine whether optical characteristics in these relatively new ecosystems are distinct from other freshwater systems. Dissolved organic carbon (DOC) concentrations ranged from 2 to 16 mg C L^{-1} across the ponds with an average value of 5.3 mg C L^{-1}. Excitation-emission matrix (EEM) spectroscopy and parallel factor analysis (PARAFAC) modelling showed urban pond DOM to be characterized by microbial-like and, less importantly, by terrestrial derived humic-like components. The relatively transparent, non-humic DOM in urban ponds was more similar to that found in open water, lake ecosystems than to rivers or wetlands. After irradiation equivalent to 1.7 days of natural solar radiation, DOC concentrations, on average, decreased by 38% and UV absorbance decreased by 25%. Irradiation decreased the relative abundances of terrestrial humic-like components and increased protein-like aspects of the DOM pool. These findings suggest that high internal production and/or prolonged exposure to sunlight exerts a distinct and significant influence on the chemistry of urban pond DOM, which likely reduces its chemical similarity with upstream sources. These properties of urban pond DOM may alter its biogeochemical role in these relatively novel aquatic ecosystems.

INTRODUCTION

In the last few decades, urbanization has been associated with widespread loss of natural (wetlands and forest) and agricultural areas [1]. This landscape conversion has been accompanied by greater imperviousness of watersheds to infiltration by surface waters and altered hydrological cycles (e.g., [2,3]). To mitigate flash flooding created by rapid drainage from highly impervious surfaces, newer urban landscapes often contain relatively shallow ponds (depth of ~2 m or less). While primarily built to retain and slow the downstream movement of stormwater, these ecosystems also potentially improve water quality through the retention of suspended sediments and dissolved nutrients [4]. Urban ponds are also thought to provide important ecological function by increasing biodiversity, serving as wildlife habitat, and altering biogeochemical cycles [5,6].

The biogeochemical role (especially for elements other than phosphorus) that urban ponds play in the developed landscape has not been well-studied, despite potential management implications. One study based on 26 urban stormwater ponds reported high rates of microbial activity and biogeochemical processes suggesting that, at current pond density, they could play an important role in regional and global carbon (C) cycles [7]. Other small, shallow freshwater systems are also typically characterized by disproportionately high rates of nutrient processing compared to larger bodies of water [8]. For example, sedimentation rates and burial of organic C can be higher in small freshwater bodies than in larger aquatic ecosystems [9-11]. In addition, allochthonous derived DOM has been shown to fuel surprising quantities of microbial production in small aquatic ecosystems and to support freshwater food webs beyond that provided by primary production alone (e.g., [12-15]).

While not particularly well-studied, urban land use has been reported to increase the loading of dissolved organic matter (DOM) above levels found in natural areas due to changes in soil and drainage conditions [16] In contrast, DOM concentrations in streams have been observed to decrease post-urbanisation due to the loss of hydrological flow paths through shallow soils [17,18]. Either way, DOM characteristics from such areas may differ from that seen in waters emerging from more natural (e.g., forested and wetland) or agricultural areas [19,20]. For example, DOM derived from urban wastewater can be distinguished from the DOM in receiving river systems through point-source increases in protein-like DOM resolved by fluorescence characterization [21]. DOM arising within urban aquatic ecosystems may have different transparency, humicity, molecular weight, and/or photoreactivity arising from limited contact with organic soils and vegetation in urban watersheds. If so, exported DOM from these urban environments could be less colored, more transparent, and/or have other, distinct fluorescent properties.

The source and chemical properties of urban pond DOM could affect the rate of DOM microbial decomposition and photodegradation. Autochthonous DOM tends to be more aliphatic and less aromatic in character compared to terrestrial derived DOM (e.g., [22]). These properties have been found to limit DOM photoreactivity (e.g., [23-25]). Photodegradation can result in more biologically available substrates (e.g., [26]) or could increase DOM humicity, reducing its bioavailability (e.g., [24,27-30]). However, how DOM in these urban ponds changes in response to exposure to solar radiation remains unknown.

In this study, we first examined the quantity and quality of DOM in urban ponds as it compares to DOM sampled from more natural aquatic ecosystems. We then examined the photochemical reactivity of urban pond DOM. Finally, we examined how the properties of DOM relate to indices of internal production (e.g., seston chlorophyll). We expected that urban pond DOM would have a distinct optical chemistry compared to that derived from more natural areas, based on either greater contributions of anthropogenic derived organic matter or from high autochthonous production. We further expected that these unique chemical properties would affect rates of *in-situ* photochemical degradation. We thus provide a thorough examination of DOM and its chemistry in stormwater retention ponds.

METHODS

Ethics Statement

All field sites were accessed by public right of way and did not require explicit permission from the municipality to collect water. Because we only collected water samples, none of our field work involved threatened or endangered species.

Sample Sites

We examined DOM properties in forty-five urban ponds in four municipalities in southern Ontario, Canada (Ottawa, Peterborough, Richmond Hill and Whitby). Selected ponds were all in residential areas and have variable catchment and drainage properties (Table 1). Additional information about these ponds and their biogeochemistry is provided in [7,31]. We compared our DOM data on urban pond water with previously published data that we compiled from studies of non-urban, aquatic environments to determine whether their DOM characteristics match that seen in natural ecosystems. Comparison data were taken from studies of different types of aquatic environments including forested and wetland-dominated streams, saline lakes, and eutrophic to hypereutrophic inland water bodies.

Table 1: Selected characteristics of urban ponds in this study

Pond Name	Municipality	Year built	Land-use	BSD (m)	$A_0(m^2)$	% I
Water St.	Peterborough	1977	R	0.75	3374	44
Carnegie	Peterborough	2005	R	1.5	5171	52
Chemong	Peterborough	2000	R	1.6	1063	64
Glenforest	Peterborough	1975	R	1.8	1657	61
Ravenwood	Peterborough	2000	R	1.5		64
Tobin Court	Peterborough		R	0.6		
White	Peterborough	2007	R	1.5		55
Loggerhead	Peterborough	2007	R	1.1	5164.6	42.6
Foxmeadow	Peterborough	2003	R	0.5		46.8
7-3	Richmond Hill	2002	R	1.2	1027	45
7-4	Richmond Hill	1998	R	0.8	2030	44.9
8-3	Richmond Hill	1999	R	2.76	1698	44.9
2-3	Richmond Hill	2000	R	1.45	1688	35.1
9-9	Richmond Hill	2004	R	1.65	4314	
16-8	Richmond Hill	1996	R	0.95	951	39.3
16-4	Richmond Hill		R	1	4620	
17-3	Richmond Hill	1987	R	0.75	7500	45
19-9	Richmond Hill	2001	R	1.5	2098	41
19-8	Richmond Hill	1997	R	1.5	1549	44.4
19-4	Richmond Hill	1997	R	2.1	3120	
9-5	Richmond Hill	2000	R	1	1129	36
9-6	Richmond Hill	2000	R	1	1698	50
8-10	Richmond Hill	2004	R	1.1		50
57-01	Whitby		R	1.8		
65-01	Whitby	2001	R	2.0		

68-02	Whitby	2000	R	2.6		
33-01	Whitby	1996	R	0.9		
34-02	Whitby	1995	R	0.7		
03-02	Whitby	1999	R	0.7		
SWF-1409	Ottawa	1999	R	1.15	21000	10.3
SWF-1410	Ottawa	1998	R	1.3	29000	13.16
SWF-1139	Ottawa	2004	M	1.45	1000	29.94
SWF-1206	Ottawa	1999	M	1.02	10000	19.82
SWF-1207	Ottawa	1995	R	1.65	11000	33.86
SWF-1211	Ottawa	1980	R	2.1	22000	42.25
SWF-1215	Ottawa	1980	R	1.51	4000	46.57
SWF-1227	Ottawa	2000	R	1.04	11000	12.35
SWF-1306	Ottawa	2000	R	1.8	5000	30.29
SWF-1309	Ottawa	2000	R	0.9	6000	16.39
SWF-1320	Ottawa	1990	R	1.15	1000	43.47
SWF-1610	Ottawa	1979	R	1.35	1000	39.55
SWF-1611	Ottawa	1980	M	1.13	3000	
SWF-1628	Ottawa	2000	M	0.65	300	40.93
SWF-1902	Ottawa	1987	I	0.5	1000	17.5
SWF-1930	Ottawa	2000	R	1.84	33000	31.31

Ponds denoted by bolded italics were included in the photo-irradiation experiment (n=25). Data provided courtesy of each municipality when they were available.

Landuse=R(residential), I(industrial), M(mixed), H(highway); BSS= Bottom sampling depth; A_0=Pond surface area; %I=impervious surface.

Ambient DOM Characteristics

We sampled water in ponds twice (June and August) during the summer of 2009. Whole water samples were collected at the deepest part of each pond at the surface (0-10 cm) and at the water-sediment interface (~0.5-2.7 m, Table 1) using a Van Dorn sampler. These samples were placed into deionized water (DI) rinsed, acid washed polypropylene bottles and held on ice in the dark during transport back to the laboratory. Within 24 hours, we filtered samples (Whatman Type polycarbonate PCTE, 0.2 μm) and placed this water into acid washed and pre-combusted glass amber bottles, which were refrigerated at 4°C until further analysis. Concentrations of DOC (mg C L^{-1}) in each sample were determined using a TOC-TN Analyser (Model 1030D OI Analytical Aurora, Texas, USA) after combustion (750°C) and acidification with 2 N hydrochloric acid. UV-visible absorbance were measured on each sample between 200 and 800 nm (Perkin Elmer Lambda 25 Spectrophotometer) and the absorption coefficient (a_{nm}) was determined by multiplying absorbance (A_{nm}) by 2.303 and dividing by the path length in meters following [32]. Molar absorptivity at 280 and 350 nm (ε_{280}, ε_{350}) was calculated as absorbance (A_{nm}) at 280 and 350 nm divided by the DOC concentration (μmol C L^{-1}; [33]). Absorbance at 440 nm (A_{440}) was used as an additional indicator of DOC color [34].

DOM Optical Characteristics

To determine DOM fluorescence characteristics of urban pond water, we selected a subset of ponds from August sampling (n=25, Table 1) for excitation emission matrix (EEM) measurements. EEM measurements on surface water samples from these ponds were made using fluorescence spectroscopy (Varian Eclipse Fluorometer, 5 nm bandwidth, integration time 0.25 s), over a range of emission (270-600 nm, 2 nm intervals) and excitation (230-500 nm, 5 nm intervals) wavelengths. The EEMs were corrected for inner filter effects [22,35], and for second-order Raman and Rayleigh scatter effects and instrument bias using manufacturer recommended instrument settings. Milli-Q water blank EEM fluorescence was subtracted from that of sample EEMs and EEMs were converted to Raman Units (RU) using the area under the Milli-Q Raman scatter peak at excitation 350 nm.

To investigate DOM fluorescence parameters, EEM data were used to calculate the humification index (HIX), fluorescence index (FI) and β:α ratio. The HIX was calculated as the ratio of peak area under each curve at emission 434-480 nm and 300-346 nm after excitation at 255 nm [36]. The FI was calculated, using an excitation wavelength of 370 nm, as the ratio of emission intensities at 470 and 520 nm after [22], and β:α ratio was calculated, using an excitation wavelength of 310 nm, from emission intensity at 380 nm (β region), divided by the emission intensity maximum observed between 420 and 435 nm (α region; [20,37]). A spectral slope ratio (Sr; [38]) was determined as the ratio of the log-transformed slope between 275 to 295 and 350 to 400 nm,

A seven component PARAFAC (Parallel Factor Analysis) model was used to examine factors of each EEM (Matlab 2012b, Mathworks; Figure 1). EEMs collected during the photo-irradiation experiment were fit to the PARAFAC model and residuals were visually examined to verify model fit. The full model validation and description is presented in detail in [7]. Of the seven PARAFAC components, C1 resembles humic-like material and occurs in most ecosystems [39,40]. C2 and C3 appear similar to humic-like DOM of terrestrial origin [41]. C4 resembles the fluorescence signatures of soil, fulvic acid-like substances [42]. C5 and C6 are similar to microbial derived humic-like substances [7,19,39]. C7 resembles protein-like materials [21,42] as well as photodegradable plant tannin-protein complexes [43]. The PARAFAC components are expressed as relative abundance (% of F_{max}).

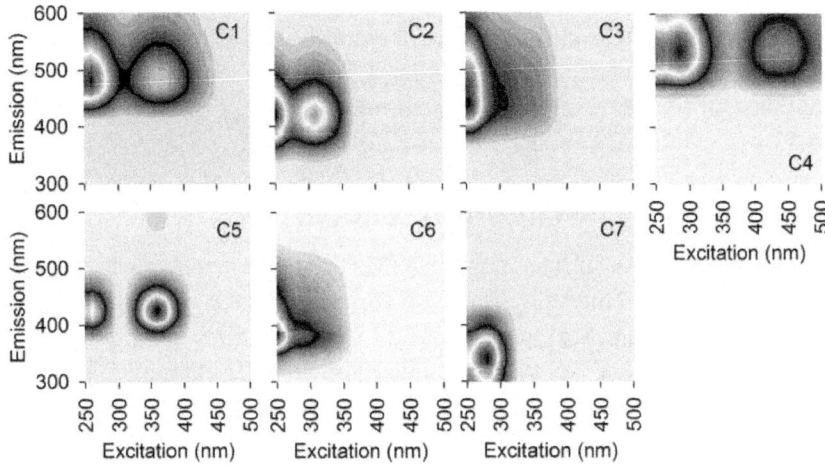

Figure 1: Countor plots of seven component PARAFAC model.

DOM Photo-Irradiation

The same samples (n=25) collected from the pond surface were included in photo-irradiation experiments. For these experiments, aliquots (25 ml) of filtered pond water were transferred into DI-water rinsed, sterile UV-transparent polyethylene bags and the samples were irradiated using a XE Arc lamp (500 W m^2 for 12 hours; Suntest XLS+ Solar Simulator, 2220W lamp, internal water bath temperature of 25°C, Atlas, Germany) approximating the solar spectral quality at sea level. Control samples treated as irradiated samples (membrane filtration and stored in polyethylene bags) were wrapped in aluminium foil and exposed to the same temperature as irradiated samples inside the solar simulator. The radiation intensity of 500 W m^{-2} is equivalent to a radiant energy dosage of 21.6 MJ m^{-2}, or to 1.7 days of natural solar radiation in June at our study location, based on estimates of the monthly averaged insolation on a horizontal surface (NASA Surface Meteorology and Solar Energy data;http://eosweb.larc.nasa.gov). The transformation of cumulative irradiation energy to the number of corresponding days of natural irradiation was done by dividing the cumulative energy by averaged insolation energy at the sampling location.

Changes in DOC concentrations during photo-irradiation are expressed against cumulative irradiation energy, which is the product of intensity of irradiation and time. The decrease in DOC with cumulative irradiation energy followed exponential decrease as observed in other studies (e.g. [44,45]). Thus the pseudo first order kinetics rate constant was used as a measure of DOC photodegradation in individual samples, following:

$$DOC = (DOC_0 - DOC_R) \cong \exp(-k_{DOC}E) + DOC_R \qquad (1)$$

Where DOC_0 is the initial DOC concentration, DOC_R is the concentration of DOC that remained after the experiment, k_{DOC} is the photochemical degradation rate constant ($m^2 MJ^{-1}$) and E is the cumulative energy of irradiation ($MJ\ m^{-2}$; see 45).

Indices of Production in Urban Ponds

We examined how urban pond DOM properties related to indices of production: seston chlorophyll, dissolved oxygen (DO) and total phosphorus (TP) concentrations. Seston chlorophyll concentrations on filtered samples (Whatman Type GF/F, 25 mm) were determined after ethanol cold extraction (24 h) with fluorometry (440 nm excitation, 660 nm emission wavelengths). Profiles of water column DO were obtained from the deepest part of each pond from the surface and every 10-20 cm depth to the water sediment interface (YSI Model 54 Oxygen Meter, Yellow Springs Instrument Co., Yellow Springs, Ohio, USA). TP concentrations were determined on unfiltered water samples stored at 4°C in acid washed polypropylene and pre-combusted (500°C) amber glass bottles, using potassium persulfate digestion (autoclave at 121°C) followed by ascorbic acid-sodium molybdate blue colorimetric analyses (Spectrophotometer, Biochrome Ultrospec Pro 500, 885 nm wavelength).

Statistical Analyses

Prior to statistical analysis, most data were transformed (log(y) or square root (y)) to meet the assumptions of normality and homoscedasticity. Pearson's Correlation analyses were used to determine significant relationships ($\alpha = 0.05$) between DOM UV absorbance and fluorescence indices. One-way Analysis of Variance (ANOVA) or Student's t-test was used to determine statistically significant differences between means, followed by the Tukey-Kramer HSD post hoc test where differences were found ($\alpha = 0.05$). Permutational Multivariate (M)ANOVA with post hoc comparison was used to determine the overall impact of photo-irradiance on fluorescent DOM. MANOVA results were visualized using Principle Component Analysis. Statistical analyses were carried out using JMP v.9 (SAS Institute, 2009), PASW Statistics 18 (SPSS Inc., 2009), and R with the VEGAN library.

RESULTS

Ambient DOM Characteristics

In the entire dataset for 45 urban ponds, including both sampling dates and depths, DOC concentrations ranged between 2.0 to 16.2 mg L^{-1} (mean ± SD; 5.3 ± 1.9 mg L^{-1}). On average, DOC concentrations were lower in June (p = 0.04, 4.9 mg C L^{-1}) compared to August (5.7 mg C L^{-1}). While molar absorptivity (ε_{280}) was significantly higher in June (p < 0.001), absorbance at 280 (a$_{280; \text{Table 2}}$) and 440 nm (a$_{440; \text{Table 1}}$) were not different between sampling dates and showed a wide range among sampled ponds.

Table 2: Range in DOC concentrations, molar absorptivity at 280 nm (ε_{280}) and corresponding absorption coefficient (a$_{280}$) for 45 urban ponds in June and August, 2009

Site	DOC (mg C L^{-1})	a$_{280}$(m^{-1})	ε_{280}(L mol C^{-1}cm^{-1})	a$_{440}$(m^{-1})	FI	HIX	β:α	Source
Stormwater Ponds								
Urban ponds, southern Ontario[June]	2.0-11.3	8.9-31.5	92-428	0.3-3.4				This study
Urban ponds, southern Ontario[Aug]	3.6-16.2	12.1-29.1	86-258	0.4-1.4	1.35-1.58	4.5-7.9	0.74-0.90	This study
Wetlands, Streams and Rivers								
Agro-Urban Streams, Australia	2.0-140.0				1.20-1.44		0.40-0.70	[40]
43 mixed watershed streams, southern Ontario, Canada	4.1-26.4	5.5-72.5	133-369		1.19-1.47	8.7-32.2	0.48-0.70	[19]
Subtropical wetland	4.6-45.0	4.8-81.9			1.28-1.47			[50]
Okavango Delta wetland, Botswana	13.3-16.6		276-324		1.45-1.50			[58]
Harp Lake watershed, south-central Ontario	10.0-35.0				1.1-1.2			[59]
Lake Superior watershed streams	3.5-34.0		138-586					[60]
Deer Creek, CO, USA	1.3-4.1				1.40-1.48			[61]
Suwannee River, GA, USA	35		509					[62,63]
Lakes								
27 Prairie lakes	13.4-328					1.4-6.4		[49]
Alpine Lake	0.5-1.7				1.33-1.71			[64]

Antarctic Lakes	2.5-22.6				1.71-2.71		[61]
Antarctic Lake (Pony)					1.5	20	[61]
30 temperate lakes	3.7-21.5			0.8-19.3			[65]
Antarctic Lake (Fryx-ell)			150				[62]

Absorbance at 440 nm is represented as a_{440}. Data is presented in comparison to other published studies.

From the subset of urban ponds selected for more detailed optical characterization (n=25), DOC concentrations ranged from 3.2 to 8.2 mg L^{-1} (mean ± SD; 5.03 ± 1.04 mg L^{-1}) and fluorescence indices were found to vary considerably (Figure 2; Table 2). DOC concentrations were positively correlated with FI and UV-visible absorbance and were negatively correlated with molar absorptivity (ε_{280}; Table 3). While PARAFAC component C6 (anthropogenic/microbial humic-like) was generally the most abundant (23 - 52%) and C4 (soil fulvic acid-like) was the least abundant (2 - 5%), the relative abundance of the individual components varied among ponds (Figure 3). These components were also related to other optical properties of the DOM. C6 correlated positively with β:α and negatively with ε_{350} (Table 3). C1 (terrestrial humic-like), C2 (terrestrial humic-like), and C4 correlated positively with HIX and ε_{350} but negatively with β:α (Table 3). C7 (protein-like) generally showed the opposite relationships from the humic-like components. C4 was the only PARAFAC component that correlated with DOC, indicating that urban pond fluorescence DOM quality varied among ponds mostly independent of bulk DOC concentration. Spectral slope (Sr) was significantly related (negatively) to one component, C7 (r=-0.72; p<0.001), and was not related to any other PARAFAC components (Table 3).

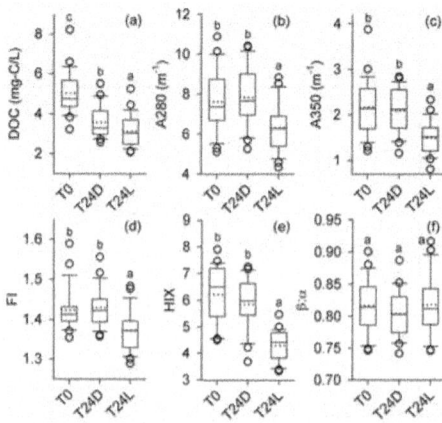

Figure 2: Effects of photo-irradiation on urban pond dissolved organic matter.

Box plots showing the median (solid line), 25th and 75th percentiles (boxes) and 10th and 90th percentiles (dots) for absorbance (A_{nm}) at (a) 280 and (b) 350 nm and for (c) DOC concentrations and measured fluorescence indices (d) FI, (e) HIX and (f) β:α for 25 urban ponds in August, prior to photo-irradiation (T0) and for light treatment (T24L), and dark treatment (T24D). Values denoted with different letters are significantly different ($p<0.05$).

Table 3: Pearson's correlation coefficient (r) between DOC concentrations and UV-Visible absorbance and fluorescence indices for 25 urban ponds in August, prior to photo-irradiation (T0)

	DOC	C1	C2	C3	C4	C5	C6	C7
DOC	1	ns	ns	Ns	-0.51	ns	ns	ns
A$_{280}$	0.65	ns	ns	Ns	ns	ns	ns	ns
A$_{350}$	0.48	ns	0.44	-0.45	ns	-0.50	ns	ns
ε$_{280}$	-0.42	ns	ns	Ns	0.62	ns	ns	-0.41
ε$_{350}$	ns	0.52	0.49	Ns	0.49	0.49	-0.41	ns
HIX	ns	0.50	0.40	Ns	0.40	ns	ns	-0.85
FI	0.53	ns	0.45	-0.82	-0.47	0.72	ns	ns
β:α	ns	-0.62	-0.62	Ns	-0.42	ns	0.60	0.45
Sr	ns	ns	ns	Ns	ns	ns	ns	-0.72

Significant correlations are indicated in bold ($p<0.01$).

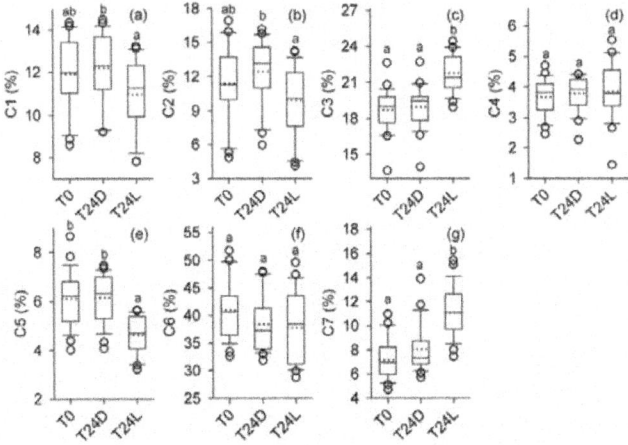

Figure 3: Responses of PARAFAC components (C1-7; %) to photo-irradiation of urban pond DOM.

Box plots show the median (solid line), mean (dotted lines), 25th and 75th percentiles (boxes), 10th to 90th percentiles (whiskers), and >90th percentiles (dots) prior to photo-irradiation (T0), after dark treatment (T24D), and after light treatment (T24L). Values denoted with different letters are significantly different ($p<0.05$).

In samples exposed to artificial light, pseudo-first order rate constants ranged between 0.003 and 0.03 m^2 MJ^{-1} after a cumulative irradiation energy exposure of 21.6 MJ m^{-2} and were significantly higher than dark treatment rate constants ($p < 0.001$; range 0.0003 to 0.02 m^2 MJ^{-1}). In 20% of samples, photo-irradiation had no effect on DOC concentrations with net irradiation rate constants ranging from 0 to 0.02 m^2 MJ^{-1}. Despite this, photo-irradiation decreased DOC concentrations, on average, by 38%. We also observed decreases in absorbance (A_{nm}) and fluorescence indices in samples exposed to light relative to the dark controls ($p < 0.05$; Figure 2). While average absorbance (A_{nm}) was reduced by 25% in the UVA and UVB regions, average molar absorptivity (ε_{280}) increased by 30% in light exposed samples relative to the dark controls.

Photo-irradiation altered characteristics of DOM quality indexed by PARAFAC, but these effects depended on the component examined (Figure 3). Among all ponds, the intensity C1 to C6 (F_{max}; RU) decreased after light exposure. The intensity of C7 (RU), however, responded non-uniformly to light exposure among ponds. C7 (RU) increased, decreased, or was similar to initial F_{max} depending on the pond water examined. The relative abundance (%) of C4 and C6 were not significantly different from initial levels after dark and light incubations ($p = 0.753$ and $p = 0.095$, respectively). C1 (%) and C2 (%) were significantly lower after light exposure relative to the dark incubation ($p = 0.039$ and $p = 0.027$, respectively). Photo-irradiation also significantly decreased C5 (%; $p < 0.001$) and increased C3 (%; $p < 0.001$) and C7 (%; $p < 0.001$) from dark incubated and initial treatments. The relative abundance of C1 consistently correlated positively with HIX and negatively with β:α across initial, light, and dark treatments (Figure 4). Correlations between C7, HIX, and β:α, however, were influenced by photo-irradiation (Figure 4). Light treated samples had higher levels of C7 and lower levels of HIX than initial and dark samples and had a similar correlation slope overall. C7 positively related to β:α for initial samples, but this relationship was not preserved after photo-irradiation. Overall, photo-irradiation significantly impacted the fluorescent DOM pool (MANOVA, $p = 0.001$), which appeared to be strongly influenced by changes in C7 (Figure 5).

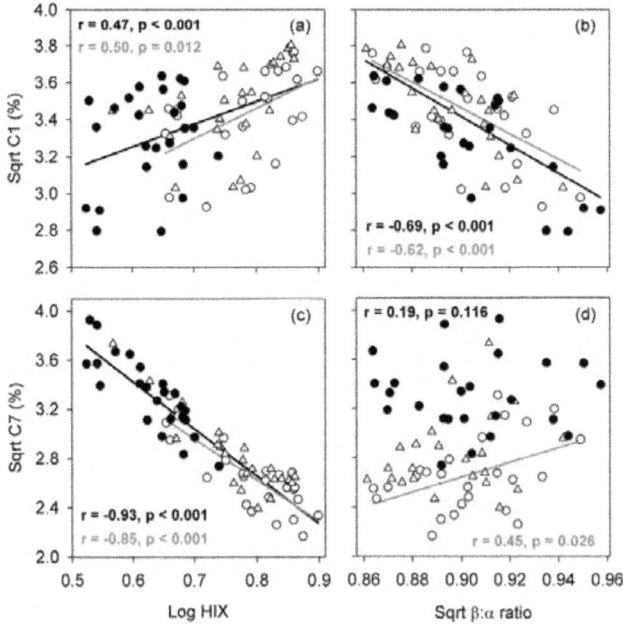

Figure 4: Photo-induced changes in DOM chemistry.

Correlations between the EEM-PARAFAC model C1 (a,b) and C7 (c,d), and the humification index (HIX; a,c) and β:α ratio (b,d) for 25 urban ponds in August (open circles=initial (T0, prior to photo-irradiation), closed circles=light treatment (T24L), triangles=dark treatment (T24D). Correlations are presented for the overall data set (black lines) and T0 (gray lines).

doi:10.1371/journal.pone.0080334.g004

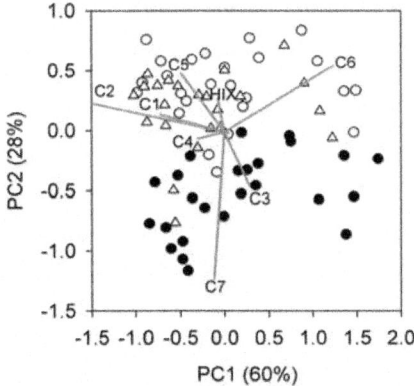

Figure 5: Principle components plot with vector loadings for the fluorescent DOM pool.

MANOVA with post hoc test indicated that the light exposed fluorescent DOM pool (closed circles) was significantly different than initial (open circles) and dark (triangles) pools, which did not differ. Note, vector loadings for β:α and FI were near the origin in the first two PCA dimensions and are not displayed in this figure.

Index of Urban Pond Production

We found that DOC concentrations (negatively) and ε_{280} (positively) were correlated with chlorophyll whereas no other optical property of the DOM was related to this index of productivity (Figure 6). Sr (negatively), HIX (positively), and ε_{280} (positively) significantly correlated with total phosphorus. Only Sr and β:α related to dissolved oxygen concentrations in the surface waters of urban ponds (Figure 6). Of the seven PARAFAC components, we found that seston chlorophyll and total phosphorus was positively correlated with C4 and C5 and negatively correlated with C3 and C7 (Figure 6). C1, C2, and C6 were not related to any of our indices of water column productivity.

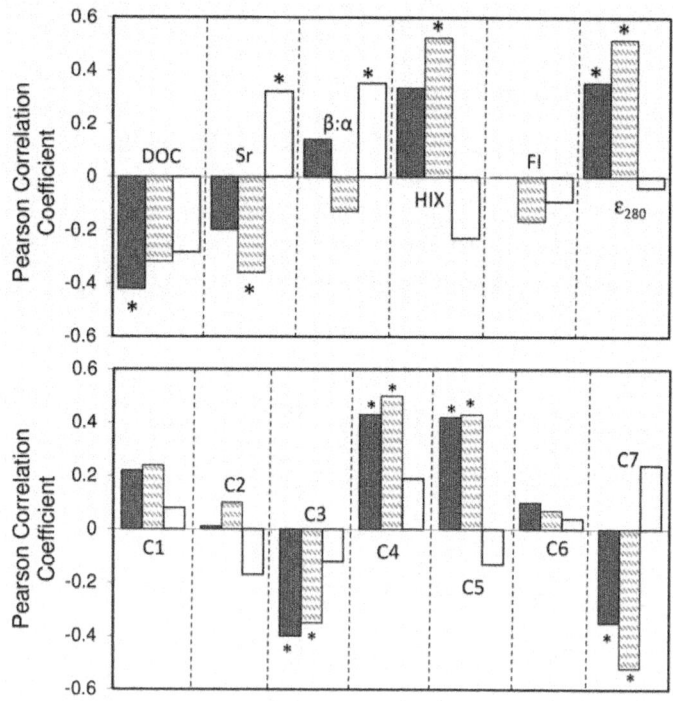

Figure 6: Correlation coefficients between DOC concentration (mg C L⁻¹), optical properties, and PARAFAC components and seston chlorophyll (dark bars), TP (light bars) and DO (open bars).

Significant correlations ($p<0.05$) are noted with *.

DISCUSSION

We examined the quantity and chemistry of DOM in ponds embedded within highly urbanized environments and report the change in its optical chemistry after irradiation. We expected that DOM in urban ponds would have a distinct optical chemistry compared to that derived from more natural areas. This predicted difference was based on our expectations of greater contributions of either anthropogenic-derived organic matter (of a distinct composition) or from high internal production of DOM. We found that urban pond DOM was relatively transparent (low ε_{280}) and uncolored (low a_{440}), had lower HIX and higher β:α ratios compared to reported values for numerous temperate lakes and streams, and some other anthropogenic impacted waters. DOM characteristics from urban ponds were most similar to that found in mixotrophic to eutrophic aquatic systems where autochthonous DOM has been shown to dominate (Table 2).

DOM in urban ponds nonetheless contained a unique mixture of microbial and terrestrial derived C, reflected in the HIX index [36], FI and β:α ratio values [39,46] (Table 1). The DOM pool was dominated by a likely algal-derived humic-like component (C6) with relatively low percentages of terrestrial-like components (Figure 3). C6 was much more abundant in urban ponds than in natural freshwaters in the same geographical region (e.g., southern Ontario [19];; C. Williams, unpublished data). It could be argued that urban ponds should not be dominated by autochthonous C, given their role in overland flood control and engineered connectivity with the watershed. The lack of terrestrial DOM influence on pond DOM may simply reflect the reduced presence of natural components (e.g. vegetation, soils, etc.) of the upstream catchment. Urban streams, however, can contain high amounts of terrestrial DOM [19,40] and we expected a stronger signal of this DOM within urban ponds. These ecosystems could either receive very different runoff or they may quickly process in-fluxing DOM. Future studies should focus on determining the relative importance of terrestrial DOM sources and internal processing rates on urban pond DOM chemistry, in part, by examining the timing and importance of external water inputs. The urban ponds in our study are downstream of stormwater drainage systems designed to channel runoff very quickly, which may further limit contact with soils and vegetation within the upstream catchment. Our results contrast with other freshwater ecosystems where allochthonous C is usually assumed to be the major fraction of DOM (e.g., [47]) due to high hydrologic connectivity between rivers and lakes and their upstream undeveloped watersheds.

The distinct character of urban pond DOM from that found in natural streams and lakes in the same region and other aquatic ecosystems largely resulted from the greater prominence of relatively transparent, internally-produced DOM. This has also been reported in other anthropogenically impacted water bodies that are relatively nutrient rich (e.g., [48] and references therein) as well as in mixotrophic to eutrophic, saline lakes (e.g., [49]). An algal-derived humic-like component (C6) was the most abundant in urban ponds and represented about 52% of the DOM fluorescence pools. C6 was higher in ponds with lower HIX scores, indicating that DOM with this chemical feature is produced internally and dilutes the allochthonous DOM pools. This component type has been documented in other human impacted areas and in highly productive ecosystems (e.g., [50]), which is further indication that there is significant internal production of DOC in urban ponds [7,41]. Urban pond DOM thus appears to have a distinct chemical signature that originates from primary production, which is fueled by high nutrient loading in these heavily urbanized catchments.

Due to its relative transparency and low aromaticity, the major DOM fraction in urban ponds should be particularly resistant to photochemical transformation. In fact, the main component of the DOM pool (C6) was not impacted significantly by light exposure. This result is similar to that reported for pelagic algal-derived DOM and DOM in some eutrophic ecosystems, which have relatively low rates of photomineralization [26]. Similarly, we also found no DOC loss with light exposure in about one fifth of our pond samples. This lack of significant total DOM photodegradation might simply reflect the chemical properties of internally produced DOM in these urban ponds. Especially transparent autochthonous material should not be photo-oxidative [25]. On the other hand, DOM in urban ponds may have already experienced prolonged UVR exposure prior to sampling, especially if there had been no recent hydrological inputs from the watershed. We would expect that rates of photochemical degradation should be higher in water samples obtained directly from in-flowing stormwater. Future comparative studies of DOM properties and photodegradation of fresh stormwater and *in situ* urban pond water would be useful to differentiate between these controls.

The less abundant terrestrial derived components of the urban pond DOM pool were the most sensitive to photochemical degradation. After photo-irradiation, DOM fluorescence shifted from humic-like to protein-like, with decreases in the most terrestrial humic-like components (C1, C2, and C5) and significant increases a protein-like component (C7; Figure 3). One terrestrial, humic-like component (C3) increased after photo-irradiation, which suggests that this component is photo-stable or photo-produced [41]. This indicates that C3 might be a product of photodegradation of the other terrestrial components. Given these changes in photo-exposed DOM samples, natural measurements of these rates of gain and losses in urban ponds over time would be a logical next step to better understand the controls of the DOM pool in these urban freshwater ecosystems. Nevertheless, DOM chemistry in urban ponds appears to be controlled by a combination of processes (i.e., DOM import, release from primary producers, and photoirradiation), which would all act in concert to shape the overall properties of this important energy source to microbial communities.

We found DOC concentration decreased in many pond samples under irradiation, but losses were also observed in some samples held in the dark. This dark-sample decrease in DOC concentration likely resulted from microbial activity [51]. While samples were filtered prior to the photo-exposure experiment, remnant microbial communities could nonetheless have consumed labile components of the DOM. The decrease in DOC concentration in dark samples was not reflected by corresponding changes in absorbance and is consistent with losses of low molecular-weight labile substrates with low absorbance. These same decreases in labile substrates with low absorbance were observed in irradiated samples, which is indicated by the increase in ε_{280} that would have been produced by a smaller decrease in absorbance compared to DOC concentration. A recent review by [52] showed variability in bacterial responses during irradiation ranging from stimulation to inhibition. However, irradiating with a full range of artificial UVA and UVB inhibited microbial growth, perhaps due to photoproduction of singlet oxygen and other reactive oxygen species [53-55]. Singlet oxygen is the primary agent of photo-oxidative stress in microorganisms [56] and high concentrations may delay consumption of the readily decomposable portions of the DOM until dark conditions prevail. Thus the microbial degradation in irradiated samples likely took place after irradiation rather than simultaneously with irradiation. The dark-sample losses nonetheless suggest that some changes observed in the photo-exposed samples resulted from the experimental conditions (i.e., non-sterility of bags) and not light per se. However, compared to the rates observed in the light exposed samples, this effect was minimal and does not significantly affect rate measurements or the observed photo-induced changes in DOM chemistry.

Chlorophyll concentrations are one surrogate index of algal biomass and should be correlated with the internal production of autochthonous DOC in freshwater ecosystems (e.g., [57]). Consequently, we expected that algal biomass in the urban ponds to be positively related to concentrations of the highly transparent DOM. Our results are not entirely consistent with this expectation as HIX and ε_{280} (both indicators of terrestrial DOM) were positively correlated with TP and chlorophyll. This result suggests that the connections between TP, productivity, and DOM chemistry are likely more complex than the simple scenario previously described. For example, greater primary production coincides with higher TP concentration in urban ponds but this may have contrasting effects on the DOM and its chemistry. While internal productivity may contribute transparent DOM and dilute terrestrial sources, it may also fuel (along with the higher nutrient concentrations) greater microbial activity that could reduce DOM concentrations. The negative relationships between chlorophyll and TP with C7 are consistent with these indirect connections. Future work should thus better examine the how DOM chemistry relates to primary producers and their external nutrient controls in urban environments.

We found a prominent signature of internally derived DOM in urban ponds of this study. This unique chemical signature likely reflects a dominance of internally produced C, the lack of external humic sources, and considerable microbial and photo-processing. While this unique chemical signature differentiates urban pond DOM from other aquatic ecosystems, how it affects the pond physical (light penetration), chemical (metal binding), or biological (microbial production) processes remains largely to be seen. Given their growing abundance and important role in urban water cycles, these ponds would appear to be potential hotspots for C processing in the urban landscape and warrant further examination of their carbon cycling.

ACKNOWLEDGMENTS

Field logistical support was provided by Environment Canada and the municipalities of Ottawa, Peterborough, Richmond Hill and Whitby. Assistance in the field and the laboratory was provided by A. McDonald, A. Scott, M. Merette, M. Lamond, and S. Nienhuis.

AUTHOR CONTRIBUTIONS

Conceived and designed the experiments: NM CW MX PP PF. Performed the experiments: NM CW PP. Analyzed the data: NM CW MX PP PF. Contributed reagents/materials/analysis tools: NM CW MX PP PF. Wrote the manuscript: NM CW MX PP PF.

REFERENCES

1. Vitousek PM, Mooney HA, Lubchenco J, Melillo JM (1997) Human domination of Earth's ecosystems. Science 277: 494-499. doi:10.1126/science.277.5325.494.

2. Lee JG, Heaney JP (2003) Estimation of urban imperviousness and its impacts on stormwater systems. J Water Resour Plan Manag 129: 419-426. doi:10.1061/(ASCE)0733-9496(2003)129:5(419).

3. Wissmar RC, Timm RK, Logsdon MG (2004) Effects of changing forest and impervious land covers on discharge characteristics of watersheds. Environ Manag 34: 91-98. PubMed: 15088123.

4. Crowe AS, Rochfort Q, Exall K, Marsalek J (2007) Controlling urban stormwater pollution by constructed wetlands: a Canadian perspective. Int J Water 3: 214-230. doi:10.1504/IJW.2007.015215.

5. Brand AB, Snodgrass JW (2010) Value of artificial habitats for amphibian reproduction in altered landscapes. Conserv Biol 24(1): 295-301. doi:10.1111/j.1523-1739.2009.01301.x. PubMed: 19681986.

6. Hamer AJ, Parris KM (2011) Local and landscape determinants of amphibian communities in urban ponds. Ecol Appl 21: 378-390. doi:10.1890/10-0390.1. PubMed:21563570.

7. Williams CJ, Frost PC, Xenopoulos MA (2013) Beyond best management practices: pelagic biogeochemical dynamics in urban stormwater ponds. Ecol Appl, 23: 1384–95. doi:101890/12-08251. PubMed: 24147410.

8. Downing JA (2010) Emerging global role of small lakes and ponds: little things mean a lot. Limnetica 29(1): 9-24.

9. Einsele G, Yan J, Hinderer M (2001) Atmospheric carbon burial in modern lake basins and its significance for the global carbon budget. Glob Planet Change 30: 167-195. doi:10.1016/S0921-8181(01)00105-9.

10. Downing JA, Prairie YT, Cole JJ, Duarte CM, Tranvik LJ et al. (2006) The global abundance and size distribution of lakes ponds and impoundments. Limnol Oceanogr 51: 2388-2397. doi:10.4319/lo.2006.51.5.2388.

11. Downing JA, Cole JJ, Middelburg J, Striegl RG, Duarte CM et al. (2008) Sediment carbon burial in agriculturally eutrophic impoundments over the last century. Global Biogoechemical Cycles 22:   GB1018 doi:10.1029/2006GB002854.

12. Caraco NF, Cole JJ (2004) When terrestrial organic matter is sent down the river: importance of allochthonous C inputs to the metabolism in lakes and rivers. In: A. PolisME Power. Food webs at the landscape level. Chicago University of Chicago Press. pp. 301-316.

13. Hanson PC, Pollard AI, Bade DL, Predick K, Carpenter SR et al. (2004) A model of carbon evasion and sedimentation in temperate lakes. Glob Change Biol 10: 1285-1298. doi:10.1111/j.1529-8817.2003.00805.x.

14. Prairie YT, Bird DF, Cole JJ (2002) The summer metabolic balance in the epilimnion of southeastern Quebec lakes. Limnol Oceanogr 47: 316-321. doi:10.4319/lo.2002.47.1.0316.

15. Kritzberg ES, Cole JJ, Pace MM, Granéli W (2005) Does autochthonous primary production drive variability in bacterial metabolism and growth efficiency in lakes dominated by terrestrial C inputs? Aquat Microb Ecol 38: 103-111. doi:10.3354/ame038103.

16. Hook AM, Yeakley A (2005) Stormflow dynamics of dissolved organic carbon and total dissolved nitrogen in a small urban watershed. Biogeochemistry 75(3): 409-431. doi:10.1007/s10533-005-1860-4.

17. Paul MJ, Meyer JL (2001) Streams in the urban landscape. Annu Rev Ecol Syst 32: 333-365. doi:10.1146/annurev.ecolsys.32.081501.114040.

18. Walsh CJ, Roy AH, Feminella JW, Cottingham PD, Groffman PM et al. (2005) The urban stream syndrome: current knowledge and search for a cure. J North Benthol Soc 24: 706-723. doi:10.1899/0887-3593(2005)024

19. Williams CJ, Yamashita Y, Wilson HF, Jaffé R, Xenopoulos MA (2010) Unraveling the role of landuse and microbial activity in shaping dissolved organic matter characteristics in streams. Limnol Oceanogr 55: 1159-1171. doi:10.4319/lo.2010.55.3.1159.

20. Wilson HF, Xenopoulos MA (2009) Effects of agricultural land use on the composition of fluvial dissolved organic matter. Nat Geosci 2: 37-41. doi:101038/ngeo391.

21. Baker A (2001) Fluorescence Excitation-Emission matrix characterization of some sewage-impacted rivers. Environ Sci Technol 35: 948-953. doi:10.1021/es000177t. PubMed: 11351540.

22. McKnight DM, Aitken GR, Smith RL (1991) Aquatic fulvic acids in microbially based ecosystems: Results from two desert lakes in Antarctica. Limnol Oceanogr 36: 998-1006. doi:10.4319/lo.1991.36.5.0998.

23. Bertilsson S, Jones JL (2003) Supply of dissolved organic matter to aquatic ecosystems: Autochthonous sources In: SEG FindlayRL Sinsabaugh. Aquatic Ecosystems: Interactivity of Dissolved Organic Matter. Academic Press. pp. 3-25.

24. Tranvik LJ, Bertilsson S (2001) Contrasting effects of solar UV radiation on dissolved organic sources for bacterial growth. Ecol Lett 4: 458-463. doi:10.1046/j.1461-0248.2001.00245.x.

25. Anesio AM, Li WG, Aiken GR, Kieber DJ, Mopper K (2005) Effect of humic substance photodegradation on bacterial growth and respiration in lake water. Appl Environ Microbiol 71: 6267-6275. doi:10.1128/ AEM.71.10.6267-6275.2005. PubMed:16204548.

26. Bertilsson S, Tranvik LJ (2000) Photochemical transformation of dissolved organic matter in lakes. Limnol Oceanogr 45: 753-762. doi:10.4319/lo.2000.45.4.0753.

27. Obernosterer I, Reitner B, Herndl GJ (1999) Contrasting effects of solar radiation on dissolved organic matter and its bioavailability to marine bacterioplankton. Limnol Oceanogr 44: 1645- 1654. doi:10.4319/ lo.1999.44.7.1645.

28. Obernosterer I, Sempere R, Herndl GJ (2001) Ultraviolet radiation induces reversal of the bioavailability of DOM to marine bacterioplankton. Aquat Microbiol 24: 61-68. doi:10.3354/ame024061.

29. Kaiser E, Sulzberger B (2004) Phototransformation of riverine dissolved organic matter (DOM) in the presence of abundant iron: Effect on DOM bioavailability. Limnol Oceanogr 49: 540-555. doi:10.4319/ lo.2004.49.2.0540.

30. Mayer LM, Schick LL, Hardy KR, Estapa ML (2009) Photodissolution and other photochemical chances upon irradiation of algal detritus. Limnol Oceanogr 54: 1688-1698. doi:10.4319/lo.2009.54.5.1688.

31. McEnroe NA, Buttle JM, Pick FR, Xenopoulos MA, Frost PC (2012) Thermal and chemical stratification of urban ponds: Are they 'completely mixed reactors'? Urban Ecosyst 16: 327-339. doi: 10.1007/s11252-012-0258-z

32. Kirk JT (1994) Light and Photosynthesis in Aquatic Ecosystems. Cambridge University Press.

33. Maurice PA, Cabaniss SE, Drummond J, Ito E (2002) Hydrogeochemical controls on the variations in chemical characteristics of natural organic matter at a small freshwater wetland. Chem Geol 18 /: 59-77. doi:10.1016/ S0009-2541(02)00016-5.

34. Cuthbert ID, del Giorgio PA (1992) Toward a standard method of measuring color in freshwater. Limnol Oceanogr 37: 1319-1326. doi:10.4319/lo.1992.37.6.1319.

35. Mobed JJ, Hemmingsen SL, Autry JL, McGown LB (1996) Fluorescence characterisation of IHSS humic substances: Total luminescence spectra with absorbance correction. Environ Sci Technol 30: 3061-3065. doi:10.1021/es9601321.

36. Zsolnay A, Baigar E, Jimenez M, Steinweg B, Saccomandi F (1999) Differentiating with fluorescence spectroscopy the sources of dissolved organic matter in soils subjected to drying. Chemosphere 38: 45-50. doi:10.1016/S0045-6535(98)00166-0. PubMed:10903090.

37. Parlanti E, Worz K, Geoffroy L, Lamotte M (2000) Dissolved organic matter fluorescence spectroscopy as a tool to estimate biological activity in a coastal zone submitted to anthropogenic inputs. Org Geochem 31: 1765-1781. doi:10.1016/S0146-6380(00)00124-8.

38. Helms JR, Stubbins A, Ritchie JD, Minor EC, Kieber DJ et al. (2008) Absorption spectral slopes and slope ratios as indicators of molecular weight source and photobleaching of chromophoric dissolved organic matter. Limnol Oceanogr 53(3): 955-969. doi:10.4319/lo.2008.53.3.0955.

39. Cory RM, McKnight DM (2005) Fluorescence spectroscopy reveals ubiquitous presence of oxidised and reduced quinones in dissolved organic matter. Environ Sci Technol 39: 8142-8149. doi:10.1021/es0506962. PubMed: 16294847.

40. Petrone JB, Fellman JB, Hood E, Donn MJ, Grierson PF (2011) The origin and function of dissolved organic matter in agro-urban coastal streams. J Geophys Res 116: G01028. doi:101029/2010JG001537.

41. Stedmon CA, Markager S (2005) Resolving the variability in dissolved organic matter fluorescence in a temperate estuary and its catchment using PARAFAC analysis. Limnol Oceanogr 50: 686-697. doi:10.4319/lo.2005.50.2.0686.

42. Coble PG, Green SA, Blough NV, Gagosian RB (1990) Characterisation of dissolved organic matter in the Black Sea by fluorescence spectroscopy. Nature 348: 432-435. doi:10.1038/348432a0.

43. Maie N, Pisani O, Jaffé R (2008) Mangrove tannins in aquatic ecosystems: their fate and possible influence on dissolved organic carbon and nitrogen cycling. Limnol Oceanogr 53: 160-171. doi:10.4319/lo.2008.53.1.0160.

44. Del Vecchio R, Blough NV (2002) Photobleaching of chromophoric dissolved organic matter in natural waters: kinetics and modelling. Mar Chem 78 (4): 231–253. doi:10.1016/S0304-4203(02)00036-1.

45. Porcal P, Amirbahman A, Kopáček J, Novák F, Norton SA (2009) Photochemical release of humic and fulvic acid-bound metals from simulated soil and streamwater. J Environ Monit 11: 1064-1071. doi:10.1039/b812330f. PubMed: 19436866.

46. Huguet A, Vacher L, Relexans S, Saubusse S, Froidefond JM et al. (2009) Properties of fluorescent dissolved organic matter in the Gironde Estuary. Org Geochem 40: 706-719. doi:10.1016/j.orggeochem.2009.03.002.

47. Carpenter SR, Cole JJ, Pace ML, Van de Bogert M, Bade DL et al. (2005) Ecosystems subsidies: Terrestrial support of aquatic food webs from 13C addition to contrasting lakes. Ecology 86: 2737-2750. doi:10.1890/04-1282.

48. Stanley EH, Powers SM, Lottig NR, Buffam I, Crawford JT (2012) Contemporary changes in dissolved organic carbon (DOC) in human-dominated rivers: is there a role for DOC management? Freshw Biol 57: 26-42. doi:10.1111/j.1365-2427.2011.02613.x.

49. Osburn CL, Wigdahl CR, Frotz SC, Saros E (2011) Dissolved organic matter composition and photoreactivity in prairie lakes of the US Great Plains. Limnol Oceanogr 56(6): 2371-2390. doi:10.4319/lo.2011.56.6.2371.

50. Yamashita Y, Scinto LJ, Maie N, Jaffé R (2010) Dissolved organic matter characteristics across a subtropical wetland's landscape: Application of optical properties in the assessment of environmental dynamics. Ecosystems 13: 1006–1019. doi:10.1007/s10021-010-9370-1.

51. Porcal P, Hejzlar J, Kopacek J (2004) Seasonal and photochemical changes of DOM in an acidified forest lake and its tributaries. Aquat Sci 66: 211-222. doi:10.1007/s00027-004-0701-1.

52. Ruiz-Gonzáles C, Simó R, Sommaruga R, Gasol JM (2013) Away from darkness. A review on the effects of solar radiation on heterotrophic bacterioplankron activity. Frontiers in Microbiol. p. 4. doi:10.3389/fmicb.2013.00131.

53. Zepp RG, Wolfe NL, Baughman GL, Hollis RC (1977) Singlet oxygen in natural waters. Nature 267: 421-423. doi:10.1038/267421a0.

54. Xenopoulos MA, Bird DF (1997) Effect of acute exposure to hydrogen peroxide on the production of phytoplankton and bacterioplankton in a mesohumic lake. Photochem Photobiol 66: 471-478. doi:10.1111/j.1751-1097.1997.tb03175.x.

55. Scully NM, Cooper WJ, Tranvik LJ (2003) Photochemical effects on microbial activity in natural waters: the interaction of reactive oxygen species and dissolved organic matter. FEMS Microbiol Ecol 46: 353-357. doi:10.1016/S0168-6496(03)00198-3. PubMed:19719565.

56. Glaeser J, Nuss AM, Berghoff BA, Klug G (2011) Singlet oxygen stress in microorganisms. Adv Microb Physiol 58: 141-173. doi:10.1016/B978-0-12-381043-4.00004-0. PubMed: 21722793.

57. Kritzberg ES, Cole JJ, Pace MM, Granéli W (2006) Bacterial growth on allochthonous carbon in humic and nutrient enriched lakes: Results from

whole lake 13C additions. Ecosystems 9: 489-499. doi:10.1007/s10021-005-0115-5.

58. Mladenov N, Huntsman-Mapila P, Wolski P, Masamba WRL, McKnight DM (2008) Dissolved organic matter accumulation reactivity and redox state in ground water of a recharge wetland. Wetlands 28: 747-759. doi:10.1672/07-140.1.

59. Wu FC, Kothawala DN, Evans RD, Dillon PJ, Cai YR (2007) Relationship between DOC concentration molecular size and fluorescence properties of DOM in a stream. Appl Geochem 22: 1659-1667. doi:10.1016/j.apgeochem.2007.03.024.

60. Frost PC, Larson JH, Johnston CA, Young KC, Maurice PA et al. (2006) Landscape predictor s of stream dissolved organic matter concentration and physicochemistry in a Lake Superior river watershed. Aquat Sci 68: 40-51. doi:10.1007/BF02508824.

61. McKnight DM, Boyer EW, Westerhoff PK, Doran PT, Kulbe T (2001) Spectroflurometric characterisation of dissolved organic matter for indication of precursor organic material and aromaticity. Limnol Oceanogr 46(1): 38-48. doi:10.4319/lo.2001.46.1.0038.

62. Chin YP, Aiken G, O'Loughlin E (1994) Molecular weight polydispersity and spectroscopic properties of aquatic humic substances. Environ Sci Technol 28: 1853-1858. doi:10.1021/es00060a015. PubMed: 22175925.

63. Chin Y-P, Aiken GR, Danielsen KM (1997) Binding of pyrene to aquatic and commercial humic substances: the role of molecular weight and aromaticity. Environ Sci Technol 31: 1630-1635. doi:10.1021/es960404k.

64. Miller MP, McKnight DM, Chapra SC, Williams MW (2009) A model of degradation and production of three pools of dissolved organic matter in an alpine lake. Limnol Oceanogr 54(6): 2213-2227. doi:10.4319/lo.2009.54.6.2213.

65. Reche I, Pace ML, Cole JJ (1999) Relationship of trophic and chemical conditions to photobleaching of dissolved organic matter in lake ecosystems. Biogeochemistry 44: 259-280. doi:10.1007/BF00996993.

Chapter 10

MESOFLUIDIC DEVICES FOR DNA-PROGRAMMED COMBINATORIAL CHEMISTRY

Rebecca M. Weisinger[1,2], Robert J. Marinelli[2], S. Jarrett Wrenn[2], Pehr B. Harbury[2]

[1]Department of Chemistry, Stanford University, Stanford, California, United States of America,

[2]Department of Biochemistry, Stanford University, Stanford, California,United States of America

ABSTRACT

Hybrid combinatorial chemistry strategies that use DNA as an information-carrying medium are proving to be powerful tools for molecular discovery. In order to extend these efforts, we present a highly parallel format for DNA-programmed chemical library synthesis. The new format uses a standard microwell plate footprint and is compatible with commercially available automation technology. It can accommodate a wide variety of combinatorial synthetic schemes with up to 384 different building blocks per chemical step. We demonstrate that fluidic routing of DNA populations in the highly parallel format occurs with excellent specificity, and that chemistry on DNA arrayed into 384 well plates proceeds robustly, two requirements for the high-fidelity translation and efficient *in vitro* evolution of small molecules.

INTRODUCTION

Natural evolution consists of iterated cycles of gene diversification, gene expression, functional selection and reproductive amplification. These cycles can be re-enacted in a test tube using populations of random biopolymer sequences as the genetic units. Functional selection is imposed by requiring individual molecules to bind to a target, or to catalyze coupling to an affinity handle, in order to survive. Remarkably, novel snippets of nucleic acid and protein with the selected functional property (binding or catalytic proficiency) emerge. The test-tube evolution paradigm can be extended to small-molecule genetic units through DNA-programmed combinatorial chemistry. [1], [2], [3], [4] Ribosomal translation is replaced with "chemical translation," wherein a DNA gene sequence programs the chemical synthesis of a covalently attached small molecule.[5]–[6] DNA-programming enables the propagation and breeding of small-molecule populations over multiple generations.

By analogy to *in vitro* biopolymer evolution, it has been suggested that evolving small-molecule libraries of more than ten billion compounds for binding to a protein target should yield ligands with dissociation constants in the nanomolar range.[1],[7] There are a number of ways to construct chemical libraries of such high complexity. One strategy would be to create synthetic decamers from an alphabet of ~10 chemical building blocks. This strategy produces high molecular-weight compounds that do not resemble small-molecule drugs, like those in the World Drug Index.[8] Alternatively, one could construct molecules in four steps using an alphabet of 384 distinct building blocks at each synthetic step. This large-alphabet strategy minimizes the molecular weight of the individual molecules that make up the population. In order to create a large-alphabet library using DNA-programmed combinatorial chemistry, some technical innovations are required.

Here, we report tools that facilitate the construction of highly complex libraries with the possibility for hundreds of diversity elements at each position. These tools build on a previously described approach to chemical translation that involves spatial partitioning of a DNA population by hybridization followed by spatially determined chemical coupling steps (Figure 1a).[6] A read of a single coding position is illustrated in Figure 1b. A degenerate library of single-stranded DNA genes is split by hybridization into different wells of a cassette holding 384 distinct oligonucleotide-conjugated resins, the "anticodon array." Following hybridization, the DNA sequences are transferred in a one-to-one fashion onto a 384-feature anion-exchange array for execution of a chemistry step on solid-supported DNA. The solid support allows reactions to be driven to completion with excess reagents, and allows reactions to be performed under conditions that are incompatible with DNA hybridization and DNA solubility. After the chemical coupling step, the library is pooled and split again by hybridization at the next coding position. Additional reads are performed until all of the coding positions have been translated.

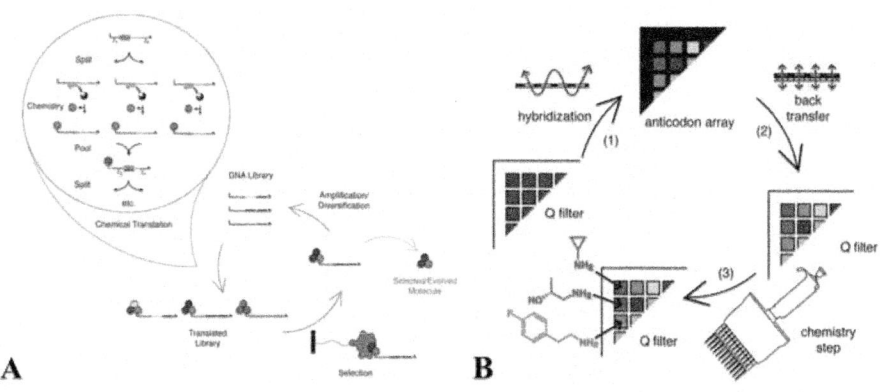

Figure 1: Small molecule evolution by DNA-programmed combinatorial chemistry.]

(A) A degenerate library of DNA "genes" is chemically translated into small molecule-DNA conjugates. The covalently attached small molecule corresponds to the structure encoded by the DNA gene. The small molecule-DNA conjugates are then selected for a desired trait, such as binding to a protein of interest. The encoding DNA is amplified and diversified. The cycle is iterated to yield selected/evolved small molecules with the desired trait. (B) One step of a highly parallel DNA-programmed chemical translation. (1) A degenerate DNA library (in which the identity at the first coding position is depicted by a range of colors) on an ion-exchange filter is split by hybridization to complementary oligonucleotides on an anticodon array. (2) The separated pools (with distinct colors representing isolated codon sequences) are transferred to a fresh anion-exchange chemistry array. (3) The encoded building blocks are incorporated by chemical coupling. These steps are repeated for each coding position until the DNA "gene" is fully translated.

doi:10.1371/journal.pone.0032299.g001

RESULTS

Our original implementation of DNA-programmed combinatorial chemistry used commercially available oligonucleotide synthesis columns to house the anticodon resins required for library splitting and the anion-exchange resins required for chemical synthesis steps.[9] This approach was inexpensive and convenient for libraries with small alphabets, but becomes unwieldy with large numbers of building blocks. Consequently, we set out to develop arrayed formats to facilitate the synthesis of large-alphabet libraries. We focused on planar substrates with a standard microplate footprint that could exploit the tools developed for high-throughput chemistry and biology, including multi-well plates, plate centrifuges, multi-channel pipetters, and pipetting robots.

We first created a chemistry array for carrying out reactions in parallel. Our efforts built on the SPOT synthesis literature in which cellulose paper is used as a stable support for the synthesis of covalently attached chemical libraries. [10],[11] To use cellulose supports for DNA-programmed combinatorial chemistry, the membrane must act as a strong anion-exchange material that can bind reversibly to DNA. We therefore derivatized the surface of the cellulose paper with quaternary amines using a process described by Genentech for the production of charged filtration membranes.[12] Typically, SPOT synthesis is performed on dense arrays of synthetic sites with no chemical or physical barriers between the sites, relying on small reagent quantities to prevent mixing of reactants at neighboring positions. Chemical transformations on solid support, however, often require the use of a reagent excess to push reactions to completion. Reagent excesses cause reactant mixing between spots on a cellulose array (Figure 2). We therefore investigated how to separate the reaction sites with a chemically resistant polymer. We solved the problem by imbedding a photo-curable fluoroelastomer[13]into the quaternary amine-modified cellulose membrane. Precise patterning was accomplished by using masks to photo-polymerize a fluoropolymer border around a field of open wells (Figure 2a) in an approach similar to the paper microzone plates reported by Carrilho and coworkers.[14] The chemically resistant gasketing prevents the contents of adjacent wells from mixing (Figure 2b and 2c).

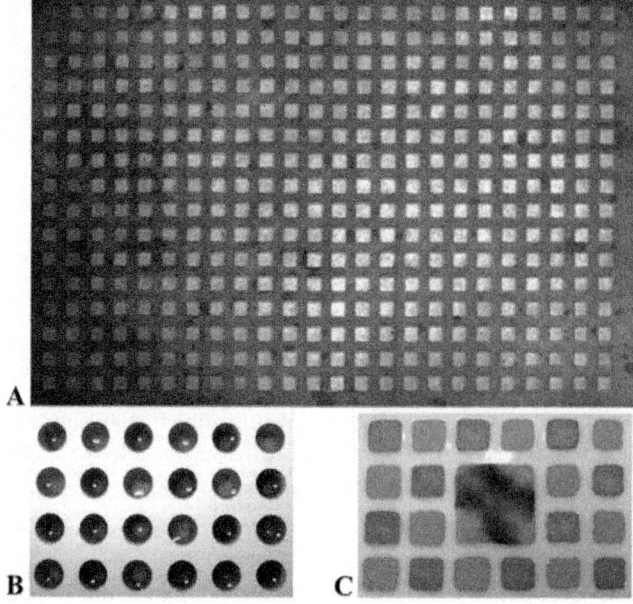

Figure 2: Anion-exchange chemistry arrays.

(A) A cellulose filter patterned with photocurable fluoropolymer. (B) The filter clamped between the two plates of the chemistry apparatus. The wells are filled in a checkerboard pattern with solutions of Orange G and xylene cyanol. (C) After evacuating the dye solutions and washing on a vacuum manifold, the filter is removed from the chemistry apparatus. The dyes mix only in the central squares, which lack fluoropolymer barriers between neighboring wells.

doi:10.1371/journal.pone.0032299.g002

To create reagent reservoirs above the wells of the chemistry array, we designed plates with 384 through holes and grooves to receive in-plane rubber gaskets (Figure 3a). When sealed on either side of the array, the plates form 384 isolated reaction vessels, which allow for different reactions to occur in adjacent wells without measurable bleed over (Figure 4a). Because the arrays have the same footprint as commercial 384-well plates, wash steps can be performed using a standard microplate vacuum manifold.

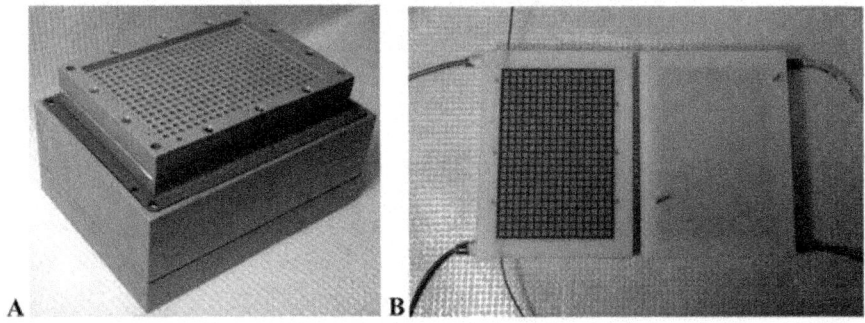

Figure 3: Mesofluidic devices for DNA-programmed synthesis.

(A) The chemistry apparatus on top of a Qiagen multi-well plate manifold. The top half of the device has a large trough to hold fluid for wash steps, and the bottom half has a raised surface that acts as an array of drip directors to facilitate the collection of material. The interior surfaces of the plates are grooved to accommodate silicone gaskets. (B) The two halves of the mesofluidic pump prior to assembly. An anticodon array is loaded onto the right plate of the pump. Throughout the temperature cycle, the DNA library is circulated over all 384 features of the anticodon array. The thick colored tubing at the top and bottom of each plate delivers compressed air or vacuum to a subset of the diaphragm features in the plate. The thin green tubing at the top and bottom of the left plate is the liquid inlet/outlet.

doi:10.1371/journal.pone.0032299.g003

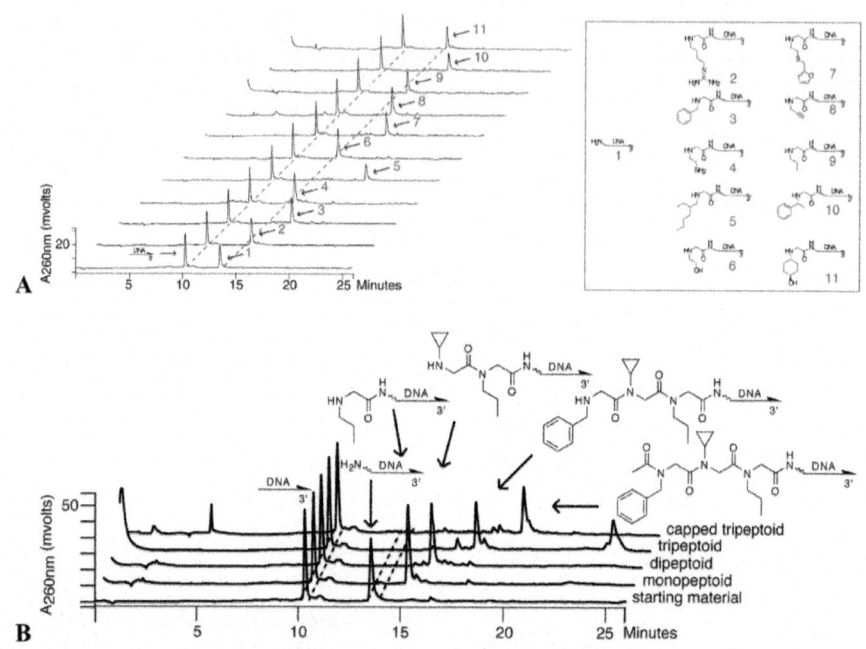

Figure 4: Parallel peptoid synthesis reactions on anion-exchange arrays.

(A) A control 10-mer and a 5′ aminated 20-mer oligonucleotide were loaded in adjacent wells of a gasketed Q cellulose chemistry array and subjected to a chloroacetylation step followed by an alkylation step with one of 10 primary amines. After elution, the products were analyzed by reverse-phase HPLC. The peaks corresponding to the peptoid products were isolated, digested with P1 nuclease, and analyzed by LC-MS to confirm the identity of the product. The masses obtained by LC-MS are listed in the supporting materials . (B) A control 10-mer and a 20-mer oligonucleotide with a 5′ primary amine were loaded in adjacent wells of a gasketed Q cellulose chemistry array (starting material) and subjected to one (monopeptoid), two (dipeptoid), or three (tripeptoid) peptoid coupling steps. Additionally, a synthesis consisting of three peptoid coupling steps and a subsequent acetylation step (capped tripeptoid) was performed. After elution, the products were analyzed by reverse-phase HPLC. The peak for the capped tripeptoid synthesis was isolated, digested with nuclease P1, and submitted for analysis by LC-MS. The identity of the major peak (approximately 88% yield) was confirmed ([M-H]⁻ observed, 898.99; expected, 898.47).

doi:10.1371/journal.pone.0032299.g004

Our DNA-programmed combinatorial chemistry approach also requires a means to split DNA by hybridization into 384 different sub-pools arranged in a planar format. One strategy would be to immobilize oligonucleotides onto a filter, but extensive studies with filter immobilization led us to the conclusion that filters cannot provide sufficient hybridization capacity. As an alternative, we used oligonucleotide-derivatized Sepharose resins housed in an array of 384 microcolumns. The microcolumns were constructed by laser cutting 3 mm square holes into a 380 µm-thick Delrin sheet, and then adhering polypropylene filters to either side. The polypropylene filters act as frits to hold Sepharose resin within each 3 mm wide and 380 µm long "column housing" (Figure 5). We call these planar structures "anticodon arrays" because each feature of the array contains a unique DNA sequence complementary to one of the codons used for DNA programming. The microcolumns hold microliter quantities of resin and remain stable to multiple rounds of hybridization and denaturation.

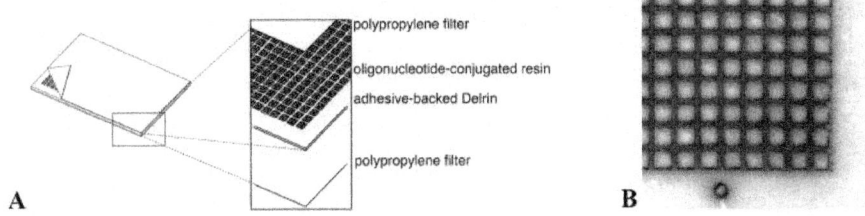

Figure 5: Construction of an anticodon array.

(A) Oblique and exploded views of an anticodon array. The arrays are constructed by sealing one side of an adhesive-backed, lasercut Delrin sheet with a polypropylene filter, filling the resulting wells with oligonucleotide-conjugated resin, and then sealing the opposite side of the array with a second polypropylene filter. (B) Photograph of the lower right corner of an anticodon array.

doi:10.1371/journal.pone.0032299.g005

During a programmed chemical step, a DNA library is partitioned by hybridization to the 384 features of the anticodon array. The entire DNA library must be distributed over all of the features on a short timescale (defined by the rate of change of the hybridization temperature). Manually passing the library through every feature of the anticodon array by vacuum filtration or by centrifugation would require at least 384 manipulations of the pooled material and was deemed too laborious. Alternatively, one could use passive mixing or capillary action to route fluid.[15] Passive hybridization proved to be slow and provided many surfaces for the nonspecific binding of library material. Ideally, the library would be pumped cyclically over all of the features of the anticodon array, with minimal surface area and dead volume. If the system were heated and then cooled slowly, it would provide multiple passes of the whole library through each microcolumn of the anticodon array in the temperature window over which hybridization occurs.

To realize the latter scheme, we constructed a mesofluidic pump (Figure 3b) that pushes liquid through the 384 features of an anticodon array in a serpentine path. As outlined in Figure 6a, application of alternating compressed air and vacuum to elastic diaphragms above and below the array produces liquid flow. This mesofluidic device is a positive displacement pump, and it operates by the same principles as the peristaltic micro-pumps used by the microfluidics community.[16]

Following spatial partitioning by hybridization, the physically separated sub-pools of DNA must be transferred from the anticodon array onto a cellulose chemistry array. To perform this transfer efficiently and conveniently, we constructed a mesofluidic Southern blotter (Figure 6). The blotter consists of two plates, each housing 384 isolated liquid columns loaded with a 10 mM sodium hydroxide solution. A chemistry array and anticodon array are clamped between the two plates. Application of alternating compressed air and vacuum to elastic diaphragms at the top and bottom of each column forces liquid to move up and down through the stacked arrays. Perfusion of the anticodon array with a denaturing solution causes release of DNA, and advection of the liquid column carries the DNA to the anion-exchange chemistry array where it rebinds. Isolation of the 384 independent liquid columns ensures a faithful one-to-one transfer of DNA between the features of the two arrays.

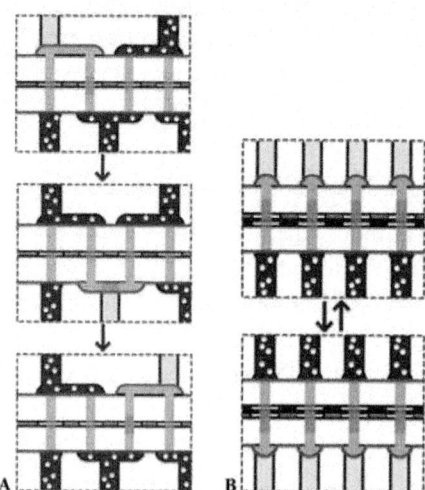

Figure 6: Schematic of hybridization pump and mesofluidic Southern blotter.

(A) The hybridization pump is shown in cross-section with the plane of the elastomeric diaphragm colored red. One volume element of fluid is depicted as an orange rectangle. Alternating application of 15 psi of air pressure (blue with white spots) and house vacuum (yellow) directs the path of fluid flow. (B) A cross-section of the mesofluidic Southern blotter illustrating the movement of liquid through the stacked anion-exchange chemistry array and the anticodon array as air and vacuum are applied.

doi:10.1371/journal.pone.0032299.g006

Collectively, the procedures described above complete one full read of a DNA-programmed synthesis. For a multi-step combinatorial library synthesis, the cycle of splitting, blotting and chemical modification is repeated multiple times.

Fidelity of DNA-programmed chemistry

The fidelity of DNA-programmed combinatorial chemistry is dependent on two things: the accuracy with which the DNA population is split into spatially separated sub-pools, and the efficiency of the subsequent sub-pool specific chemistry steps. To characterize the arrays and mesofluidic devices described above, we measured these two parameters.

First, we tested the accuracy of the DNA splitting onto anticodon arrays. For these experiments, an anticodon array was constructed using two different oligonucleotide-conjugated resins. One of the two resins was added to wells that form the shape of the letter S, and the second resin was used to fill the remaining wells as shown in the schematic in Figure 7a. Radiolabelled 40-mers complementary to the resin used to define the "S" were hybridized to the array with the positive displacement pump for one hour at 45°C (the pump displaces the dead volume of the system in approximately five minutes). After hybridization, the anticodon array was imaged (Figure 7b), and the flowthrough was analyzed using a scintillation counter. More than than 95 percent of the radioactivity had bound. The 40-mers on the anticodon array were then transferred to an anion-exchange chemistry array using the mesofluidic Southern blotter, and the arrays were imaged again (Figure 7c). The labeled DNA partitioned to the "S" wells as expected, and the intensity of signal on the spots complementary to the 40-mer probe oligonucleotide were at least 29-fold higher than the background on adjacent non-complementary wells, showing that these devices can accurately divide DNA populations by sequence identity into spatially patterned sub-pools.

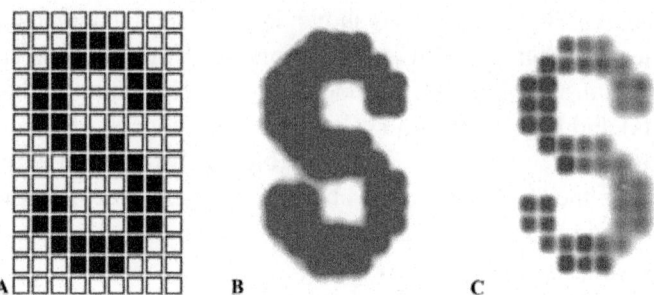

Figure 7: Patterned hybridization on an anticodon array and subsequent transfer to a chemistry array.

(A) An anticodon array was assembled with two distinct oligonucleotide-coupled resins in the pattern of the letter S as shown in this schematic. (B) The anticodon array was then probed using a radiolabelled 40-mer complementary to one of the two oligonucleotides. Hybridization was performed with the mesofluidic pump. The array was imaged on a phosphor screen. (C) The hybridized 40-mer was transferred from the anticodon array onto an anion-exchange chemistry array using the mesofluidic Southern blotter, and imaged on a phosphor screen. Image C has finer resolution than image B because the chemistry array is much thinner than the anticodon array.

doi:10.1371/journal.pone.0032299.g007

Second, we verified by HPLC that DNA (a 20-mer oligonucleotide with a 5′ primary amine modification) could be bound, chemically modified, and eluted from the anion-exchange chemistry array. Within the accuracy of our measurements, the binding and elution steps were quantitative. In adjacent wells, we performed one to four steps of a chemical synthesis employing peptoid chemistry. The final product of the four-step procedure was an acylated tripeptoid. Peptoid chemistry was used as a test case because it is one of several synthetic schemes that are versatile enough to accommodate hundreds to thousands of monomers in a multistep synthesis. The four-step, microwave-assisted synthesis in Figure 4b, which produces a representative member of a library of capped tripeptoids, proceeds with greater than 95 percent recovery of the DNA and 88 percent conversion to the desired product, a yield comparable to previously reported syntheses of peptoid-DNA conjugates in a column format.[1]

DISCUSSION

The specific DNA hybridization and efficient chemical conversion facilitated by the arrays and mesofluidic devices reported here constitute a high-fidelity system for DNA-programmed combinatorial chemistry. The highly parallel format should prove to be valuable in accessing large libraries with hundreds of diversity elements at each step of the synthesis. In future implementations, this format could easily be adapted for use with 1536-well plates. The microplate devices should facilitate the synthesis of small-molecule libraries with complexities comparable to those of the biopolymer libraries used for *in vitro* evolution.

The system described here and our previous realization of DNA-programmed combinatorial chemistry are limited in the amount of material that can be effectively synthesized; both yield roughly 50 picomoles of DNA-small molecule conjugate corresponding to $\sim 10^{13}$ molecules. As increasingly diverse DNA populations are translated, 50 picomoles will not adequately sample all of the library members. A system that could translate orders of magnitude more material would provide better coverage, and could potentially allow for larger fold-enrichments of fit molecules per generation.

A major challenge for DNA-programmed and DNA-encoded combinatorial chemistry technologies is the development of combinatorial synthetic schemes that accommodate hundreds to thousands of diversity elements at each step. Peptoid submonomer synthesis is one scheme that can achieve this building block complexity. Another is the sequential functionalization of triazine scaffolds with nucleophilic substituents, including Fmoc-amino acid building blocks.[2] These examples represent a small subset of the structures available to medicinal chemists. Combinatorial schemes that exploit other reagent classes or scaffolds would increase the region of chemical space that DNA-programmed chemical libraries can explore. For these reasons, efforts to expand DNA-compatible combinatorial chemistry will be extremely valuable.

Finally, routing of DNA populations can facilitate multiplexed selections of affinity reagents on a proteome-wide scale. For example, the microplate hybridization device could be used to divide bar-coded aptamer or mRNA display libraries for selections against 384 different targets. If the selected genetic material were pooled after selection into a common transcription/translation mixture, and then split by hybridization prior to the next selection round, the laborious library preparation process would be vastly simplified.

The recent successes of DNA-encoded combinatorial chemistry and its development and utilization in both academic and industrial settings point to a future in which these techniques will become an increasingly important part of the lead identification and drug optimization process. The tools we report will be useful in those efforts as they represent one straightforward and robust option for the programmed synthesis of diverse chemical libraries.

MATERIALS AND METHODS

Materials

Chemicals and solvents were purchased from Acros (Geel, Belgium), Alfa Aesar (Ward Hill, MA, USA), Novabiochem (La Jolla, CA, USA), Oakwood Chemical (West Columbia, SC, USA), Sigma-Aldrich (St. Louis, MO, USA), TCI America (Portland, OR, USA), VWR International (West Chester, PA, USA), or from the supplier indicated.

5′ pentynyl oligonucleotides used for hybridization were purchased from Bioneer (Alameda, CA, USA). All other oligonucleotides, including sequences with reactive primary amine functionalities (Glen Research, Sterling, VA, USA; 5′-Amino Modifier C12 and Spacer Phosphoramidite with 5′-Amino Modifier 5), were purchased from the Stanford PAN Facility (Stanford, CA, USA).

Equipment

Small parts and bulk materials were purchased from McMaster-Carr (Aurora, OH, USA) and Fisher Scientific. The ultraviolet flood lamp (#38100) used to cure the fluoropolymer gasketing material was purchased from Dymax (Torrington, CT, USA). Microwave-assisted chemical couplings were performed in a Panasonic NN-H965WFX 1250 W microwave. The pressure/vacuum manifold (M5A-0404-10) and valves (V3A-C231-AE1 and V3A-C231-BE1) were purchased from Mead USA (Chicago, IL, USA). The Stamp PLC (30064) and BASIC Stamp 2 module (BS2-IC) were purchased from Parallax (Rocklin, CA, USA). The laser cutter used to produce arrays and gaskets was a Legend 36EXT from Epilog Laser (Golden, CO, USA). The components for the routing devices were machined by Enviro-Tech (Boise, ID, USA), Patai Quality Machining (Santa Clara, CA, USA), and CCT Plastics (Grapevine, TX, USA). Engineering drawings are provided for these components. All 384-well vacuum filtration procedures were performed using a vacuum manifold (#9014579) from Qiagen (Venlo, Netherlands).

General methods

HPLC analysis of peptoid reactions was performed on a Microsorb reverse-phase C18 analytical column (Varian; Palo Alto, CA, USA) heated to 50°C and monitored at 260 and 280 nm using a UV detector (Spectra Focus, Spectra-Physics; Irvine, CA, USA). Linear gradients between 100 mM triethylammonium acetate pH 5.5 and 100 mM triethylammonium acetate pH 5.5, 90 percent acetonitrile were used. DNA-peptoid conjugates were digested with P1 nuclease and analyzed on a Micromass ZQ LC-MS at the Vincent Coates Foundation Mass Spectrometry Laboratory (Stanford, CA, USA).

Patterned quaternary ammonium (Q) cellulose

Rectangles measuring 11 by 15 cm were cut from sheets of cellulose filter paper (Whatman 542). Each rectangle was incubated in 15 mL of 1 M (3-bromopropyl) trimethylammonium bromide and 0.1 M sodium hydroxide at 37°C for 16 to 18 hours. Following derivatization, the filters were washed with water and acetone, dried, and flattened.

The rectangular cellulose filters were then patterned using photolithography. Each filter was immersed in 5 mL of a solvent-resistant photocurable liquid fluoropolymer, synthesized as described in reference 14 with the following modifications: Fluorolink D4000 (Solvay Solexis) was substituted for ZDOL, dichloropentafluoropropane (DCPFP) (SynQuest; Alachua, FL, USA) was substituted for Freon 113, and a silica plug was used to purify the polymer instead of an alumina column. The polymer-impregnated filter was placed between two masks of transparency film laser-printed with the negative image of a 384-well array, and the fluorinated material was polymerized by exposure to a UV flood lamp for approximately 20 s on each side. The unpolymerized material was removed by washing with DCPFP and acetone.

Generation of azido-Sepharose

An amino azide PEG400 linker was synthesized following reference [17]. The linker (20 µmol) in 1 mL 200 mM diisopropylethylamine in N,N-dimethylformamide (DMF) was coupled to 100 mg NHS-activated Sepharose that had been previously washed with DMF. The reaction was incubated overnight at room temperature, washed with DMF, and incubated with 1 mL of 1 M ethanolamine in DMF for twelve hours at room temperature to cap unreacted sites. The resin was then washed with DMF and water and stored at 4°C.

Synthesis of anticodon resin

A 5' alkyne-modified phosphoramidite was prepared as previously reported by Duckworth and coworkers,[18] and two 20-mer oligonucleotides with the sequences 5'- GTGAT TAAG TCTGC TTCG GC-3' and 5'-CCCAG TGC TGAC ATCTAT GA-3' were synthesized by Bioneer (Alameda, CA, USA) using that material. 47 µl of 860 µM Cu(I) tris-(benzyltriazolylmethyl)amine (TBTA)[19] in DMSO were added to 54 µl of an aqueous solution of 20 µM crude alkyne-terminated oligonucleotide and 1 mM sodium ascorbate. The solution was incubated with the azide substrate for thirty minutes at room temperature. The remaining azide groups were capped by repeating the reaction with 1 M propargyl alcohol in place of the oligonucleotide.

Anticodon arrays

Six inch segments were cut from a 0.015″ thick and 4″ wide strip of Delrin. Each piece was coated on both sides with double-sided tape (Scapa 702 Double Coated Silicone Tape). A 16 by 24 array of 3 mm-square holes on 4.5 mm centers was created using a laser cutter. To form a reservoir for the oligonucleotide-conjugated resin, the tape liner on one side of the array was removed, and a polypropylene filter with a 10 μm pore size (#60342) from Pall Corporation (Port Washington, NY, USA) was adhered to the adhesive surface. The array was inverted, the tape liner was removed from the opposite face, and the wells were filled with 10 μL of a 50:50 water:resin slurry. Excess fluid was removed using a vacuum manifold, and a second polypropylene filter was used to seal the array.

Hybridization

The displacement pump was assembled so that an anticodon array was sandwiched between the internal plates, and the entire apparatus was clamped between two aluminum plates. The inlet and outlet tubes were connected to a buffer reservoir containing a radiolabeled 40-mer oligonucleotide (5′-ATGGTATCAAGCTTGCCAC   AGCCGAAGCAGACTTAATCAC-3′). Air and vacuum pressure were used to pump buffer cyclically through the 384 features of the anticodon array. The application of air and vacuum was alternated using a valve array controlled by a BASIC stamp processor. After one hour in a water bath at 70°C, the temperature was lowered to 45°C for one hour and then to room temperature. Finally, the array was washed with 5 mL of hybridization buffer, and the system was disassembled. The array was incubated with a storage phosphor screen (Molecular Dynamics; Sunnyvale, CA, USA) for one hour, and the screen was imaged using a Typhoon 9400 (General Electric; Fairfield, CT, USA).

DNA transfer

The two halves of the mesofluidic Southern blotter were filled with transfer buffer (10 mM sodium hydroxide, 0.005% Triton X-100, 1 mM EDTA). An anticodon array and a fresh chemistry array were stacked on one plate, then covered with the second plate, and secured with an aluminum clamp. The device was submerged in a water bath heated to 80°C. The air and vacuum lines were opened to approximately 10 psi of air pressure and house vacuum and connected to a valve array that alternated air and vacuum on each side of the arrays. After one hour, the assembly was removed from the water bath, unclamped, and the arrays were removed. Both the anticodon array and the ion-exchange array were incubated with a storage phosphor screen for five minutes. The screen was imaged as above.

Peptoid coupling

The chemistry array with bound, 5′ amino modified oligonucleotides was secured between two gasketed plates used for chemistry. The array was then washed with 40 mL of methanol using a vacuum manifold. 40 µL of 150 mM 4-(4,6-dimethoxy-1,3,5-triazin-2-yl)-4-methylmorpholinium chloride and 100 mM sodium chloroacetate in distilled methanol were pipetted into each well and the reaction was incubated for 10 min at room temperature. The solvent was removed using a vacuum manifold, the array was washed with distilled methanol, and the acylation step was repeated twice more. After the third acylation step, the array was washed with 40 mL of 1 M propylamine in methanol and 40 mL of DMSO. Each well of the chemistry array was subsequently incubated with 40 µL of a 2 M solution of a primary amine in DMSO, and the entire assembly was microwaved for 13 s at 100% power six times over 30 min. The reactions were allowed to cool for approximately 5 min after each microwave step. Following alkylation, the arrays were washed with 40 mL each of DMSO and water. Finally, the chemistry plates were disassembled, and the DNA was eluted using Elute Buffer (50 mM Tris pH 8, 1.5 M NaCl, 0.005% Triton X-100).

Supporting Information

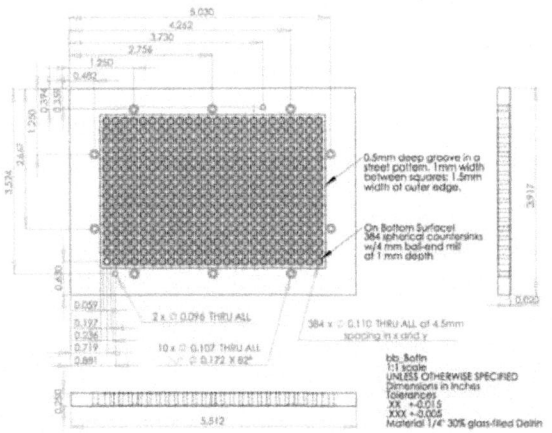

Engineering drawing of internal plate for backtransfer device.

ACKNOWLEDGMENTS

We thank F.E. Boas, G.F. Burkhard, N.K. Devaraj, E.L. Girvin, D.R. Halpin, H. Jung, C.J. Krusemark, C.J. Layton, R.S. Mathew, W.L. Martin, G.P. Miller, K.L. Schmidt, N.P. Tilmans, and P.A. Walker for helpful discussions.

REFERENCES

1. Wrenn SJ, Weisinger RM, Halpin DR, Harbury PB (2007) Synthetic ligands discovered by in vitro selection. Journal of the American Chemical Society 129: 13137–13143.

2. Clark MA, Acharya RA, Arico-Muendel CC, Belyanskaya SL, Benjamin DR, et al. (2009) Design, synthesis and selection of DNA-encoded small-molecule libraries. Nature chemical biology 5: 647–654.

3. Kleiner RE, Dumelin CE, Tiu GC, Sakurai K, Liu DR (2010) In vitro selection of a DNA-templated small-molecule library reveals a class of macrocyclic kinase inhibitors. Journal of the American Chemical Society 132: 11779–11791.

4. Mannocci L, Melkko S, Buller F, Molnar I, Bianke JP, et al. (2010) Isolation of potent and specific trypsin inhibitors from a DNA-encoded chemical library. Bioconjugate chemistry 21: 1836–1841.

5. Gartner ZJ, Liu DR (2001) The generality of DNA-templated synthesis as a basis for evolving non-natural small molecules. Journal of the American Chemical Society 123: 6961–6963.

6. Halpin DR, Harbury PB (2004) DNA display II. Genetic manipulation of combinatorial chemistry libraries for small-molecule evolution. PLoS biology 2: E174.

7. Clark MA (2010) Selecting chemicals: the emerging utility of DNA-encoded libraries. Current opinion in chemical biology 14: 396–403.

8. Lipinski CA, Lombardo F, Dominy BW, Feeney PJ (2001) Experimental and computational approaches to estimate solubility and permeability in drug discovery and development settings. Advanced drug delivery reviews 46: 3–26.

9. Halpin DR, Harbury PB (2004) DNA display I. Sequence-encoded routing of DNA populations. PLoS biology 2: E173.

10. Heine N, Ast T, Schneider-Mergener J, Reineke U, Germeroth L, et al. (2003) Synthesis and screening of peptoid arrays on cellulose membranes. Tetrahedron 59: 9919–9930.

11. Heine N, Germeroth L, Schneider-Mergener J, Wenschuh H (2001) A modular approach to the SPOT synthesis of 1,3,5-trisubstituted

hydantoins on cellulose membranes. Tetrahedron Letters 42: 227–230.

12. van Reis RD (2006) Charged filtration membranes and uses therefor. United States: Genentech, Inc.

13. Rolland JP, Van Dam RM, Schorzman DA, Quake SR, DeSimone JM (2004) Solvent-resistant photocurable liquid fluoropolymers for microfluidic device fabrication [corrected]. Journal of the American Chemical Society 126: 2322–2323.

14. Carrilho E, Phillips ST, Vella SJ, Martinez AW, Whitesides GM (2009) Paper microzone plates. Analytical chemistry 81: 5990–5998.

15. Martinez AW, Phillips ST, Whitesides GM (2008) Three-dimensional microfluidic devices fabricated in layered paper and tape. Proceedings of the National Academy of Sciences of the United States of America 105: 19606–19611.

16. Unger MA, Chou HP, Thorsen T, Scherer A, Quake SR (2000) Monolithic microfabricated valves and pumps by multilayer soft lithography. Science 288: 113–116.

17. Schwabacher AW, Lane JW, Schiesher MW, Leigh KM, Johnson CW (1998) Desymmetrization reactions: efficient preparation of unsymmetrically substituted linked

18. Duckworth BP, Chen Y, Wollack JW, Sham Y, Mueller JD, et al. (2007) A universal method for the preparation of covalent protein-DNA conjugates for use in creating protein nanostructures. Angewandte Chemie 46: 8819–8822.

19. Chan TR, Hilgraf R, Sharpless KB, Fokin VV (2004) Polytriazoles as copper(I)-stabilizing ligands in catalysis. Organic letters 6: 2853–2855.

Chapter 11

GUEST-HOST CHEMISTRY WITH DENDRIMERS—BINDING OF CARBOXYLATES IN AQUEOUS SOLUTION

Mario Ficker[1] , Johannes F. Petersen[1] , Jon S. Hansen[1] , Jørn B. Christensen[1]

[1] Department of Chemistry, University of Copenhagen, Thorvaldsensvej 40, DK-1871 Frederiksberg C, Denmark

ABSTRACT

Recognition and binding of anions in water is difficult due to the ability of water molecules to form strong hydrogen bonds and to solvate the anions. The complexation of two different carboxylates with 1-(4-carbomethoxypyrrolidone)-terminated PAMAM dendrimers was studied in aqueous solution using NMR and ITC binding models. Sodium 2-naphthoate and sodium 3-hydroxy-2-naphthoate were chosen as carboxylate model compounds, since they carry structural similarities to many non-steroidal anti-inflammatory drugs and they possess only a limited number of functional groups, making them ideal to study the carboxylate-dendrimer interaction selectively. The binding stoichiometry for 3-hydroxy-2-naphthoate was found to be two strongly bound guest molecules per dendrimer and an additional 40 molecules with weak binding affinity. The NOESY NMR showed a clear binding correlation of sodium 3-hydroxy-2-naphthoate with the lyophilic dendrimer core, possibly with the two high affinity guest molecules. In comparison, sodium 2-naphthoate showed a weaker binding strength and had a stoichiometry of two guests per dendrimer with no additional weakly bound guests. This stronger dendrimer interaction with sodium 3-hydroxy-2-naphthoate is possibly a result of the additional interactions of the dendrimer with the extra hydroxyl group and an internal stabilization of the negative charge due to the hydroxyl group. These findings illustrate the potential of the G4 1-(4-carbomethoxy) pyrrolidone dendrimer to complex carboxylate guests in water and act as a possible carrier of such molecules.

INTRODUCTION

Dendrimers are well-defined nano-scale macromolecules formed by repetitive branching from a core. Depending on the branch-cell unit, dendrimers can have cavities capable of hosting smaller molecules. Guest-host chemistry in dendrimers is divided into endo- or exo-complexation which is determined by whether the guest molecule is bound in the interior or to the surface of the dendrimer. Both types of guest-host chemistry have been a popular topic due to the potential applications in drug-delivery.[1–4] 1-(4-Carbomethoxy) pyrrolidone coated PAMAM dendrimers are especially promising candidates for the complexation and release of drug molecules, since they have unique and favorable solubility properties in both organic solvents and aqueous solutions[5] and have a benign toxicity profile.[6–8]

We recently reported a study of endo-complexation of the γ-lactam antibiotic oxacillin in a G4 1,4-diaminobutane-core 1-(4-carbomethoxy) pyrrolidone functionalized PAMAM-dendrimer, where it was found that the stoichiometry of the guest-host complexes showed solvent dependency.[9]

However, oxacillin and the other penicillins are sold as alkali metal salts due to the low stability of the free carboxylic acids; this raised the question of whether it could be possible to have binding of carboxylate anions to the pyrrolidone-terminated dendrimer in water.

Recognition and binding of anions in water is difficult because of waters ability to form strong hydrogen bonds and to solvate the anions. Many of the best examples of anion receptors are pre-organized macromolecules with suitable cavities such as cryptands,[10] calixarenes,[11] or curcubiturils.[12] Guest-host chemistry with dendrimers in water is much less investigated, but there are examples of binding of pharmaceutically interesting compounds such as cis-Platin,[13] Nadifloxacine[14] and Prulifloxacine,[15] Campthotecin,[16] Dexamethasone phosphate,[17] anti-inflammatoric drugs (NSAIDs)[18,19] and of course DNA and siRNA.[20–23]

Initially, we tried the sodium salt of oxacillin, but because the results were inconclusive, we decided to look at the more simple molecules such as sodium 2-naphthoate and sodium 3-hydroxy-2-napthoate. These two carboxylates were chosen as model guests, since they possess similar structural features and water solubility as many antibiotics and non-steroidal anti-inflammatory drugs. The G4 1-(4-carbomethoxy) pyrrolidone dendrimer and the two encapsulated guest molecules are illustrated in Fig 1.

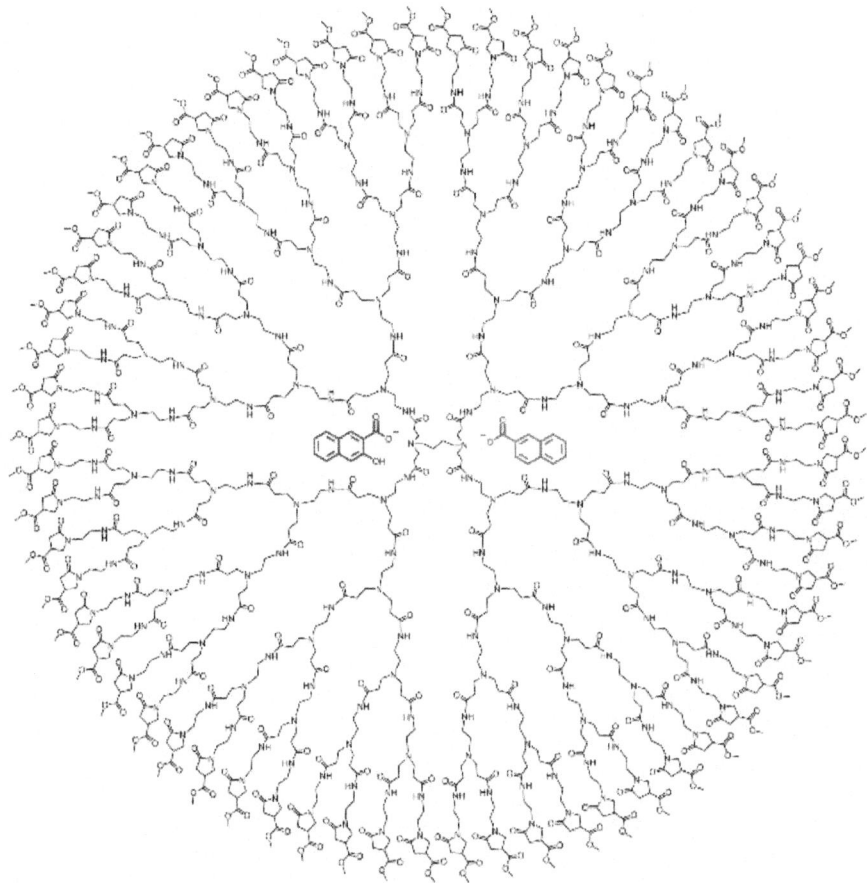

Figure 1:. The two model guests illustrated within a G4 1-(4-carbomethoxy) pyrrolidone dendrimer.

doi:10.1371/journal.pone.0138706.g001

MATERIALS AND METHODS

Unless otherwise stated, all starting materials were obtained from commercial suppliers and used as received. Solvents were HPLC grade and used as received. ^{1}H-NMR spectra were recorded on a 500 MHz NMR (Bruker) apparatus. Chemical shifts are reported in ppm downfield of TMS (tetramethylsilane) using the resonance of the deuterated solvent as internal standard (s = singlet, d = doublet, t = triplet, q = quartet, m = multiplet). ^{1}H-NMR titration data were fitted with Origin 9.0. The employed ITC apparatus was a NanoITC Model 5300, TA Instruments, Lindon, UT, USA, with a cell volume of 1038 μl. All ITC data were fitted in NanoAnalyze.

Preparation of dendrimers

The PAMAM dendrimer of the generation 4 was synthesized by published procedures,[24] starting from 1,4-diaminobutane as the core. The 1-(4-carbomethoxy-pyrrolidone) surface functionalization was done by reacting the amino terminated dendrimers with dimethyl itaconat. The reaction was monitored by performing Kaiser-tests until completion.

Preparation of the corresponding sodium salt of the naphthoic acids

Sodium 3-hydroxy-2-naphthoate and sodium 2-naphthoate were prepared by the slow addition of the corresponding carboxylic acid (2.9 mmol) to an aqueous solution (10 mL) containing an equivalent amount of NaOH (2.9 mmol). Excess water was removed by freeze drying and the carboxylates were gained in a quantitative yield (2.9 mmol).

Sodium 3-hydroxy-2-naphthoate: ^{1}H-NMR: δ = 8.33 (s, 1H); 7.85–7.88 (m, 1H); 7.69–7.72 (m, 1H); 7.50–7.53 (m, 1H); 7.35–7.38 (m, 1H); 7.18 (s, 1H). Sodium 2-naphthoate: ^{1}H-NMR: δ = 7.97–8.02 (m, 1H); 7.74–7.80 (m, 2 H); 7.32–7.44 (m, 4 H).

Preparation of NMR-Titration samples and dissociation constant determination

5 ml stock solutions in D_2O were prepared, containing 1 mM G4 1-(4-carbomethoxy pyrrolidone) dendrimer and (2) containing 1 mM of G4 1-(4-carbomethoxy pyrrolidone) dendrimer and 150 mM of the respective carboxylate. The dendrimer concentration was kept constant, while the carboxylate concentration was varied from 0 to 100 mM. For the determination of the dissociation constant of sodium 2-naphthoate, a NMR fitting model was used as described in an earlier paper.[4] Binding saturation curves were fitted employing *Origin 9.0.*

Preparation of samples for 2D-NOESY-experiments

NOESY experiments were conducted on a 500 MHz NMR (Bruker) apparatus that was equipped with a cryo-probe. The concentration of G4 1-(4-carbomethoxypyrrolidone) PAMAM-dendrimer was 5.2 mM and the concentration of sodium 3-hydroxy-2-naphthoate was 35.7 mM in D_2O. The dendrimer sample was incubated at room temperature with sodium 3-hydroxy-2-naphthoate for 2 hours prior to starting the experiment. The concentration of sodium 2-naphthoate incubated with dendrimer in D_2O was the same. The experiments were performed at 25°C with a 2 s relaxation delay, 205 ms acquisition time, 300 ms mixing time and a 8.2 μs¹H 90° pulse width. Eight transients were averaged for 1024 *t1* increments.

Preparation of NMR-samples for Job-plot

5 ml stock solutions in D$_2$O were prepared with 10 mM of G4 1-(4-carbomethoxy pyrrolidone) dendrimer, and 10 mM of the respective carboxylate. The samples were prepared by injecting a total volume of 500 μl in 12 NMR tubes for each Job-plot, keeping a constant total concentration of 10 mM ([*den*] + [*carboxylate*]), where the ratio [*den*]/[*carboxylate*] was varied.

ITC-titration experiments.

ITC measurements have been performed following an ITC best practice guideline by Freyer and Lewis [26]. For each compound there have been 42 individual heat signals collected. The computational fitting of the data was performed using the *NanoAnalyze* standard software for ITC measurements by TA Instruments (manufacturing company of the NanoITC Model 5300). Before and after the measurements the accuracy of the instrument was tested by standardized blank titration of water into water. Control experiments for baseline determination were performed (blank titration of the carboxylate into water and a blank titration of water into the dendrimer solution). The error values of the obtained values (fitting by the *NanoAnalyze*standard instrument software) are reported after each value.

Titration of G4 4-carbomethoxy pyrrolidone terminated PAMAM-dendrimer with sodium 3-hydroxy-2-naphthoate

A stock solution of 0.1 mM G4 4-carbomethoxy pyrrolidone terminated PAMAM-dendrimer in MQ water was prepared (2 ml), **A**. 40 mM sodium 3-hydroxy-2-naphthoate was prepared (5 ml), **B**. Both solutions **A** and **B** were carefully degassed by ultra sonification in order to remove the air content. The pH of the aqueous carboxylate solutions was adjusted to match the pH of the 0.1 mM dendrimer solution (pH 7.8). The ITC cell was filled with solution **A** (1038 μl). The ITC syringe (250 μl) was filled with solution **B**. The temperature was maintained at 25°C throughout the experiment. Addition of solution **B** into solution **A** was carried out by injection of 6 μl for each titration increment, giving rise to 42 heat signals.

Baseline Subtraction Blank ITC-titration experiments

The ITC cell was filled with solution **A** (1038 μl). The ITC syringe (250 μl) was filled with MQ water, which was carefully degassed by ultra sonification.

Addition of MQ water into solution **A** was carried out by injection of 6 μl for each titration increment, giving rise to 42 heat signals.

Blank titration of solution **B** into MQ water was performed by filling the ITC cell with 1038 μl of degassed MQ water followed by the addition of 250 μl of solution **B**. 6 μl of solution **B** was added for each titration increment, giving rise to 42 heat signals. The same blank titration experiment was conducted with sodium 2-naphthoate as described for sodium 3-hydroxy-2-naphthoate. Another blank experiment was conducted by addition of water to the dendrimer solution. This did not result in a significant heat signal and was thus neglected.

RESULTS AND DISCUSSION

The complexation of the two carboxylate compounds was studied by means of ^{1}H-NMR- and ITC-titrations, these two techniques have been applied previously in similar studies.[27–30] Both analytical techniques showed guest encapsulation in aqueous solution. The dissociation constants were calculated from the obtained ITC data and a recently used ^{1}H-NMR binding model was also applied to study the binding strength.[9]

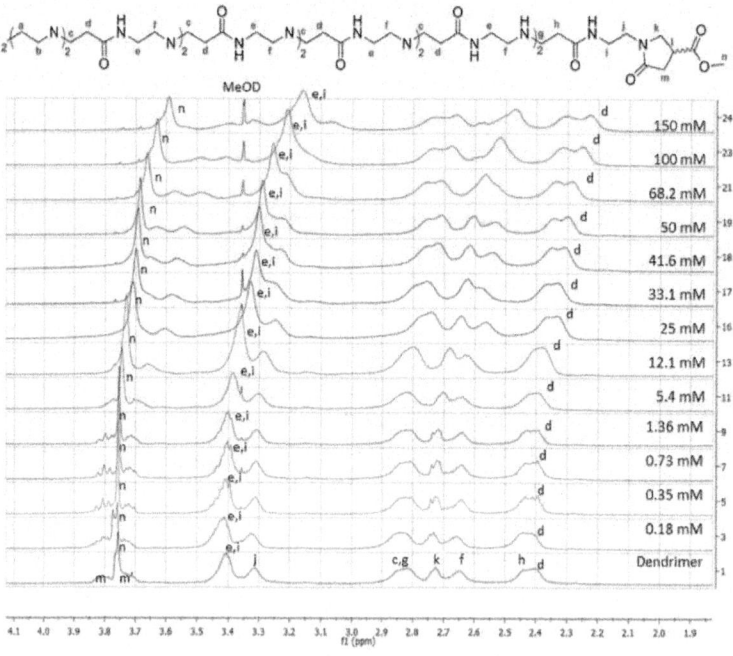

Fig 2: Stacked ^{1}H-NMR spectra of different ratios of 3-hydroxy-2-naphthoate incubated with a 1 mM G4 1-(4-carbomethoxy) pyrrolidone dendrimer in D$_2$O.

doi:10.1371/journal.pone.0138706.g002

The complex formation of these carboxylates within the dendrimer was elucidated by a titration series of different concentrations of sodium 3-hydroxy-2-naphthoate and sodium 2-naphthoate incubated with a 1 mM aqueous solution of G4 1-(4-carbomethoxy) pyrrolidone dendrimer. The dendrimer signals underwent a significant change in chemical shift and line broadening as a consequence of guest encapsulation, as illustrated in Fig 2. The full titration series for both carboxylates, where the change in chemical shift for the host and guest molecules are shown, can be found in the supporting information.

The inner dendrimer signals, d and e, experience a particularly large shift (0.1 ppm and 0.25 ppm respectively) due to the possible close proximity of the corresponding guest molecule. At increased guest concentration, the surface protons m and n also show a change in chemical shift. This binding mode is most likely caused by a combination of lipophilic and electrostatic interactions between the carboxylate and the dendrimer. The electrostatic interaction presumably occurs between the negatively charged carboxylate and the partially protonated interior of the dendrimer, i.e. the tertiary amine focal points.

Job´s method was used to calculate the respective binding stoichiometries,[9] where sodium 2-naphthoate and sodium 3-hydroxy-2-naphthoate were both calculated to form 1:2 dendrimer-carboxylate ratios. The Job-plot experiments are shown in the supporting information . It should be noted that the maxima calculated from the Job-plots only provide the stoichiometry that contributes most to the observed chemical shift. The found ratio is thus the predominant host-guest interaction, even though there might be different guest-host stoichiometries present in low concentrations. Two dimensional NOE experiments were conducted to obtain a deeper understanding of the binding phenomenon within the dendrimer cavity. The 3-hydroxy-2-naphthoate interacts with the core and interior branching of the dendrimer, as can be seen in Fig 3. In particular, the dendrimer proton signal a correlates with the aromatic guest protons, presumably due to a favorable lipophilic interaction with the butyl core. The dendrimer surface protons m and n did not give NOE correlations to the complexed carboxylate guests, this observation indicates that the binding motif is situated in the interior of the dendrimer. The full spectra for both carboxylates and a graphic illustration of the presumed binding site can be found in the supporting information (S7, S8 and S9 Figs).

The association constants and binding enthalpies by complexation of the two carboxylates within the dendrimer were determined by ITC-experiments. Titration of a 40 mM solution of the respective carboxylate guest to a 0.1 mM solution of G4 1-(4-carbomethoxy) pyrrolidone dendrimer in water was performed in order to obtain the titration series. A blank experiment for baseline subtraction was performed by gradual addition of the guest molecule solution into a dendrimer free aqueous solution. Another blank experiment was conducted adding water to a dendrimer solution. This did not result in a significant heat signal and was thus neglected in the determination methodology.

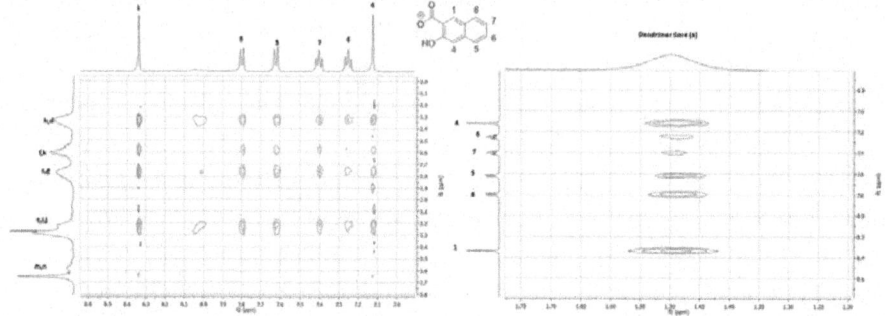

Fig 3. 2D-NOE-spectrum showing a significant correlation between sodium 3-hydroxy-2-naphthoate and the G4 1-(4-carbomethoxypyrrolidone) PAMAM-dendrimer.

doi:10.1371/journal.pone.0138706.g003

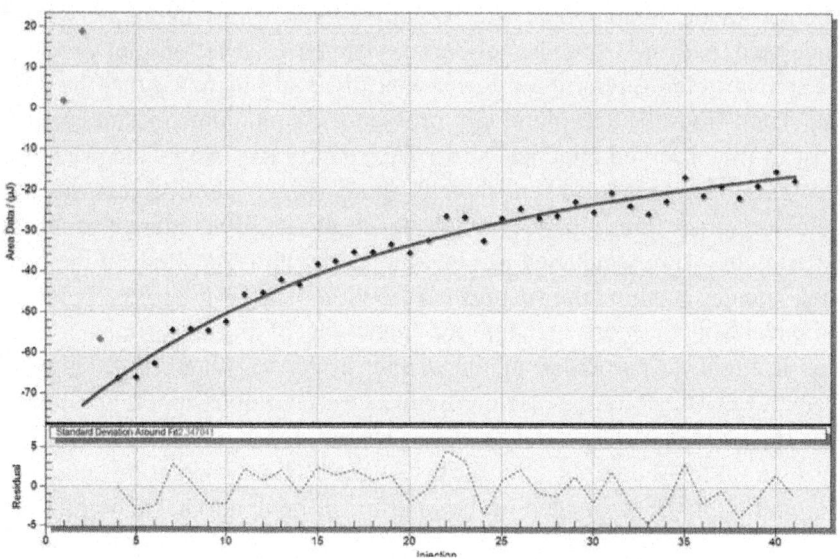

Fig 4: ITC-binding curve of 2-naphthoate, showing the best fit.

The ITC raw data was treated by blank subtraction (titration of guest into water).

doi:10.1371/journal.pone.0138706.g004

The ITC-results (Table 1) show a difference in binding behavior of the two model guests. The sodium 2-naphthoate exhibits a weak binding interaction (K_a = 88.2 ± 8.4 M^{-1}) with the dendrimer, showing a 2:1 stoichiometry. This ratio is in accordance with the determined ratio from the ^{1}H-NMR Job-plot experiment. In comparison, the sodium 3-hydroxy-2-naphthoate exhibits a more complex binding interaction. The dendrimer shows a strong binding (K_{a1} =2369 ± 927 M^{-1}) with a ratio of 2:1 carboxylate to dendrimer. Again, this number of strong bound guest molecules is in accordance with the ^{1}H-NMR determined stoichiometry.

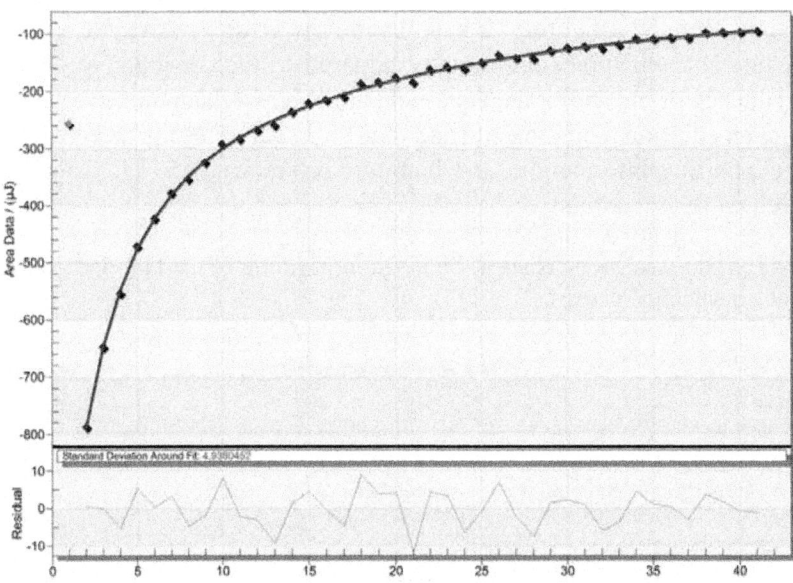

Figure 5: ITC-binding curve of 3-hydroxy-2-naphthoate, showing the best fit.

The ITC raw data was treated by blank subtraction (titration of guest into water).

doi:10.1371/journal.pone.0138706.g005

Additionally to this very strong binding interaction of two guest molecules, presumably at the lipophilic dendrimer core, an additional weak binding contribution was found during the ITC studies. Fitting of the heat signal indicated that two guest molecules alone were not the only contribution to the system; besides these two strongly bound guests, a larger number (app. 40) of guest molecules showed an additional low affinity binding (K_{a2} = 259 ± 101 M^{-1}) to the dendrimer. This weak binding is visible in the long tailing of the heat signals over a broad range of high concentration ratios of guest to dendrimer. While the high affinity binding side is quickly occupied with

2 guest molecules, the weak binding takes places over a very broad range of concentration; and even at a dendrimer to guest ratio of 1:100, the dendrimer binding sites were not fully occupied, which indicates a very low binding affinity of the second guest molecule. Due to the low K_{a2} value, compared to the K_{a1}, which is around 10 times stronger, it can be justified why the ^{1}H-NMR Job plot only resulted in a 1:2 ratio of dendrimer to guest, since the Job-plot gives an out-read of the most dominant contribution to the system, which is in this case the strong K_{a1}. This demonstrated, how important it is to apply multiple analytical techniques to get a full understanding of the binding behavior to a complex system like a dendrimer. Due to the more complex binding scenario in the case of the 3-hydroxy-2-naphthoate, the error bars for the association constants and enthalpies are larger compared to the 2-naphthoate. This is a consequence of the additional equation parameters, which are necessary when having an additional binding site and thus more variables that need to be fitted (see supporting information and literature reference[26,31] for more details about the ITC equations).

Table 1: The association constants obtained from fitting of the ITC-data, along with calculated enthalpy increases

	3-hydroxy-2-naphthoate	2-naphthoate
K_{a1} [M^{-1}]	2369 ± 927	88.2 ± 8.4
K_{a2} [M^{-1}]	259 ± 101	-
ΔH_1 [kJ/mol]	-37 ± 8	-18.0 ± 1.7
ΔH_2 [kJ/mol]	-87 ± 17	-
n_1	2 ± 1	2.1 ± 0.2
n_2	41 ± 24	-

doi:10.1371/journal.pone.0138706.t001

The stronger binding of 3-hydroxy-2-naphthoate compared to 2-naphthoate may be explained by the higher lipophilicity of 3-hydroxy-2-naphthoate. The carboxylate unit in 3-hydroxy-2-naphthoate is capable of forming an intramolecular hydrogen bond with the adjacent hydroxyl group, resulting in a higher degree of negative charge delocalization and thus decreasing the overall hydrophilicity of the carboxylate. This intramolecular stabilization correlates well to the association of 3-hydroxy-2-naphthoate to the lipophilic interior of the dendrimer at low guest concentration, as illustrated by the conducted NOE experiments. At higher carboxylate concentration these preferred binding sites are already occupied, forcing additional carboxylate guests to associate with the more hydrophilic branches of the outer parts of the dendrimer.

Recently, we applied a binding model for ^{1}H-NMR titrations in order to determine association constants between guest molecules and dendrimers.[9]

In this binding model it is assumed that each dendrimer has a defined number of equal and independent binding sites in order for the algorithm to be applied to the system. For further information about this correlation of chemical shifts to the association constant we refer to this previously published study. Due to the requirement of equal binding sites this model could not be applied to the 3-hydroxy-2-naphthoate study, since the assumption of equal binding sites is not coherent with the previously discussed results.

However, the fitting algorithm could successfully be employed for the 2-naphthoate. The obtained association constant K_a = 5.35±0.7 M^{-1} (for n = 2 carboxylate molecules) is approximately in the same order of magnitude as the one determined by the ITC experiment. The best fit for the titration series is shown in Fig 6. The difference between the determined binding constant by ITC and NMR is most likely due to the assumptions made in the NMR model, e.g. equal contribution for first and second bound guest. In comparison, the ITC measures the actual heat output of the complexation, which include all contributions, e.g. conformational changes in the dendrimer structure, replacement of water with guest molecules etc. In contrast the NMR model correlates the chemical shift to a ratio between bound and unbound guest molecules, it neglects to take into account any structural changes incurred by the dendrimer during the binding process. Both in the ITC and the NMR study, the 2-naphthoate was found to exhibit weak binding to the G4 1-(4-carbomethoxy) pyrrolidone dendrimer.

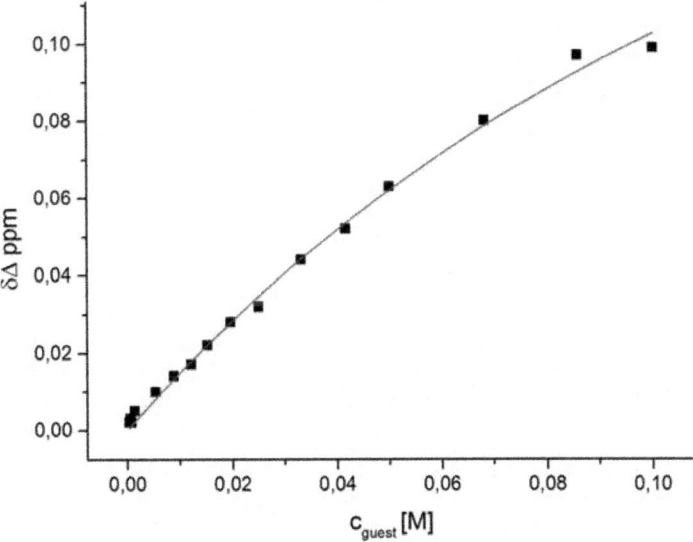

Fig 6: The best fit obtained for the ^{1}H-NMR-titration of sodium 2-naphthoate into the PyrG4 dendrimer in aqueous solution.

The calculated binding constant corresponds to $K_a = 5.35 \pm 0.7$ M^{-1} for $n = 2$ carboxylates.

doi:10.1371/journal.pone.0138706.g006

CONCLUSION

The G4 1-(4-carbomethoxy) pyrrolidone terminated PAMAM-dendrimer was found to form dendrimer-carboxylate complexes with both of the model compounds, sodium 2-naphthoate and sodium 3-hydroxy-2-naphthoate, in water. This was elucidated by NMR- and ITC-experiments. Complexation of the guest was found to occur in the interior of the dendrimer, possibly attributed to favorable lipophilic interactions. A difference in binding stoichiometry and binding strength was found for the two different model compounds. Both carboxylates had a primary stoichiometry of two guest molecules per dendrimer, presumably in close proximity to the dendrimer core. The 3-hydroxy-2-naphthoate derivative was found to associate more tightly with this binding site than the 2-naphthoate compound. The conducted ITC-experiments suggest an additional weak binding of app. 40 units of 3-hydroxy-2-naphthoate molecules to the interior branching of the G4 1-(4-carbomethoxy) pyrrolidone dendrimer.

These findings illustrate the potential of the G4 1-(4-carbomethoxy) pyrrolidone dendrimer to complex carboxylate guests and act as a possible nano-carrier of such molecules. Future studies on the potential of this dendrimer family as hosts for biologically active carboxylates, e.g. antibiotic and non-steroidal anti-inflammatory drugs, are currently in progress. This future work will also focus on the role of the cation and whether it too is complexed within the dendrimer cavities. Finally, the release of the bound carboxylate guests is also being investigated. The 1-(4-carbomethoxy) pyrrolidone surface also has synthetic "handles", which can be utilized to link targeting units to the dendrimer.

ACKNOWLEDGMENTS

The authors would like to thank Christian G. Tortzen for his professional guidance in recording of the 2D-NOE-spectra. We would also like to acknowledge Louise S. Holm for her enthusiastic guidance in the recording of the ITC-spectra, and Hanne Moerck Nielsen for providing access to the ITC-apparatus at the Department of Pharmacy. We thank Nathaniel George Breffni King for proof reading of the manuscript language and grammar. For financial support we would like to thank the University of Copenhagen.

REFERENCES

1. Hu J, Xu T, Cheng Y (2012) NMR Insights into Dendrimer-Based Host–Guest Systems. Chemical Reviews 112: 3856–3891. doi: 10.1021/cr200333h. pmid:22486250

2. Zeng F, Zimmerman SC (1997) Dendrimers in Supramolecular Chemistry: From Molecular Recognition to Self-Assembly. Chemical Reviews 97: 1681–1712. pmid:11851463 doi: 10.1021/cr9603892

3. Wu L-p, Ficker M, Christensen JB, Trohopoulos PN, Moghimi SM (2015) Dendrimers in Medicine: Therapeutic Concepts and Pharmaceutical Challenges. Bioconjugate Chemistry. doi: 10.1021/acs.bioconjchem.5b00031

4. Donald A. Tomalia JBC, Boas Ulrik (2012) Dendrimers, Dendrons, and Dendritic Polymers: Discovery, Applications, and the Future. Cambridge: Cambridge University Press.

5. Ficker M, Petersen JF, Gschneidtner T, Rasmussen A-L, Purdy T, Hansen JS et al. (2015) Being two is better than one-catalytic reductions with dendrimer encapsulated copper- and copper-cobalt-subnanoparticles. Chemical Communications 51: 9957–9960. doi: 10.1039/c5cc00347d. pmid:25997569

6. Ciolkowski M, Petersen JF, Ficker M, Janaszewska A, Christensen JB, Klajnert B et al. (2012) Surface modification of PAMAM dendrimer improves its biocompatibility. Nanomedicine 8: 815–817. doi: 10.1016/j.nano.2012.03.009. pmid:22542820

7. Janaszewska A, Ciolkowski M, Wrobel D, Petersen JF, Ficker M, Christensen JB et al. (2013) Modified PAMAM dendrimer with 4-carbomethoxypyrrolidone surface groups reveals negligible toxicity against three rodent cell-lines. Nanomedicine 9: 461–464. doi: 10.1016/j.nano.2013.01.010. pmid:23434674

8. Janaszewska A, Studzian M, Petersen JF, Ficker M, Christensen JB, Klajnert-Maculewicz B (2015) PAMAM dendrimer with 4-carbomethoxypyrrolidone—In vitro assessment of neurotoxicity. Nanomedicine: Nanotechnology, Biology and Medicine 11: 409–411. doi: 10.1016/j.nano.2014.09.011

9. Hansen JS, Ficker M, Petersen JF, Nielsen BE, Gohar S, Christensen JB (2013) Study of the Complexation of Oxacillin in 1-(4-Carbomethoxypyrrolidone)-Terminated PAMAM Dendrimers. The Journal of Physical Chemistry B 117: 14865–14874. doi: 10.1021/jp408613z. pmid:24219418

10. Kang SO, Llinares JM, Day VW, Bowman-James K (2010) Cryptand-like anion receptors. Chemical Society Reviews 39: 3980–4003. doi: 10.1039/c0cs00083c. pmid:20820597

11. Kim SK, Sessler JL (2010) Ion pair receptors. Chemical Society Reviews 39: 3784–3809. doi: 10.1039/c002694h. pmid:20737073

12. Gramage-Doria R (2013) Metallocyclodextrins: Combining Cavitands with Metal Centres. ChemistryOpen 2: 176–179. doi: 10.1002/open.201300033. pmid:24551563

13. Tarazona-Vasquez F, Balbuena PB (2007) Dendrimer-Tetrachloroplatinate Precursor Interactions. 2. Noncovalent Binding in PAMAM Outer Pockets. The Journal of Physical Chemistry A 111: 945–953. pmid:17266236 doi: 10.1021/jp065016b

14. Arnusch CJ, Branderhorst H, de Kruijff B, Liskamp RMJ, Breukink E, Pieters RJ (2007) Enhanced Membrane Pore Formation by Multimeric/Oligomeric Antimicrobial Peptides†. Biochemistry 46: 13437–13442. pmid:17944489 doi: 10.1021/bi7015553

15. Cheng Y, Qu H, Ma M, Xu Z, Xu P, Fang Y et al. (2007) Polyamidoamine (PAMAM) dendrimers as biocompatible carriers of quinolone antimicrobials: An in vitro study. European Journal of Medicinal Chemistry 42: 1032–1038. pmid:17336426 doi: 10.1016/j.ejmech.2006.12.035

16. Morgan MT, Nakanishi Y, Kroll DJ, Griset AP, Carnahan MA, Wathier M et al. (2006) Dendrimer-encapsulated camptothecins: increased solubility, cellular uptake, and cellular retention affords enhanced anticancer activity in vitro. Cancer Res 66: 11913–11921. pmid:17178889 doi: 10.1158/0008-5472.can-06-2066

17. Yang K, Weng L, Cheng Y, Zhang H, Zhang J, Wu Q et al. (2011) Host–Guest Chemistry of Dendrimer–Drug Complexes. 6. Fully Acetylated Dendrimers as Biocompatible Drug Vehicles Using Dexamethasone 21-Phosphate as a Model Drug. The Journal of Physical Chemistry B 115: 2185–2195. doi: 10.1021/jp111044k. pmid:21338144

18. Avila-Salas F, Sandoval C, Caballero J, Guiñez-Molinos S, Santos LS, Cachau RE et al. (2012) Study of Interaction Energies between the PAMAM Dendrimer and Nonsteroidal Anti-Inflammatory Drug Using a Distributed Computational Strategy and Experimental Analysis by ESI-MS/MS. The Journal of Physical Chemistry B 116: 2031–2039. doi: 10.1021/jp2069122. pmid:22324459

19. Gupta U, Agashe HB, Jain NK (2007) Polypropylene imine dendrimer mediated solubility enhancement: effect of pH and functional groups of hydrophobes. Journal of pharmacy & pharmaceutical sciences: a

publication of the Canadian Society for Pharmaceutical Sciences, Societe canadienne des sciences pharmaceutiques 10: 358–367. pmid:17727799

20. Lee CC, MacKay JA, Frechet JMJ, Szoka FC (2005) Designing dendrimers for biological applications. Nat Biotech 23: 1517–1526. doi: 10.1038/nbt1171

21. Caminade A- M, Turrin C- O, Majoral J-P (2008) Dendrimers and DNA: Combinations of Two Special Topologies for Nanomaterials and Biology. Chemistry–A European Journal 14: 7422–7432. doi: 10.1002/chem.200800584

22. Liu X-x, Rocchi P, Qu F-q, Zheng S-q, Liang Z-c, Gleave M et al. (2009) PAMAM Dendrimers Mediate siRNA Delivery to Target Hsp27 and Produce Potent Antiproliferative Effects on Prostate Cancer Cells. ChemMedChem 4: 1302–1310. doi: 10.1002/cmdc.200900076. pmid:19533723

23. Pavan GM, Posocco P, Tagliabue A, Maly M, Malek A, Danani A et al. (2010) PAMAM Dendrimers for siRNA Delivery: Computational and Experimental Insights. Chemistry–A European Journal 16: 7781–7795. doi: 10.1002/chem.200903258

24. Fréchet JMJ, Tomalia DA (2001) Dendrimers and other Dendritic Compounds. Chichester: John Wiley & Sons.

25. Huang B, Swanson DR, Tomalia DA (2005) Heterocycle functionalized dendritic polymers. Google Patents.

26. Freyer MW, Lewis EA (2008) Isothermal titration calorimetry: experimental design, data analysis, and probing macromolecule/ligand binding and kinetic interactions. Methods in cell biology 84: 79–113. pmid:17964929 doi: 10.1016/s0091-679x(07)84004-0

27. Fielding L (2003) NMR methods for the determination of protein-ligand dissociation constants. Curr Top Med Chem 3: 39–53. pmid:12577990 doi: 10.2174/1568026033392705

28. Leavitt S, Freire E (2001) Direct measurement of protein binding energetics by isothermal titration calorimetry. Curr Opin Struct Biol 11: 560–566. pmid:11785756 doi: 10.1016/s0959-440x(00)00248-7

29. Turnbull WB, Daranas AH (2003) On the Value of c: Can Low Affinity Systems Be Studied by Isothermal Titration Calorimetry? Journal of the American Chemical Society 125: 14859–14866. pmid:14640663 doi: 10.1021/ja036166s

30. Biswas T, Tsodikov OV (2010) An easy-to-use tool for planning and modeling a calorimetric titration. Anal Biochem 406: 91–93. doi: 10.1016/j.ab.2010.06.050. pmid:20615384

31. Freire E, Mayorga OL, Straume M (1990) Isothermal titration calorimetry. Analytical Chemistry 62: 950A–959A. doi: 10.1021/ac00217a002

Chapter 12

UNIQUE REACTIVITY OF TRANSITION METAL ATOMS EMBEDDED IN GRAPHENE TO CO, NO, O$_2$ AND O ADSORPTION: A FIRST-PRINCIPLES INVESTIGATION

Minmin Chu, Xin Liu, Yanhui Sui, Jie Luo and Changgong Meng

School of Chemistry and State Key Laboratory of Fine Chemicals, Dalian University of Technology, Dalian 116024, China

ABSTRACT

Taking the adsorption of CO, NO, O$_2$ and O as probes, we investigated the electronic structure of transition metal atoms (TM, TM = Fe, Co, Ni, Cu and Zn) embedded in graphene by first-principles-based calculations. We showed that these TM atoms can be effectively stabilized on monovacancy defects on graphene by forming plausible interactions with the C atoms associated with dangling bonds. These interactions not only give rise to high energy barriers for the diffusion and aggregation of the embedded TM atoms to withstand the interference of reaction environments, but also shift the energy levels of TM-d states and regulate the reactivity of the embedded TM atoms. The adsorption of CO, NO, O$_2$ and O correlates well with the weight averaged energy level of TM-d states, showing the crucial role of interfacial TM-C interactions on manipulating the reactivity of embedded TM atoms. These findings pave the way for the developments of effective monodispersed atomic TM composites with high stability and desired performance for gas sensing and catalytic applications.

INTRODUCTION

In graphene, the newest allotrope of carbon, all the C atoms are sp^2 hybridized and interconnected in a two-dimensional honeycomb lattice through σ bonds, while the remaining C-p$_z$ states that are not involved in the hybridization stand vertically to the lattices and conjugate strongly for formation of the delocalized π bonds. This special bonding feature among C atoms makes graphene a semimetal Dirac fermion system with a zero electronic band gap, good chemical stability, excellent electrical and thermal conductivity, as well as the remarkably high mechanical strength [1,2]. Owing to these unique properties, graphene has been extensively investigated for fabrication of electronic devices [3,4], supercapacitors [5,6], electrodes and sensors [7,8], *etc.*, as well as support material for dispersion of TM nanostructures for gas sensing and catalytic applications [9,10,11,12].

For TM composites for catalytic and gas sensing applications, high density of reaction sites with outstanding reactivity and stability are required to achieve superior performance [13]. One solution to this is to downsize the TM NPs to sub-nano scale or even single atoms to expose more unsaturated TM atoms that are the real reaction sites as determined by the delocalized nature of TM-d states [13]. The interfacial interaction may also promote the reactivity of the composites [14]. However, due to the delocalized π bonds, pristine graphene can only interact with ultrafine TM nanostructures through σ-π bonding and interaction of this type is normally below than −3 eV, which can hardly influence the electronic structure of TM [15]. Furthermore, the large surface energy is one of the key problems in fabrication of single TM atom or ultrafine TM structures, driven by which atomic diffusion and aggregation through Ostwald ripening may take place and result in formation of large TM structures with lowered performance and utilization of TMs. The energy barriers for TM atoms diffusion are typically lower than 1 eV which makes graphene hard to stabilize the deposited TM nanostructures [14,16,17]. In this sense, pristine graphene is not a good support material for fabrication of TM-graphene composites for gas sensing and catalytic applications.

The graphene materials used for chemical applications are generally made by oxidative exfoliation followed by reduction [18]. Driven by the large exothermicity for formation of CO and CO_2, the exfoliation always generates various types of localized defects on the as synthesized graphene, whose evolution can be regulated with additional treatment such as electron beam radiation, chemical functionalization, *etc.* [19,20]. The existence and evolution of these defects provide a new platform to manipulate the reactivity of the deposited TM nanostructures through defect engineering [21,22,23,24]. The recent investigations on the electronic structure of sub-nanosized TM particles/reduced graphene oxide composites showed that these localized defect structures, such as monovacancies and multivacancies on graphene can not only act as strong trapping sites for TM nanostructures and inhibit their aggregation [25], but also determine the electronic structure as well as the reactivity of the composites through TM-graphene interfacial interaction [24,26]. As the size of TM nanostructures goes down, the impact of this interaction would in principle become more significant and this has been examplified by the recent successful invention of single TM atom catalysts for various reactions including CO oxidation [27,28,29]. Stabilization of TM atoms and sub-nanosized nanostructures has been realized experimentally [30,31,32], the electronic structure of TM atoms on graphene has been investigated before and predicted to be effective for hydrogen storage [33,34], while monodispersed Pt [27], Au [35], Cu [36] and Fe [37] atoms stabilized by vacancy defects over graphene have also been proposed to be efficient for CO oxidation. In this sense, it would be of great significance to investigate the electronic structure of these monodispersed TM atoms to generalize the role of interfacial TM-defect interaction in determining the reactivity of the formed composites to rationalize the design of single TM atom and subnanosized TM structures for gas sensing and catalytic applications.

In this work, we investigated the electronic structure of TM atoms (TM = Fe, Co, Ni, Cu and Zn) embedded in graphene and the adsorption of CO, NO, O_2 and O on them by first-principles-based calculations. We showed that the plausible interactions between TM atoms and the defects on the graphene significantly enhance the binding of TM atoms and exclude the possibility for embedded TM atoms to aggregate and form large particles. These interactions also shift the energy levels of TM-d states and tune the reactivity of these atomic composites to small molecules. Further investigation shows that the adsorption of CO, NO, O_2 and O correlates well the energy level of d states of the corresponding TM atoms. These findings pave the way for the developments of effective monodispersed atomic TM composites with high stability and desired performance for gas sensing and catalytic applications. The rest of the paper is organized as the following: the results are presented and discussed in Section 2, the theoretical methods and computational details are described in Section 3, and our conclusions are presented in Section 4.

RESULTS AND DISCUSSION

We began with the atomic deposition of TM atoms on pristine graphene (PG) and the main results are summarized in Table 1. According to the hexagonal symmetry of the PG lattice, TM atoms can be deposited on the top of a C atom (Atop), on the middle of two adjacent C atoms (Brg) and above the center of the C_6 ring (Hol). The calculated E_b for deposition of TM atoms on PG are in the range from 0.02 to 1.51 eV [33]. According to the E_b and deposition configurations, chemical bonds are formed upon atomic deposition of Fe, Co and Ni on PG, while Cu and Zn only bind PG through van der Waals interaction. We reinvestigated these cases using GGA-PBE functional with Grimme scheme for DFT-D correction and found that the contribution of van der Waals interaction is within 0.3 eV and will not alter their relative stability [33]. We also calculated the diffusion barriers of deposited TM atoms and found that the barriers on PG fall in the range of 0.01 to 0.44 eV, which indicates that the deposited TM atoms should be highly mobile even at moderate temperatures [16,38]. The fast diffusion of TM atoms makes PG less eligible for fabrication of atomic TM/graphene composites for sensing and catalytic applications.

Table 1: The most plausible structures, E_{ad} and diffusion barriers of TM atoms deposition on PG.

TM	E_b (eV) [a]	E_a (eV) [b]	d_{TM-C} (Å) [c]	h (Å) [d]
Fe	−0.92 (Hol)	0.42	2.12	1.56
Co	−1.44 (Hol)	0.44	2.10	1.53
Ni	−1.51 (Hol)	0.21	2.12	1.56
Cu	−0.23 (Atop)	0.03	2.20	2.04
Zn	−0.02 (Hol)	0.01	3.02	2.78

Stabilization of these single TM atoms is one of the outstanding problems in fabrication of single TM atoms for chemical sensing and catalysis, as their ultrahigh surface energy will drive them to aggregate on the surface of the substrate resulting in formation of large TM NPs and low utilization of TMs [14]. One possible solution to this is to enhance the stability of the atomic TM-GN composites by strengthening the interaction at the TM-graphene interface that rises the diffusion barriers and tunes the atomic diffusion endothermic [25]. The graphene materials used for chemical applications are generally made through Hummer's method. The exfoliation always generates various types of localized defects on the as synthesized graphene [19]. The existence and evolution of these defects provides a new platform to manipulate the reactivity of the deposited TM nanostructures. We previously studied the interaction of Ru atom with various defects observed on graphene sample synthesized through wet chemistry method and found that monovacancy (MG) is typical and these defects can effectively stabilize the monondispersed Ru atoms [22]. Inspired by previous findings, we then investigated the electronic structure of TM-MG composites with the structural and energetic information summarized in Figure 1.

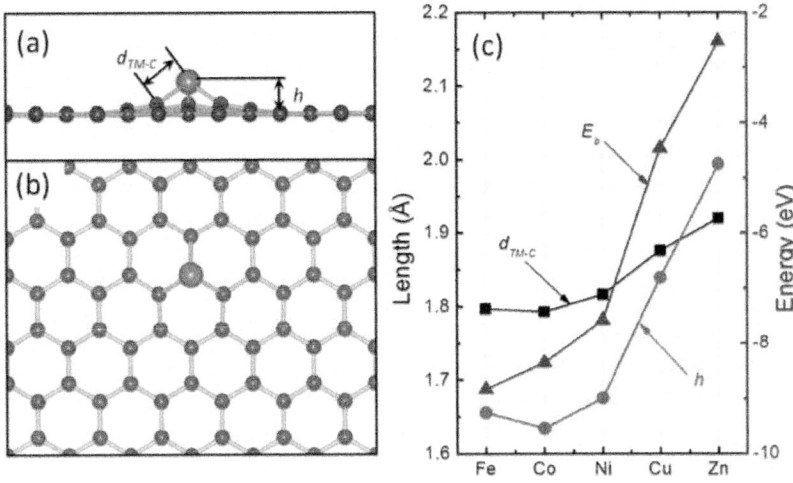

Figure 1: The side view (**a**) and top view (**b**) of the optimized structures of a TM atom embedded in graphene, and the evolution of the corrugation (h); the nearest TM-C distance (d_{TM-C}) and binding energy (E_b) when TM varies in Fe, Co, Ni, Cu and Zn (**c**).

Due to the large size of TM atoms as compared with C atom, the TM atoms will stand out of the basal plane of graphene and the embedment will cause strong corrugation within the graphene (Figure 1a,b). We characterized the stability of these composites with the E_b calculated as the difference between the TM atom in the substitution position and the energy of a reconstructed naked MG plus the energy of an isolated TM atom according to Equation (1) (Figure 1c). The TM-d states are screening the interaction between C-sp states and the TM states. Starting from Fe to Ni and with the increase of the population in TM-d states, the interaction C-sp states with TM states are weakened and so as the stability of the atomic composites, from −8.83 eV for Fe to −7.58 eV for Ni. As for Cu and Zn that have a closed d-shell, the involvement of TM-d states in TM-C bonding is quite limited due to the poor compatibility between these states and defect states of MG in energy space. The calculated E_b are −4.46 and −2.52 eV for Cu and Zn, respectively and are significantly unstable than those TM with a partially occupied d-shell. Resulting from these, the out-of-plane displacement (h, Figure 1a,c) is increasing from 1.63 (Fe) to 1.99 Å (Zn). Following this, the nearest TM-C distance (d_{TM-C}, Figure 1a,c) also changes from 1.79 Å for Co to 1.92 Å for Zn. The small d_{TM-C} and h in these TM-MG composites provide direct evidence for the formation of chemical bonding between the TM states and the dangling bonds localized on C atoms around the vacancy. In Figure 1c, the small d_{TM-C} at Co can be attributed to the highly spin polarized nature of Co-d states due to the limited coordination in Co-MG [39].

Comparing with TM atomic deposition on PG, the binding energies are significantly enhanced. The E_b for Fe-MG is -8.83 eV and is enhanced by about 9 times as compared with that on PG (-0.92 eV). Even for Cu, the E_b of Cu-MG is -4.46 eV and is enhanced by more than 10 times than that over the PG (-0.23 eV). The significant enhanced E_b on MG makes the diffusion of TM atoms endothermic. After embedment, Zn diffusion from the defect site becomes endothermic by ~2.50 eV, implying that the barrier for Zn atomic diffusion would be at least 2.50 eV. The endothermicity for diffusion of embedded Fe atoms is even enhanced to 7.91 eV. The significant enlarged diffusion barriers suggest that the atomic diffusion of embedded TM atoms cannot take place in conventional condition and thus ensure the chemical stability of these TM-MG composites for sensing and catalytic applications. The density of states (DOS) analysis was performed to understand the interfacial TM-C interactions, as shown in Figure 2a.

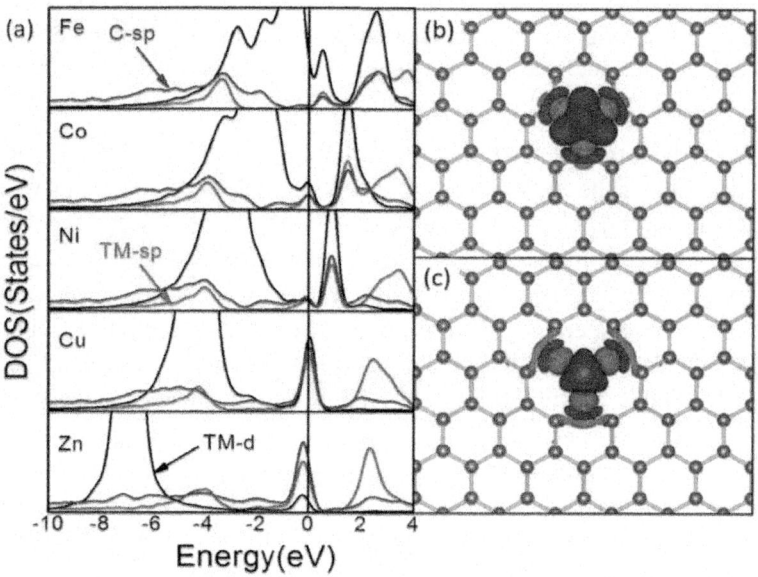

Figure 2: Density of states curves for TM-MG composites (**a**) and contour plots of differential charge density for Fe-MG (**b**) and Zn-MG (**c**) composites. In (**a**), the DOS curves of TM-d, TM-sp and C-sp states are in black, red and blue, respectively. The DOS curves are aligned with the calculated Fermi level (E_F) which is set to zero. In (**b,c**), the contour value is $\pm 3 \times 10^{-7}$ a.u. The charge accumulation regions are in red and the charge depletion regions are in blue.

The DOS of MG is characterized by the sharp spikes around the E_F which are associated with the dangling bonds localized on the C atoms around the defect site [22]. After the embedment of TM atoms, the DOS peaks of these dangling bonds and valence states of TMs are significantly downshifted, get broadened and strong resonance among these states is obvious. This is a sign that these TM atoms are using their valence states to interact with the dangling bonds of MG. The degree of hybridization among TM-sp, TM-d and C-sp states shows a strong dependence on the population in TM-d states. This hybridization gives rise to the shoulder on the DOS curves of those TMs with a partially occupied d-shell at −4.0 eV where TM-d, TM-sp and C-sp have contributions. When the d-shell is fully occupied, the TM-d and TM-sp states are no longer compatible as shown by the large difference in energy levels of these states and the narrow distribution of TM-d state, and the interfacial interaction is mainly contributed by the TM-sp and C-sp. These hold for the relatively weak stability of embedded Cu and Zn atoms. We also plotted the differential charge density to visualize the charge transfer (Figure 2b,c). There are large charge accumulation regions at the TM-C interface while the charge depletion regions are on the TM and C atoms, showing that these TM-C interactions are of partially covalent nature. The shape of those charge depletion regions on both TM and C atoms confirms that the interfacial interaction is the contribution of TM valence states and the dangling bonds and is in line with the DOS analysis.

Due to the joint effect of interfacial interaction and the characteristics of the embedded TM atoms, the energy distribution of TM-d states varies with the TM. There is also a clear trend that with the increase of the population, the main peaks of TM-d states are shifting to lower energy. For all the TMs, the TM-d states are split into two parts due to the TM-MG interaction, one part is around the E_F and is partially occupied or unoccupied, while another part is occupied. The split is the smallest for Fe with a partially occupied d-shell and largest for Zn with a filled d-shell. It should be noted that for some cases, such as Cu and Fe, the TM-C interaction also shifts some of the TM-d states to the E_F [27,37]. The energy levels of TM-d states are known vital for the TM-adsorbate bonding and account for the superior reactivity of TM atoms at the edge and corners of TM nanostructures. The different distribution of TM-d states implies that these TM-MG composites would exhibit distinct reactivity to the adsorbates as compared with the bulk TMs.

We then used O_2, O, CO and NO as adsorbates to probe the reactivity of these TM-MG composites. These adsorbates are also important reaction species for some widely investigated reactions including CO oxidation, oxygen reduction and conversion of nitrogen oxides [40]. The calculated O-O, C-O and N-O distances in freestanding O_2, CO and NO are 1.23 Å, 1.14 Å, and 1.16 Å, respectively. On PG, O_2 lying parallel to the graphene surface with the O-O axis parallel to the axis of two opposite C atoms in the same C_6 ring is the most plausible. The calculated E_{ad} is less than −0.10 eV and the O-O bond length almost remains the same as the free O_2 molecule. The stacking between the π states of O_2 and PG accounts for this weak interaction. Due to the dominant role of π-π stacking in stabilization of CO and NO on PG, the most plausible adsorption configuration of CO and NO on PG are similar to that of O_2. The calculated adsorption energies are less than 0.10 eV and the nearest distances between CO, NO and PG are all above 3.20 Å [41,42]. As for the adsorption of O on PG, the O atom lying above the middle of two adjacent C atoms is plausible and the calculated E_{ad} is −2.07 eV [42,43,44].

The most plausible adsorption configurations for O_2 and O adsorption on TM-MGs are listed in Table 2. As the interfacial interaction split the TM-d states, the O_2 can use its antibonding π states to interact with either the in-plane TM-d states of t_{2g} symmetry in a distorted octahedral (Octa) or to interact with the TM-d states of t_2 symmetry in a tetrahedral (Tetr). This results in formation of two different types of adsorption configurations, as shown in Figure 3. O_2 lies parallel to the substrate plane and two O-TM bonds are formed in-plane with and in the reverse direction of two C-TM bonds in Octa, while two O-TM bonds are formed in the plane vertical the substrate plane with the O_2 standing immediately on top of the TM atom in Tetr. Due to the TM-C interactions and different hybridization between the TM-sp and TM-d states originated from the population in TM-d states and the TM-C interfacial interactions, the O_2 adsorption is plausible in Octa for Co-MG, Ni-MG, Cu-MG and Zn-MG, and in Tetr for Fe-MG. As for O atom, it prefers to act as a ligand to interact directly with the TM atoms in a tetrahedral coordination environment.

Table 2. The structural and energetic parameters for O_2 and O adsorption on TM-MG composites.

TM	O_2					O		
	E_{ad} [a] (eV)	d_{O-TM} [b] (Å)	d_{O-O} [c] (Å)	d_{C-TM} [d] (Å)	C.E. [e]	E_{ad} [a] (eV)	d_{O-TM} [b] (Å)	d_{C-TM} [d] (Å)
Fe	−2.05	1.85	1.39	1.82	Tetr	−2.09	1.62	1.82
Co	−1.74	1.90	1.37	1.80	Octa	−1.80	1.64	1.80
Ni	−1.52	1.94	1.37	1.84	Octa	−1.66	1.66	1.82
Cu	−1.30	1.92	1.36	1.89	Octa	−0.94	1.72	1.92
Zn	−0.71	2.02	1.35	1.97	Octa	−0.23	1.79	2.00

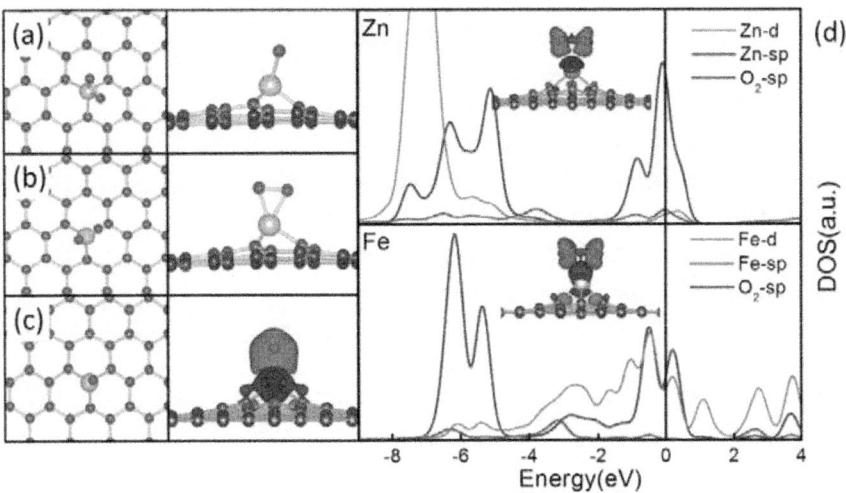

Figure 3: The top view and side view of Octa (**a**) and Tetr (**b**) configurations for O_2 adsorption on TM-MGs, the top view of the configuration for O adsorption on Fe-MG and the corresponding contour plot of differential charge density (**c**); and the DOS curves of O_2 adsorption on Fe-MG and Zn-MG (**d**) in Octa. In structures, the C atoms are in brown, the TM atoms are in silver and the O atoms are in red. The contour plots of differential charge density for O_2 adsorption are shown as insets in (**d**). The charge accumulation regions are in red and the charge depletion regions are in blue. In (**d**), the DOS curves were aligned by the calculated E_F that was set to 0.

The most plausible adsorption configurations of NO and CO are listed in Table 3 with the typical adsorption structures shown in Figure 4a,b. They prefer to interact with the TM atoms with both their filled σ states associated with the lone pair on C or N for formation of σ-donation and their unfilled π states of antibonding character to accommodate the back-donated TM electrons for formation of the π-back donation (Figure 4c). These two interactions account for the stabilization of NO and CO on the embedded TMs. However, due to the Jahn-Teller distortion originated from the population in TM-d states, ligands of this type can interact with either axial components with e_g symmetry of the TM-d states in a distorted octahedral with CO or NO along the direction reverse to a TM-C bond (Octa', Figure 4b) or with the TM-d states with t_2 symmetry in a distorted tetrahedral with adsorbates in the axial position (Tetr', Figure 4a). The stability preference for CO and NO in these 2 configurations are also dependent on the population in TM-d states and their hybridization with the TM-sp states and C-sp state of MG which is regulated by the TM-C interfacial interaction. As the Jahn-Teller distortion is significant for TMs with a more than half-occupied TM-d states, the adsorbed CO is stable in Octa' for Co, Ni and Cu. Different from CO, NO has an unpaired electron and the electron pairing upon adsorption will introduce additional stability and distortion to the coordination around the embedded TM atom. NO adsorption over Fe-MG and Ni-MG is plausible in the Octa', while the Tetr' is preferred for other TM-MGs.

Table 3: The structural and energetic parameters for CO and NO adsorption on TM-MG composites

TM	CO			NO		
	E_{ad}^{a} (eV)	d_{C-TM}^{b} (Å)	d_{C-O}^{c} (Å)	E_{ad}^{a} (eV)	d_{N-TM}^{d} (Å)	d_{N-O}^{e} (Å)
Fe	−1.23	1.91	1.16	−1.30	1.77	1.19
Co	−1.10	1.89	1.16	−1.18	1.77	1.19
Ni	−1.12	1.87	1.16	−1.13	1.74	1.19
Cu	−1.13	1.87	1.16	−1.00	1.79	1.18
Zn	−0.94	1.94	1.16	−0.69	1.99	1.19

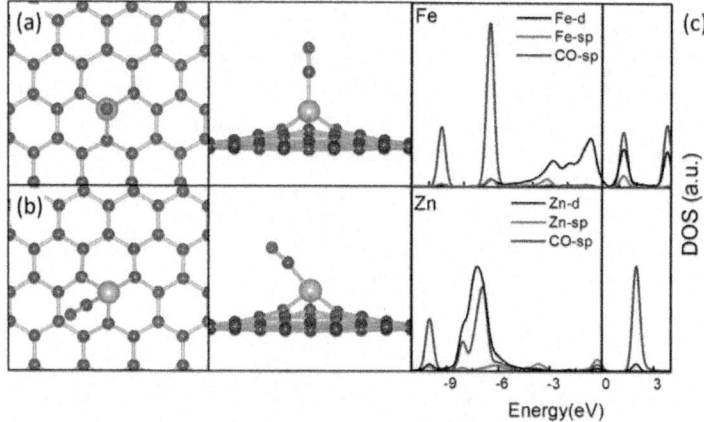

Figure 4: The top view (left panel) and side view (middle panel) of the Tetr' (**a**) and Octa' (**b**) configurations for CO and NO adsorption on TM-MGs, and the DOS curves for CO adsorption on Fe-MG and Zn-MG (**c**). In (a,b), the C atoms are in brown, the TM atoms are in silver and the O atoms are in red. In (**c**), the DOS curves were aligned by the calculated E_F.

According to Table 2 and Table 3, it is apparent that the E_{ad} of all these adsorbates are dramatically enhanced as compared with their adsorption over PG. Furthermore, there are significant elongation of the O-O, C-O and N-O bonds in the plausible adsorption configurations, proving both the direction of the charge transfer is from the TM-MGs to the adsorbs and the activation of the these adsorbates. The variations of the TM-C distance after adsorption are within 0.05 Å, showing that due to the interfacial TM-C interactions, these TM-MG composites can survive in existence of these adsorbates.

We also noticed that for these adsorbates, there is a correlation between the E_{ad} of the adsorbates and the electronic structure of embedded TM. Adsorption of gaseous molecules and associated charge transfer are vital for various chemical processes including chemical sensing and catalysis. If we consider the adsorption of a gaseous molecule as a simple chemical reaction, according to the Frontier Molecular Orbital (FMO) theory proposed by Fuki *et al.*, the E_{ad} and the amount of charge transfered onto the TM composite would be determined by the energetic and spacial compatibility among the orbitals of the gas molecule and the TM-MG composite [45]. To simplify the discussion, we adapted the "d-band model" of proposed by Norskov and Hammer. In this model, the interaction of the adsorbate with the TM assembly is described as the joint effect of both the hybridization of adsorbate states with the TM-s states and the interaction with the TM-d states for production of states of either bonding or antibonding character. This theory has been successfully applied in extended TM systems ranging from bulk truncated surface of various TMs and TM alloys to even small TM clusters for reaction mechanism studies and catalyst design [46]. We aligned the energy levels of TM-d states of these TM-MG composites by the vacuum level and calculated the weight-averaged d-state centers (ε_d). The adsorption energies of O_2, O, CO and NO are plotted *vs.* the calculated ε_d in Figure 5, demonstrating a clear linear relationship. In this sense, the variation of E_{ad} of these adsorbates over TM-MG composites can be directly correlated with the ε_d determined by both the TM-C interfacial interactions and the population in the TM-d states.

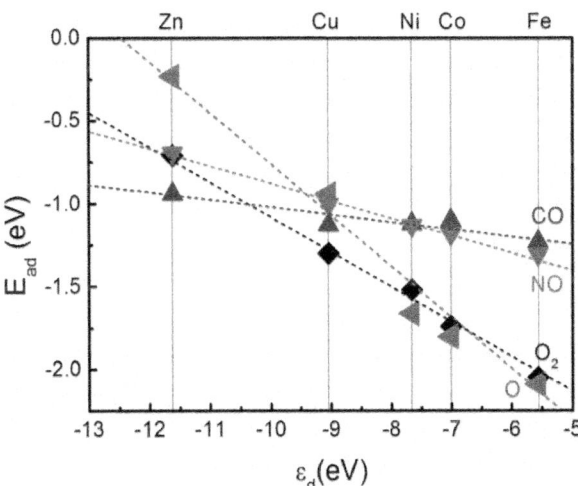

Figure 5: The calculated adsorption energy, E_{ad}, plotted *vs.* calculated weight-average of TM-d states with respect to vacuum, ε_d. The calculated E_{ad} correlates well with ε_d. The dashed lines are to guide the eye.

THEORETICAL METHODS

The first-principles based spin-polarized calculations were done using GGA-PBE functional and DSPP pseudopotentials with DND basis sets as implemented in the DMol^3 package [39,47,48,49,50]. A hexagonal 9×9 supercell of pristine graphene was used to mimic the graphene and the TM-graphene composites. TM atoms were introduced by substituting C atoms. The minimum distance between the graphene sheet and its mirror images is set to be larger than 20 Å to avoid the interactions among the periodic images. We used empirical potential to preoptimize the substrate and the embedded composites and then had them fully relaxed using DMol^3 until the residue forces were below 1×10^{-2} eV/Å [51,52]. A Γ centered $4 \times 4 \times 1$ k-point grid was used for the Brillouin zone sampling during geometric optimization and a $20 \times 20 \times 1$ k-point grid was used for electronic structure analysis [53]. All calculations were performed with a convergence criterion of 2×10^{-4} eV on the total energy and a real-space global orbital cutoff of 4.5 Å. The reaction barriers were calculated with the synchronous method with conjugated gradient refinements [54]. With the above setup, the minimum C-C distance in pristine graphene is 1.42 Å [1,55].

The binding energy (E_b) of TM atom onto grapheme (GN) is calculated as the energy difference between the TM atomically deposited graphene (TMGN) and the separated graphene plus the freestanding TM atom, following Equation (1):

$$E_b = E_{TMGN} - (E_{TM} + E_{GN})$$

(1)

For the study concerning adsorption of CO, NO, O_2, O, *etc.*, the adsorption energy (E_{ad}) is calculated as the energy difference between the species absorbed TM deposited graphene and the gaseous species plus the bare TMGN, following Equation (2):

$$E_{ad} = E_{adsorbate + TMGN} - (E_{adsorbate} + E_{TMGN})$$

(2)

CONCLUSIONS

We addressed the electronic structure of transition metal (TM, TM = Fe, Co, Ni, Cu and Zn)-atoms embedded in graphene and the adsorption of CO, NO, O_2 and O on them by first-principles-based calculations. Due to the plausible interaction between TM atoms and the defects on the graphene, the binding of TM atoms onto MG are significantly enhanced, which not only ensure the high stability of the embedding, but also dramatically rise the barrier for embedded TM atoms to diffuse and aggregate. This interaction also shifts the energy levels of TM-d states and regulates the reactivity of these TM-MG composites to small molecules. Further investigation shows that the adsorption of CO, NO, O_2 and O correlates well the weight averaged energy level of TM-d states of the corresponding TM-MG. As the interfacial C-TM interactions play an important role in regulating the reactivity of the embedded TM atoms, any effect that involves the interactions with the TM atoms, such as doping or co-doping with N, B, P, S and other heteroelements, controlled fabrication of defects on the graphenic support, introduction of co-adsorbents, *etc.*, may help to optimize the reactivity of these embedded TM atoms to the adsorption of gaseous molecules. These findings pave the way for the development of monodispersed TM atomic composites for gas sensing and catalytic applications.

ACKNOWLEDGMENTS

This work was supported by NSFC (21373036, 21573034, 21103015 and 21271037), the Fundamental Research Funds for the Central Universities (DUT15LK18, DUT14LK09 and DUT12LK14).

REFERENCES

1.   Castro Neto, A.H.; Guinea, F.; Peres, N.M.R.; Novoselov, K.S.; Geim, A.K. The electronic properties of graphene. *Rev. Mod. Phys.* 2009, *81*, 109–162.

2.   Novoselov, K.S.; Geim, A.K.; Morozov, S.V.; Jiang, D.; Zhang, Y.; Dubonos, S.V.; Grigorieva, I.V.; Firsov, A.A. Electric field effect in atomically thin carbon films. *Science* 2004, *306*, 666–669.

3.   Schwierz, F. Graphene transistors. *Nat. Nanotechnol.* 2010, *5*, 487–496.

4.   Avouris, P.; Chen, Z.H.; Perebeinos, V. Carbon-based electronics. *Nat. Nanotechnol.* 2007, *2*, 605–615.

5.   Zhu, Y.W.; Murali, S.; Stoller, M.D.; Ganesh, K.J.; Cai, W.W.; Ferreira, P.J.; Pirkle, A.; Wallace, R.M.; Cychosz, K.A.; Thommes, M.; *et al.* Carbon-based supercapacitors produced by activation of graphene. *Science* 2011, *332*, 1537–1541.

6.   Wang, Y.; Shi, Z.Q.; Huang, Y.; Ma, Y.F.; Wang, C.Y.; Chen, M.M.; Chen, Y.S. Supercapacitor devices based on graphene materials. *J. Phys. Chem. C* 2009, *113*, 13103–13107.

7.   Shao, Y.Y.; Wang, J.; Wu, H.; Liu, J.; Aksay, I.A.; Lin, Y.H. Graphene based electrochemical sensors and biosensors: A review. *Electroanalysis* 2010, *22*, 1027–1036.

8. Huang, X.; Zeng, Z.Y.; Fan, Z.X.; Liu, J.Q.; Zhang, H. Graphene-based electrodes. *Adv. Mater.* 2012, *24*, 5979–6004.

9. Huang, X.; Yin, Z.Y.; Wu, S.X.; Qi, X.Y.; He, Q.Y.; Zhang, Q.C.; Yan, Q.Y.; Boey, F.; Zhang, H. Graphene-based materials: Synthesis, characterization, properties, and applications. *Small* 2011, *7*, 1876–1902.

10. Julkapli, N.M.; Bagheri, S. Graphene supported heterogeneous catalysts: An overview. *Int. J. Hydrog. Energy* 2015, *40*, 948–979.

11. Cheng, Y.; Fan, Y.; Pei, Y.; Qiao, M. Graphene-supported metal/metal oxide nanohybrids: Synthesis and applications in heterogeneous catalysis. *Catal. Sci. Technol.* 2015, *5*, 3903–3916.

12. Fan, X.B.; Zhang, G.L.; Zhang, F.B. Multiple roles of graphene in heterogeneous catalysis. *Chem. Soc. Rev.* 2015, *44*, 3023–3035.

13. Yang, F.; Deng, D.H.; Pan, X.L.; Fu, Q.; Bao, X.H. Understanding nano effects in catalysis. *Natl. Sci. Rev.* 2015, *2*, 183–201.

14. Yang, X.F.; Wang, A.Q.; Qiao, B.T.; Li, J.; Liu, J.Y.; Zhang, T. Single-atom catalysts: A new frontier in heterogeneous catalysis. *Acc. Chem. Res.* 2013, *46*, 1740–1748.

15. Tang, Q.; Zhou, Z.; Chen, Z.F. Graphene-related nanomaterials: Tuning properties by functionalization. *Nanoscale* 2013, *5*, 4541–4583.

16. Chan, K.T.; Neaton, J.B.; Cohen, M.L. First-principles study of metal adatom adsorption on graphene. *Phys. Rev. B* 2008, *77*.

17. Krasheninnikov, A.V.; Lehtinen, P.O.; Foster, A.S.; Pyykko, P.; Nieminen, R.M. Embedding transition-metal atoms in graphene: Structure, bonding, and magnetism. *Phys. Rev. Lett.* 2009, *102*.

18. Hummers, W.S.; Offeman, R.E. Preparation of graphitic oxide. *J. Am. Chem. Soc.* 1958, *80*, 1339–1340.

19. Meyer, J.C.; Kisielowski, C.; Erni, R.; Rossell, M.D.; Crommie, M.F.; Zettl, A. Direct imaging of lattice atoms and topological defects in graphene membranes. *Nano Lett.* 2008, *8*, 3582–3586.

20. Wang, Z.; Zhou, Y.G.; Bang, J.; Prange, M.P.; Zhang, S.B.; Gao, F. Modification of defect structures in graphene by electron irradiation: *Ab initio* molecular dynamics simulations. *J. Phys. Chem. C* 2012, *116*, 16070–16079.

21. Liu, X.; Yao, K.X.; Meng, C.G.; Han, Y. Graphene substrate-mediated catalytic performance enhancement of Ru nanoparticles: A first-principles study. *Dalton Trans.* 2012, *41*, 1289–1296.

22. Liu, X.; Sui, Y.; Meng, C.; Han, Y. Tuning the reactivity of Ru nanoparticles by defect engineering of the reduced graphene oxide support. *RSC Adv.* 2014, *4*, 22230–22240.

23. Liu, X.; Li, L.; Meng, C.; Han, Y. Palladium Nanoparticles/defective graphene composites as oxygen reduction electrocatalysts: A first-principles study. *J. Phys. Chem. C* 2012, *116*, 2710–2719.

24. Liu, X.; Meng, C.G.; Han, Y. Defective graphene supported MPd12 (M = Fe, Co, Ni, Cu, Zn, Pd) nanoparticles as potential oxygen reduction electrocatalysts: A first-principles study. *J. Phys. Chem. C* 2013, *117*, 1350–1357.

25. Liu, X.; Meng, C.G.; Han, Y. Substrate-mediated enhanced activity of Ru nanoparticles in catalytic hydrogenation of benzene. *Nanoscale* 2012, *4*, 2288–2295.

26. Liu, X.; Meng, C.; Han, Y. Unique reactivity of Fe nanoparticles-defective graphene composites toward NHx (x = 0, 1, 2, 3) adsorption: A first-principles study. *Phys. Chem. Chem. Phys.* 2012, *14*, 15036–15045.

27. Liu, X.; Sui, Y.; Duan, T.; Meng, C.; Han, Y. CO oxidation catalyzed by Pt-embedded graphene: A first-principles investigation. *Phys. Chem. Chem. Phys.* 2014, *16*, 23584–23593.

28. Yao, K.X.; Liu, X.; Li, Z.; Li, C.C.; Zeng, H.C.; Han, Y. Preparation of Ru nanoparticles/defective graphene composite as a highly efficient arene hydrogenation catalyst. *ChemCatChem* 2012, *4*, 1938–1942.

29. Liu, X.; Sui, Y.; Duan, T.; Meng, C.; Han, Y. Monodispersed Pt atoms anchored on *N*-doped graphene as efficient catalysts for CO oxidation: A first-principles investigation. *Catal. Sci. Technol.* 2015, *5*, 1658–1667.

30. Qiao, B.; Wang, A.; Yang, X.; Allard, L.F.; Jiang, Z.; Cui, Y.; Liu, J.; Li, J.; Zhang, T. Single-atom catalysis of CO oxidation using Pt1/FeOx. *Nat. Chem.* 2011, *3*, 634–641.

31. Wei, H.S.; Liu, X.Y.; Wang, A.Q.; Zhang, L.L.; Qiao, B.T.; Yang, X.F.; Huang, Y.Q.; Miao, S.; Liu, J.Y.; Zhang, T. FeOx-supported platinum single-atom and pseudo-single-atom catalysts for chemoselective hydrogenation of functionalized nitroarenes. *Nat. Commun.* 2014, *5*.

32. Guo, X.; Fang, G.; Li, G.; Ma, H.; Fan, H.; Yu, L.; Ma, C.; Wu, X.; Deng, D.; Wei, M.; *et al.* Direct, nonoxidative conversion of methane to ethylene, aromatics, and hydrogen. *Science* 2014, *344*, 616–619.

33. Manadé, M.; Viñes, F.; Illas, F. Transition metal adatoms on graphene: A systematic density functional study. *Carbon* 2015, *95*, 525–534.

34. Valencia, H.; Gil, A.; Frapper, G. Trends in the hydrogen activation and storage by adsorbed 3d transition metal atoms onto graphene and nanotube surfaces: A DFT study and molecular orbital analysis. *J. Phys. Chem. C* 2015, *119*, 5506–5522.

35. Lu, Y.H.; Zhou, M.; Zhang, C.; Feng, Y.P. Metal-embedded graphene: A possible catalyst with high activity. *J. Phys. Chem. C* 2009, *113*, 20156–20160.

36. Song, E.H.; Wen, Z.; Jiang, Q. CO catalytic oxidation on copper-embedded graphene. *J. Phys. Chem. C* 2011, *115*, 3678–3683.

37. Li, Y.F.; Zhou, Z.; Yu, G.T.; Chen, W.; Chen, Z.F. CO catalytic oxidation on iron-embedded graphene: Computational quest for low-cost nanocatalysts. *J. Phys. Chem. C* 2010, *114*, 6250–6254.

38. Sevincli, H.; Topsakal, M.; Durgun, E.; Ciraci, S. Electronic and magnetic properties of 3d transition-metal atom adsorbed graphene and graphene nanoribbons. *Phys. Rev. B* 2008, *77*.

39. Valencia, H.; Gil, A.; Frapper, G. Trends in the adsorption of 3d transition metal atoms onto graphene and nanotube surfaces: A DFT study and molecular orbital analysis. *J. Phys. Chem. C* 2010, *114*, 14141–14153.

40. Liu, X.; Duan, T.; Meng, C.; Han, Y. Pt atoms stabilized on hexagonal boron nitride as efficient single-atom catalysts for CO oxidation: A first-principles investigation. *RSC Adv.* 2015, *5*, 10452–10459.

41. Zhou, M.; Lu, Y.H.; Cai, Y.Q.; Zhang, C.; Feng, Y.P. Adsorption of gas molecules on transition metal embedded graphene: A search for high-performance graphene-based catalysts and gas sensors. *Nanotechnology* 2011, *22*.

42. Leenaerts, O.; Partoens, B.; Peeters, F.M. Adsorption of H_2O, NH_3, CO, NO_2, and NO on graphene: A first-principles study. *Phys. Rev. B* 2008, *77*.

43. Zhang, Y.-H.; Chen, Y.-B.; Zhou, K.-G.; Liu, C.-H.; Zeng, J.; Zhang, H.-L.; Peng, Y. Improving gas sensing properties of graphene by introducing dopants and defects: A first-principles study. *Nanotechnology* 2009, *20*.

44. Wu, M.; Liu, E.Z.; Jiang, J.Z. Magnetic behavior of graphene absorbed with N, O, and F atoms: A first-principles study. *Appl. Phys. Lett.* 2008, *93*.

45. Fukui, K.; Yonezawa, T.; Shingu, H. A molecular orbital theory of reactivity in aromatic hydrocarbons. *J. Chem. Phys.* 1952, *20*, 722–725.

46. Hammer, B.; Norskov, J.K. Electronic factors determining the reactivity of metal surfaces. *Surf. Sci.* 1995, *343*, 211–220. [Google Scholar]

47. Delley, B. An all-electron numerical-method for solving the local density functional for polyatomic-molecules. *J. Chem. Phys.* 1990, *92*, 508–517.

48. Delley, B. From molecules to solids with the DMol(3) approach. *J. Chem. Phys.* 2000, *113*, 7756–7764.

49. Perdew, J.P.; Burke, K.; Ernzerhof, M. Generalized gradient approximation made simple. *Phys. Rev. Lett.* 1996, *77*, 3865–3868.

50. Delley, B. Hardness conserving semilocal pseudopotentials. *Phys. Rev. B* 2002, *66*.

51. Liu, X.; Meng, C.G.; Liu, C.H. Molecular dynamics study on superheating of Pd at high heating rates. *Phase Transit.* 2006, *79*, 249–259.

52. Liu, X.; Meng, C.G.; Liu, C.H. Melting and superheating of Ag at high heating rate. *Acta Phys. -Chim. Sin.* 2004, *20*, 280–284. [Google Scholar]

53. Monkhorst, H.J.; Pack, J.D. Special points for Brillouin-zone integrations. *Phys. Rev. B* 1976, *13*, 5188–5192.

54. Govind, N.; Petersen, M.; Fitzgerald, G.; King-Smith, D.; Andzelm, J. A generalized synchronous transit method for transition state location. *Comput. Mater. Sci.* 2003, *28*, 250–258.

55. Gajdos, M.; Eichler, A.; Hafner, J. CO adsorption on close-packed transition and noble metal surfaces: Trends from ab initio calculations. *J. Phys. -Condens. Matter* 2004, *16*, 1141–1164.

Chapter 13

FIRST-PRINCIPLES ELUCIDATION OF THE SURFACE CHEMISTRY OF THE C_2H_x (X = 0–6) ADSORBATE SERIES ON FE(100)

Ashriti Govender [1], Daniel Curulla-Ferré [2], Manuel Pérez-Jigato [3] and Hans Niemantsverdriet [3]

[1]Sasol Technology R&D, PO Box 1, Sasolburg 1947, South Africa

[2]Gaz & Energies Nouvelles, Total S.A., Paris La Defense 6, France

[3]Physical Chemistry of Surfaces, Eindhoven University of Technology, PO Box 513, 5600 MB, Eindhoven, The Netherlands

ABSTRACT

Ab initio total-energy calculations of the elementary reaction steps leading to acetylene, ethylene and ethane formation and their decomposition on Fe(100) are described. Alongside the endothermicity of all the formation reactions, the crucial role played by adsorbed ethyl as main precursor towards both ethylene and ethane formation, characterises Fe(100) surface reactivity towards C_2H_x (x = 0–6) hydrocarbon formation in the low coverage limit. A comprehensive scheme based on three viable mechanisms towards ethyl formation on Fe(100), including methyl/methylene coupling, methyl/methylidyne coupling followed by one hydrogenation and methyl/carbon coupling followed by two hydrogenations, is the main result of this article.

INTRODUCTION

Aiming at the remediation of the current world fuel shortage and spearheaded by heterogeneous catalysis, hydrocarbon production from natural gas, coal or biomass via synthesis gas ($CO+H_2$) and the Fischer-Tropsch synthesis is at the heart of some of the main efforts currently taking place within the chemical energy industry. Furthermore, the investigation of fundamental adsorption and reaction properties of hydrocarbons on metal single crystal surfaces under ultra-high-vacuum conditions has become part of the foundation as well as a significant avenue towards the progress of modern surface science and catalysis [1].

The Fischer-Tropsch synthesis (FTS) is a versatile process that can be tuned to alkanes in the diesel and wax range using high-pressure–low-temperature conditions and cobalt or iron catalysts, or to shorter hydrocarbons, including small olefins, using high-pressure–high-temperature conditions and iron catalysts. Understanding the fundamental surface chemistry of paraffinic and olefinic hydrocarbons on the metals involved is an essential requirement for obtaining molecular level insight in (details of) the FTS mechanism.

There are typically three popular FTS mechanisms [2], although much debate on the subject still lingers on. The most accepted is the *carbide mechanism* which entails CO adsorption and dissociation towards adsorbed carbon and oxygen atoms, as well as successive hydrogenation of surface carbon atoms towards CH_x fragments, and insertion of CH_x monomers into the metal-carbon bond of an adsorbed alkyl chain. With methyl as chain initiator, methylene is considered the frame-building monomer. Chain termination takes place via the dehydrogenation of the resulting alkyl adsorbates towards an α-olefin, or the incorporation of an OH group to form n-alcohols.

The *enol mechanism* involves the partial hydrogenation of adsorbed CO to an enol, an oxygen-containing surface species, its chain growth occurring via the condensation of pairs of -CHOH species, via water elimination towards adsorbed -CHROH. Termination steps lead to oxygenates and α-olefins, with alkanes forming in a secondary step.

The *CO insertion mechanism* proceeds via the insertion of adsorbed CO into the metal-alkyl bond, leading to a surface acyl species, its chain termination taking place by either hydrogenation or dehydrogenation towards olefins, paraffins, aldehydes or alcohols.

Understanding the composition of the resulting complex mixture in FTS, has become a crucial information source for its mechanism, the Anderson-Schulz-Flory product distribution comprising the main description in terms of chain-length [3], on the basis of a stepwise full-polymerisation model reaction. There exist many FTS mechanistic studies [4,5,6,7,8,9,10,11,12,13,14,15,16,17,18,19,20,21,22,23,24] that have been reported in the literature, mainly on Fe, Ni, Co and Ru. Some of them involve the investigation of relevant CH_x species and include the possibility of two different active sites driving independent reactions. Considering the large number of adsorbed species present during FTS on the catalyst surface, it does seem likely that more than one mechanism is at play.

Several *ab initio* studies of the chain-growth process of linear hydrocarbons on metal surfaces appear in the literature, probably the most detailed being the *carbide mechanism* based on reports by van Santen and co-workers [25,26,27], as well as those by Hu and co-workers [28,29,30,31]. Studies addressing higher level of molecular complexity for chain growth (including CO insertion, CHO insertion and synchronic C-C coupling/CO dissociation) have appeared in the literature [32,33,34], some on the iron carbide surfaces [35,36].

Lo and Ziegler [37] recently reported two-carbon adsorbate formation on Fe(100) from *ab initio* calculations. They found that most two-carbon species favour adsorption at the hollow site and that ethylene prefers to be π-bonded on Fe(100). On the other hand, they identify adsorbed ethyl on the bridge site, on Fe(100), and ethane in the physisorbed state, on the same surface. In that study, ethynyl and vinylidene are the most thermodynamically stable (ethylidyne is next in stability). Lee *et al.* [38] studied the decomposition of acetylene on Fe(001) from planewave-pseudopotential calculations. They concluded that the decomposition of acetylene by C-H bond breakage: CHCH → CCH → C+CH → C+C was just as likely as C-C bond breakage: CHCH → CH+CH → C+CH → C+C. The full dissociation of acetylene into atomic carbon is energetically favourable and exothermic. Anderson and Mehandru [39,40] calculated the adsorption of acetylene on Fe(100), (110) and (111) clusters, using the atom superposition and electron delocalisation molecular orbital theory (ASED-MO) method. They found that the high coordination hollow sites are favoured on (100) and (110) surfaces and the bridging site is favoured on Fe(111).

Cheng *et al.* [41] report the promotion effect of transition metal adsorbed atoms on ethylene chemisorption on Co(0001), from linear-combination of atomic-orbitals *ab-initio* calculations, and emphasize the role of Pd and Cu in decreasing ethylene adsorption energy, with the ensuing improvement in α-olefin selectivity. Xu *et al.* [42] present photoemission evidence for the coverage dependent ethylene decomposition on Co(0001) via acetylene, and identify several adsorbed carbon species, including dicarbon. The latter is identified as a stable species on fcc-Co(111), from *ab-initio* calculations, by Ciobîca and co-workers [43,44,45,46].

Several *ab initio* reports on the adsorption of ethylene on Ni(100) [47,48], Ni(110) [49,50,51,52] and Ni(111) [53] are known from the literature. A comparative investigation of ethylene dissociation on both Ni(111) and Ni(211) by Vang *et al.* [53] concludes that the C-C dissociation barrier is lowered at steps much more than the dehydrogenation barrier does.

Our group has published on important steps in the formation of hydrocarbons on the (100) surface of iron, being the direct dissociation of CO [54,55], the H-assisted dissociation of CO [56], the subsequent formation of water from adsorbed oxygen and hydrogen [57], and the formation of CH4 from adsorbed carbon and hydrogen [58]. This article completes the series with the ab-initio characterisation of the surface reactivity of Fe(100) towards C2 hydrocarbon formation in the low coverage regime. We believe our mechanism to provide understanding towards the process of linear hydrocarbon formation conditions in general.

A short justification for using the Fe(100) surface as a substrate for the reactions is in order. It is well known that metallic iron is unstable under FTS conditions and that iron is converted to carbides during early stages of the reaction [59,60]. We see the studies on Fe(100) as a simplified reference case for those on the much more complicated iron carbide surfaces, among which several have relatively similar surface free energy and stability [61]. We already published a preliminary study on the methane formation on Fe_5C_2 [62] and further investigations are in progress.

This paper comprises the Results and Discussion inSection 2, the computational methodology as largely covered by references [63,64,65,66,67,68,69,70] in Section 3 and the conclusions in Section 4. Within Section 2, we first discuss the adsorption behaviour of C_2H_x (x = 0–6) species, followed by co-adsorption configurations and minimum energy pathways. After determining the viability of the elementary steps in terms of barrier, we present the potential energy surfaces for acetylene, ethylene and ethane. After identifying three likely individual mechanisms, we bring these together into a comprehensive scheme describing a selection of FT reactions.

RESULTS AND DISCUSSION

The stability of adsorbed hydrocarbon molecules and intermediates is hereby considered by means of two different thermodynamic properties, adsorption energy and chemical potential. Whereas the absolute stability scale is given by the adsorption energies of the surface species, the relative stability scale of chemical potentials provides enthalpies under a constant supply of adsorbed atomic hydrogen atoms, therefore allowing the comparison of the energetics of two adsorbed moieties that possess a different number of hydrogen atoms in their chemical formulas, as required by the study of chemical reactions.

The build-up of chemical potential profiles for the reactions of formation of adsorbed acetylene, ethylene and ethane involves the full study of all possible intermediates, starting from adsorbed atomic hydrogen, carbon and C_xH_y moieties, and systematically progressing by chemical reaction towards the three original adsorbed molecules. In order to scan the ensemble of involved intermediates, the following reactions are studied; carbon-carbon coupling, hydrogenations, isomerisations and dehydrogenations. The methanation reaction has already been reported [58], and is not to be considered any further.

Acetylene, Ethylene, Ethane and Intermediates: Stable Adsorption States

In order to identify all minima for acetylene, ethylene and ethane as well as all intermediates in between, geometry optimisations and (vibrational) normal-mode analysis are carried out. A careful sampling of optimal geometry enthalpy lets us choose the lowest energy configurations, and the normal-mode analysis lets us identify true minima by ensuring that all vibrational frequencies are real. Table 1 describes the lowest energy configuration for all the adsorbed species considered in terms of geometries, adsorption energies and vibrational frequencies, including *dicarbon, ethynyl, vinylidene, acetylene, vinyl, ethylidyne, ethylidene, ethylene, ethyl* and *ethane*.

The complete set of optimal adsorbate geometries and vibrational frequency data is available in a PhD thesis [70]. C_2 adsorbates with no hydrogen atoms on the carbon bound to the metal, - C_α-, tend to have the highest adsorption energy, and as the α-carbon gets more and more hydrogenated (until the α-carbon is fully hydrogenated), the adsorption bond becomes destabilised. The most stable moieties are dicarbon (adsorption energy −8.78 eV), ethynyl (adsorption energy −6.28 eV), and ethylidyne (adsorption energy −6.26 eV), although the dicarbon formation reaction is shown below not to be viable in the low coverage regime hereby considered. On the other hand, the molecule with the lowest adsorption energy is ethane, which is physisorbed, and will probably leave the surface as soon as it forms.

The species involved in acetylene, ethylene and ethane formation (C_2H_x, x = 0–6), adsorb on fourfold hollow sites, except for ethyl that sits on the bridge site, and ethane, which does not exhibit any preference for any particular site on Fe(100). Whereas dicarbon, acetylene, ethylene and ethane adopt a flat configuration on Fe(100) in their most stable adsorption geometry, ethylidyne is the only species that stands completely upright on the surface. Ethynyl, vinyl, ethyl, vinylidene and ethylidene all exhibit a tilted geometry, with vinyl adopting an asymmetric (with the two β-hydrogens skewed) configuration on Fe(100), see Table 1.

The resulting dicarbon geometry is at variance to a previous *ab-initio* study by Lo and Ziegler [37] that suggests tilted structures. On the other hand the ethynyl (CCH) geometry is in agreement with that reported by the same authors, and the hereby reported vibrational frequencies for CCH compare well with experimental values from HREELS [71]. Furthermore, the vibrational frequencies hereby computed for acetylene are comparable to reported experimental frequencies from RAIRS [72]. Our optimal structure for vinyl on Fe(100) differs from that reported by Lo and Ziegler, but the ethylidene geometry is in agreement with theirs. The quad–σ-bonded configuration (hollow site) of acetylene does not agree with Lo and Ziegler, who report a π-bonded ethylene geometry (ontop) as the most stable. However, our result is in agreement with the experimental geometry reported by Hung and Bernasek [72]. All structures in Table 1 are characterised by real vibrational frequencies, except for the physisorbed ethane, and a low frequency of $36i$ cm^{-1} for ethylidyne. We nevertheless accept the ethylidyne in the hollow geometry as the most stable structure as a) it exhibits the strongest adsorption of all geometries considered [70] and b) adsorbate–surface coupling is not considered in the frequency calculations, which makes very low frequencies somewhat doubtful.

Table 1: Lowest energy configuration for all intermediates in acetylene, ethylene and ethane formation and decomposition reactions, along with heats of adsorption, specific structural parameters and vibrational frequencies

			ΔH_{ads} (eV)	$d_{C\text{-}C}$ (Å)	$z_{C\alpha}$ (Å)	$z_{C\beta}$ (Å)	vibrational frequencies (cm^{-1})
dicarbon	CC		-8.78	1.32	0.85	0.85	259 277 278 319 562 1488
ethynyl	CCH		-6.28	1.39	0.64	1.43	223 266 345 352 429 608 843 1236 3096
vinylidene	CCH2		-4.71	1.44	0.63	1.77	19 240 285 340 359 574 786 931 1129 1589 2943 2999
acetylene	CHCH		-2.83	1.4	1.36	1.36	229 235 287 588 599 679 817 895 1099 1265 2907 2930
vinyl	CHCH2		-3.73	1.46	0.73	1.54	162 215 273 324 383 415 578 785 950 1072 3083 3481 / 3283 3495 2998
ethylidyne	CCH3		-6.26	1.54	0.67	2.21	36i 83 124 292 320 521 954 940 955 1151 1394 1426 / 2802 2950 2964
ethylidene	CHCH3		-5.27	1.55	0.76	2.05	50 127 184 259 319 384 818 878 965 980 1115 1151 / 1378 1414 1978 2769 2799 2979
ethylene	CH2CH3		-0.83	1.5	1.61	1.61	87 187 227 341 295 932 578 872 885 1004 1005 1144 / 1230 1317 2799 2798 2831 2840
ethyl	CH2CH2		-2.18	1.53	1.8	2.75	48 74 175 197 229 815 407 855 913 982 1121 1177 / 1325 1344 1426 1464 2775 2802 2918 2961 2989
ethane	CH3CH3		+0.23	1.53	3.74	3.74	51i 54i 26 47 60 90 306 802 805 994 1176 1179 / 1361 1379 1447 1452 1455 1458 2953 2959 3007 3007 3057 3040

Elementary Reaction Steps Connecting Reaction Intermediates and Molecules: Co-Adsorption Configurations and Minimum-Energy Paths

Geometry optimisations for different co-adsorption systems, as required by the chemical reactions hereby investigated, are followed by vibrational normal-mode analysis. After identifying minima and transition states, the selection of initial state for reactions is followed by the computation of minimum-energy paths, in order to characterise elementary reactions. Isomerisation steps do not require the study of co-adsorption systems, usually implying adsorbates are in their ground state geometry.

Table 2 describes the elementary steps linking all intermediates appearing during acetylene, ethylene and ethane formation and decomposition, in terms of the geometry of initial (IS), transition (TST) and final (FS) state, as well as forward and reverse activation energies and reaction energies. Pre-reaction and/or post-reaction steps are described in Table 2whenever relevant (see the discussion section). Figure 1 illustrates the nomenclature used in Table 2. In case more than one pre-step is involved, we include the total change in energy before the main activation barrier as ΔE_1. Table 2 represents a selection of elementary reactions starting from either the most stable or the most likely configurations. The complete set of co-adsorption geometries and minimum-energy path data is available [70].

Surface Reaction Mechanisms

In order to arrive at mechanistic descriptions of the formation or decomposition of C_2 hydrocarbons, the viability of the elementary reaction steps is further investigated by means of the individual barrier and total barrier concept. All data in Subsection 2.2 are categorised below in terms of chemical potentials for all the species (including co-adsorbates) involved in the reaction sequence up to the formation of ethane, which provide a relative stability scale. A selection is then carried out on the basis of reaction barriers (either apparent or total), which leads to our mechanistic conclusion for the reaction of formation of C_2H_x hydrocarbons.

Two separate approaches can be followed for interpreting the viability of the computed elementary reaction steps, when more than one stage is present. In the first, the concept of *individual reaction barrier/rate determining step* (RDS) is applied, and the main assumption in our energy barrier analysis is that the activation energy of the main-reaction has the highest value of all sequential steps (Figure 1). Provided a pre-reaction or post-reaction step had a higher activation energy than the main reaction, the new highest energy barrier value would be taken as the determining step. However, virtually all cases to be described in Table 2 comply with the situation sketched in Figure 1. Another assumption in this first analysis is that any pre-reaction or post-reaction stages accompanying the main reaction are all uncoupled from each other, in contrast to reactions that take place via a transient intermediate or to non-equilibrium multi-stage processes. The second analysis approach involves the complete picture provided by *the total/apparent reaction barrier*. The indicator employed now for the viability of the surface reactions under scrutiny is *the maximum enthalpy deviation*, the difference between the highest enthalpy value for all states (usually a transition state) and the lowest enthalpy value of the initial state (usually the ground state configuration for the initial state). Our work is limited by the fact that not all barriers are computed for pre- and post-reaction steps. Table 2 includes both, the rate determining step barrier corresponding to the individual thermal reaction rate (*individual* or RDS barrier); and the maximum enthalpy deviation for the total thermal reaction rate method (*total* or *apparent* barrier).

Table 2: Elementary steps for C_2H_x (x = 0–6) formation, including (**a**) acetylene formation steps, and (**b**) ethylene and ethane formation steps, along with energy parameters as defined in Figure 1

Table 2a

Reaction	IS	IS'	TST	FS'	FS	product	$E_{act}^{f,ov}$	$E_{act}^{r,ov}$	ΔE_{react}^{ov}
	E_0	ΔE_1	$E_{act}^f (E_{act}^f)$	ΔE_{react}	ΔE_2			(eV)	
C + C → CC						dicarbon	2.31	0.88	1.43
	0	0.37	1.94 (0.63)	1.31	-0.25				
CC + H → CCH						ethynyl	0.71	1.15	-0.44
	0	0.16	0.55 (1.09)	-0.54	-0.06				
CC + H → CCH						ethynyl	0.96	1.34	-0.38
	0	0.2	0.76 (1.34)	-0.58					
C + CH → CCH						ethynyl	2.13	0.85	1.28
	0	0.76	1.37 (0.85)	0.52					
C + C + H → CCH						ethynyl	1.97	0.85	1.12
	0	0.6	1.37 (0.85)	0.52					
CCH + H → CCH2						vinylidene	0.5	0.2	0.3
		0.22	0.28 (0.13)	0.15	-0.07				
CCH + H → CCH2						vinylidene	0.58	0.11	0.47
	0		0.58 (0.13)	0.45	0.02				
C + CH2 → CCH2						vinylidene	1.88	1.14	-0.74
	0	1.15	0.73 (1.07)	-0.34	-0.07				
C + CH2 → CCH2						vinylidene	2.08	1.14	0.94
	0	1.35	0.73 (1.07)	-0.34	-0.07				
CCH + H → HCCH						acetylene	0.92	0.38	0.54
	0	0.06	0.86 (0.38)	0.48					
CH + CH → HCCH						acetylene	1.89	1.25	0.64
	0	0.73	1.16 (0.66)	0.5	-0.59				
CH + C + H → HCCH						acetylene	1.7	1.25	0.45
	0	0.54	1.16 (0.66)	0.5	-0.59				
CCH2 → HCCH						acetylene	1.5	1.5	0
	0		1.5 (1.5)	0					

Table 2b

Reaction	IS	IS'	TST	FS'	FS	product	$E_{act}^{f,ov}$	$E_{act}^{r,ov}$	ΔE_{react}^{ov}
	E_0	ΔE_1	$E_{act}^f\ (E_{act}^r)$	ΔE_{react}	ΔE_2			(eV)	
CCH2 + H → CHCH2						vinyl	0.89	0.1	0.79
	0	0.24	0.65 (0.10)	0.55					
CCH3 → CHCH2						vinyl	1.6	1.22	0.38
	0		1.60 (1.22)	0.38					
H + CCH2 → CCH3						ethylidyne	0.68	0.28	0.4
	0	0.24	0.44 (0.28)	0.16					
C + CH3 → CCH3						ethylidyne	0.49	0.24	0.25
	0		0.49 (0.24)	0.25					
H + CHCH2 → CHCH3						ethylidene	0.62	0.25	0.37
	0	0.18	0.44 (0.25)	0.19					
H + CCH3 → CHCH3						ethylidene	0.75	0.02	0.73
	0		0.75 (0.02)	0.73					
CH + CH3 → CHCH3						ethylidene	0.58	0.04	0.54
	0		0.58 (0.04)	0.54					
CH2 + CH2 → H2CCH2						ethylene	2.21	1.68	0.53
	0	0.56	1.61 (1.68)	-0.07					
CH2 + CH2 → H2CCH2						ethylene	1.94	1.41	0.53
	0	1.49	0.45 (1.25)	-0.8	-0.16				
CHCH2 + H → H2CCH2						ethylene	0.85	0.38	0.47
	0		0.85 (0.38)	0.47					
CHCH3 → H2CCH2						ethylene	1.88	1.78	0.1
	0		1.88 (1.78)	0.1					
CH2 + CH3 → H2CCH3						ethyl	0.71	0.55	0.16
	0		0.71 (0.55)	0.16					
CH2CH2 + H → H2CCH3						ethyl	0.95	0.66	0.29
	0	0.33	0.62 (0.66)	-0.04					
CHCH3 + H → H2CCH3						ethyl	0.72	0.42	0.3
	0	0.03	0.69 (0.42)	0.27					
CH2CH3 + H → H3CCH3						ethane	1.03	0.8	0.23
	0	0.17	0.86 (0.80)	0.06					
CH3 + CH3 → H3CCH3						ethane	1.62	-	-
	0		1.62 (-)	-					

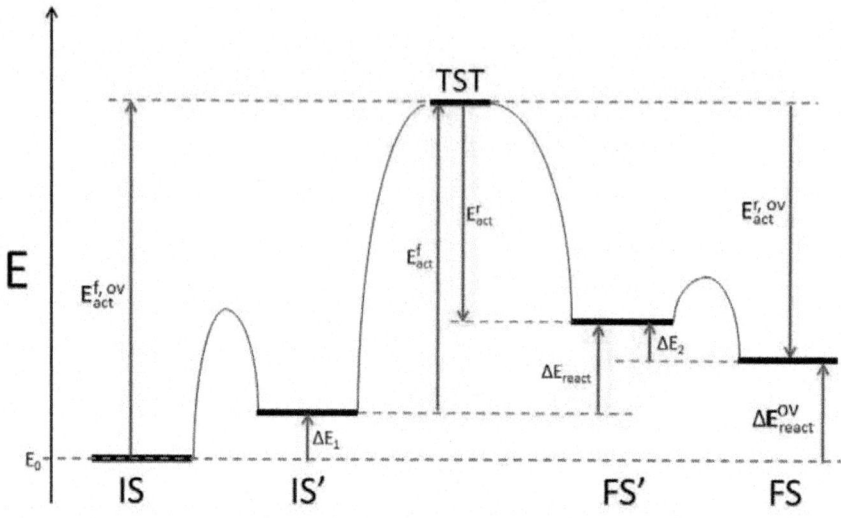

Figure 1: Schematic enthalpy diagram for a multi-step chemical reaction, as per the nomenclature of Table 2.

Chemical Potential Profiles for Acetylene, Ethylene and Ethane Formation

The information presented in Table 2 describes a complete series of elementary reaction steps connecting the adsorbed intermediates that appear on the Fe(100) surface within the process of formation of the three adsorbed hydrocarbons containing two carbon atoms; acetylene, ethylene and ethane, the steps connecting each of the formed hydrocarbons with the intermediates of the other hydrocarbons being included. The complete uninterrupted sequence of reaction steps up to the formation of adsorbed ethane provides a full description of the system reactivity, both in terms of C-H and C-C couplings. Figure 2 shows the total chemical potential profile for all adsorbed intermediates and molecules. The common scale used for Figure 2 corresponds to the ethane gas at zero energy and corresponding 2C+6H stoichiometry. On the other hand, the x-axis in Figure 2 follows a logical ordered sequence of adsorbed species complexity, starting from dicarbon and two co-adsorbed carbon atoms, and progressing step-wise via acetylene and ethylene/ethylidene up to ethane.

All elementary reaction steps connecting *dicarbon, ethynyl, vinylidene and acetylene* to acetylene formation and decomposition comprise the first section of Figure 2, including the steps linking. The sequence commences with C+C coupling to form dicarbon. The chemical potential for the reacting configuration of the co-adsorbed C/C system (initial-state) provides the lowest value in Figure 2 (neither the pre-reaction nor the post-reaction steps described in Table 2 for this reaction are included).

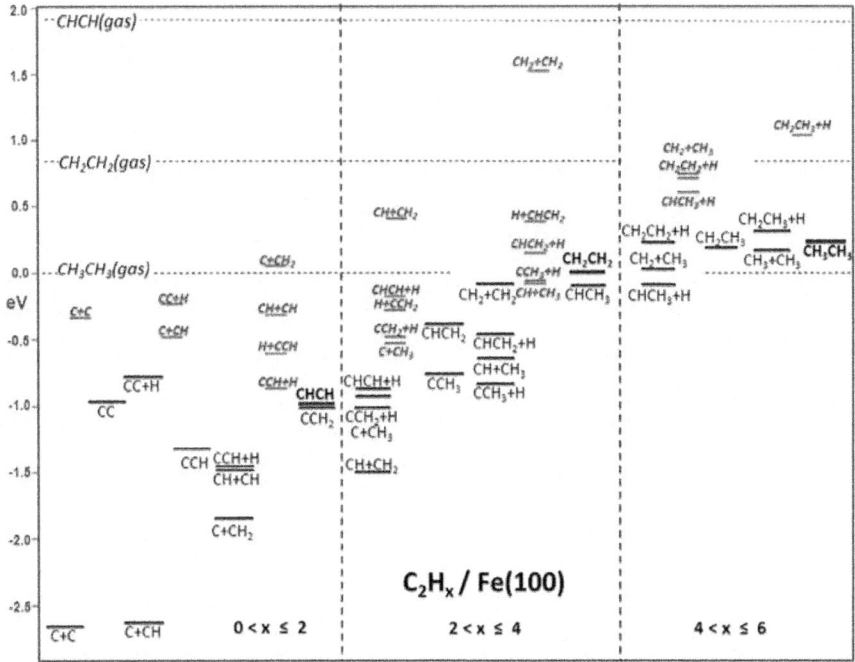

Figure 2: Compilation of chemical potential profiles of C_2H_x for x = 0–6; transition states are indicated in red. The x-axis shows the systematic H-addition while the y-axis expresses energy in electronvolts (eV).

The chemical potential of ethynyl, CCH, is even more negative than that for dicarbon, as seen from Figure 2, and it is only overcome by the more negative values of some co-adsorption systems, viz C+C, C+CH, CH+CH, CCH+H. Our relative stability of ethynyl is in agreement with the high stability found experimentally for ethynyl being abundant at low hydrogen gas pressures on Fe(100), by Bernasek and co-workers [71,72]. Vinylidene and acetylene have similar chemical potentials, between those for dicarbon and ethynyl. The least negative of the series being that for dicarbon, the most negative that for ethynyl. Still, there are two co-adsorption systems with even less negative chemical potential values than all four intermediates/molecules (they are the least negative of all IS and FS in Figure 2), viz CC+H and C+CH₂, these being very similar to each other.

The ethylene formation/decomposition sequence follows in Figure 2. The specific adsorbed species involved in this sequence of elementary steps include *vinyl, ethylidyne, ethylidene and ethylene*. The $C+CH_3$ co-adsorbate in its ground state geometry, along with the acetylene and vinylidene adsorbates within the second relative scale, have similar chemical potentials, the most negative of all chemical potential values for C_2H_x ($2 < x \leq 4$). The least stable intermediate/molecule is ethylene, a preferred product in experimental FTS on iron catalysts, and the second least stable is ethylidene. The next intermediate in the relative stability scale is vinyl, and a more negative value corresponds to ethylidyne. The involved co-adsorption systems follow the chemical potential sequence $C+CH_3 < CCH_2+H < CH+CH_2 < CHCH+H < CCH_3+H < CH_2+CH_2 < CH+CH_3 < H+CHCH_2 < CHCH_2+H$. Similarly to what happens for acetylene formation, and in the methanation chemical potential profile on Fe(100) reported by Govender *et al.* [58], all forward elementary reaction steps leading to ethylene formation, are endothermic.

The chemical potential profile describing the ethane formation reaction rounds off Figure 2. The two intermediates formed in this range of reactions are *ethyl* and *ethane*. The most stable adsorbed species now, which in turn have negative chemical potentials, are ethylidene and the ethylidene-hydrogen co-adsorption system, with similar chemical potentials. They are followed by ethylene, and then the CH_2+CH_3 co-adsorption system. The remaining chemical potential values for adsorbed intermediates-molecules as well as for co-adsorbates follow the sequence $CH_2CH_3+H < CH_2CH_3 < CH_3CH_3 < CH_2CH_2+H < CH_3+CH_3$. Adsorbed ethane appears at chemical potential very close to that for CH_2CH_2+H.

Selection of Three Viable Reaction Mechanisms

Considering that high temperature Fischer-Tropsch synthesis is carried out up to 625 K, and that this temperature corresponds to an overall kinetic barrier in the order of 1.5 eV, we will not consider reactions with barriers well above this value, and particularly not in cases where alternatives at more favourable barriers are available.

Three different pathways towards ethyl formation on Fe(100) have been identified in Figure 3 as most favourable, including CH_2+CH_3, $CH+CH_3$ (followed by one hydrogenation) and $C+CH_3$ (followed by two subsequent hydrogenations), with barriers of 0.71 eV, 0.58 eV and 0.49 eV, respectively. All other couplings of $CH_x + CH_y$ possess barriers that are substantially higher, and are unlikely. The selected C-C coupling steps have similar barrier values; therefore the parallel operation of all three is to be expected, rather than a single dominant mechanism. Figure 3 shows the details of the *three*selected mechanisms, including C-C step, and subsequent hydrogenations (when necessary) towards ethylene and ethane, via ethyl formation.

In Mechanism 1, co-adsorbed methyl and methylene would require 0.71 eV in order to couple and form adsorbed ethyl. Ethyl would then need to cross a 0.86 eV barrier in order to further hydrogenate and form ethane, or it could dissociate to ethylene, for which a lower activation energy, 0.66 eV, is required. Ethylene is thermodynamically more stable on Fe(100) and its formation from ethyl is actually exothermic, rather than the less favourable endothermic reaction to form ethane. Note that Mechanism 1 is in essence the one proposed by Biloen and Sachtler [13], which is often quoted in Fischer-Tropsch literature.

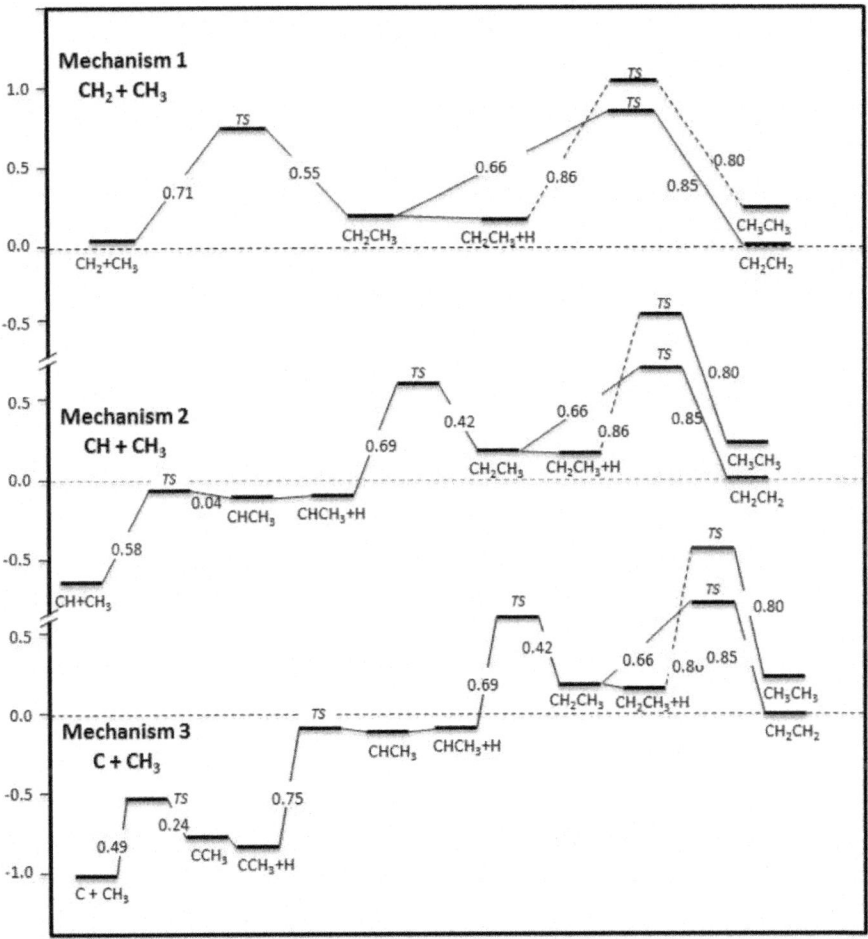

Figure 3: Detailed chemical potential profile characterising the three selected reaction mechanisms. The *y*-axis indicates energy in electronvolts (eV).

Co-adsorbed methylidyne and methyl, in Mechanism 2, would cross a barrier of 0.58 eV in order to couple and form ethylidene, which happens to be a reversible reaction. In fact, the C-C dissociation reaction of ethylidene has a low barrier of 0.04 eV. In case the forward C-C coupling reaction proceeds, ethylidene will be hydrogenated to form ethyl, a process that requires an activation energy of 0.69 eV. Ethylene and ethane formation would proceed as previously explained.

Co-adsorbed atomic carbon and methyl on Fe(100) would require climbing a barrier of 0.49 eV in order to couple as per Mechanism 3. Two subsequent hydrogenations occur with similar activation energies to those of the methanation reaction on Fe(100) [58], viz 0.75 eV to ethylidene and 0.69 eV to ethyl. The barrier for C-H dissociation of ethylidene is 0.02 eV, very similar to that for its C-C dissociation, as seen in Mechanism 2.

The *total* and *apparent forward barriers* for ethane and ethylene formation, respectively, forMechanism 1 are 1.02 eV and 0.82 eV; for Mechanism 2, 1.67 eV and 1.47 eV; and for Mechanism 3, 2.08 eV and 1.88 eV. Hence Mechanism 3 has an overall barrier which is probably not compatible with the 625 K maximum reaction temperature of the high-temperature Fischer-Tropsch reaction. Both stepwise and in total, the formation reactions for ethylene and ethane–as well as for methane [58] are endothermic, making Fe(100) particularly reactive towards decomposition.

Co-Existence of Many Surface Species: An Interconversion Scheme

The selected mechanisms exhibit substantial overlap with each other; therefore, merging all three into a single scheme would lead to a compact reactivity description of the C_2H_x series. Figure 4 shows a comprehensive scheme with selected reaction steps that lead to the formation of the hydrocarbon series C_2H_x (x = 0–6), and includes both forward and reverse energy barriers (useful when considering competing reactions). Note that we do not claim to be complete, we left out steps with high barriers, or steps for which favorable competing steps are available. Moreover, the selection has been carried out with the formation reaction in mind, rather than the decomposition of species, although several of these have been included. We realise that the present analysis is based on energy barriers only, and we acknowledge that entropic effects may exert a profound role in the rate constants of the elementary steps that we studied. In addition, distributions of reactant species participating in the reaction step as well as their mobility are factors that are expected to have profound influence. We nevertheless hope that the present analysis may contribute to insight in the complexity of Fischer-Tropsch mechanisms.

Figure 4 illustrates that it would be hard to decide on a 'most abundant surface intermediate' during C_2H_x formation on Fe(100) (microkinetic simulations may provide further insight). On the other hand, the prominent role of ethyl, C_2H_5, as a precursor to both C_2H_4 and C_2H_6 stands out in Figure 4, although a competing path to ethylene via vinyl appears feasible, albeit with a somewhat higher barrier. Note also that acetylene, not a common product in Fischer-Tropsch synthesis, cannot be formed directly, but may result by indirect reaction and dehydrogenation of species richer in hydrogen.

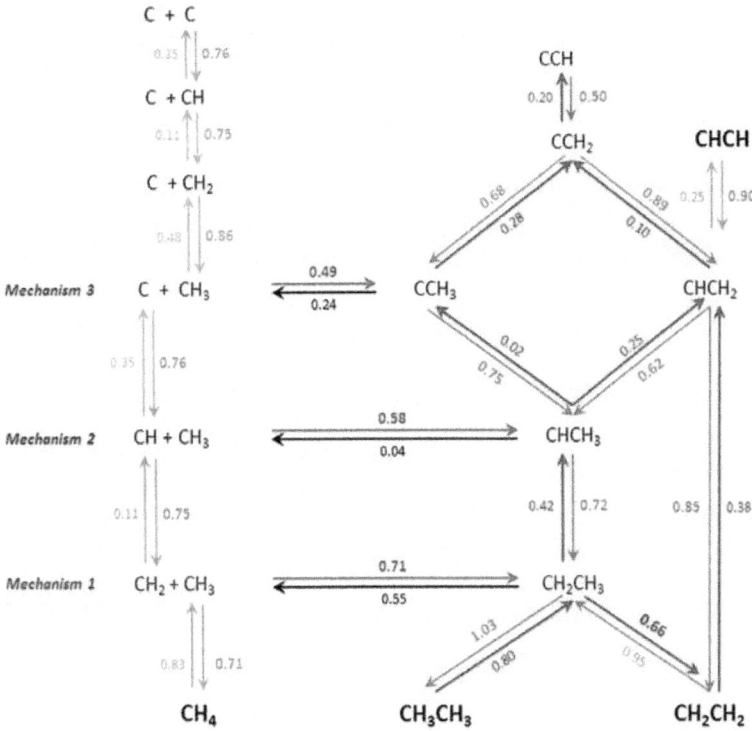

Figure 4: Comprehensive diagram describing a selection of surface reactions of C_2H_x species on Fe(100); the selection considers barriers below 1 eV for C-C bond formation reactions.

COMPUTATIONAL METHOD

Projector augmented-wave (PAW) [63,64,65] total-energy calculations, geometry optimisations with analytic forces, normal mode analysis from finite-difference force-constant computation and diagonalisation of the dynamical matrix; and*climbing-image nudged-elastic-band*[66] minimum-energy path computations are carried out by means of the Vienna *ab Initio*Simulation Package (VASP, version 4.6.31).

A kinetic energy cut-off of 400 eV is used for the plane-wave expansion of the single-particle wave-function. The Perdew-Wang 91 (PW91) [67] exchange and correlation energy-functional and potential of *spin-density functional theory* is employed for the self-consistent RMM-DIIS electronic cycles and for building the provided atomic PAW potentials for C, H, O and Fe. Fractional occupancies of the bands is ensured by the first-order Methfessel-Paxton [68] approach, with a 0.1 eV broadening parameter. The RMM-DIIS method is employed for geometry optimisations, as well as for the electronic cycles.

The Fe(100) surface model uses the slab method with a p(2x2) two-dimensional supercell and four metal layers, as well as a vacuum size equivalent to six metal layers. Adsorbates and co-adsorbates sit on one side of the slab, with the top metal layer and adsorbed species allowed to relax. A 5x5x1 Monkhorst-Pack *k*-point sampling [69] is utilised for the two-dimensional Brillouin-zone integrations of surface calculation. A convergence criterium of 0.01 eV/Å is used for the geometry optimisations, normal mode analysis and minimum-energy path computations. The resulting PW91 *bcc*-Fe bulk lattice constant is 2.8313 Å, to be compared with the experimental value of 2.8665 Å. The resulting *spin-only magnetic moment*for *bcc*-iron is 2.16 μ_B, which is in fair agreement with the experimental value, 2.22 μ_B.

Zero-point energy corrections, as described by Govender *et al.* [58], have been applied throughout. All adsorbate and co-adsorbate stable systems have been characterised as energy minima when all normal modes have real frequency values. Transition states are first order saddle points which contain just one imaginary frequency.

CONCLUSIONS

The formation of C_2 hydrocarbons, starting from adsorbed atomic hydrogen, carbon and hydrogenated C_1 adsorbates on Fe(100), has been described as sequentially endothermic as a function of adsorbate complexity, viz as a new hydrogen or a new carbon is sequentially added. Adsorbed ethyl plays the role of main precursor towards the formation of ethylene and ethane, the products of our model reaction.

A systematic first-principles investigation of elementary reaction steps lead to the build-up of a comprehensive scheme comprising three independent mechanisms, nominally methyl and methylene direct coupling, methyl and methylidyne direct coupling followed by one hydrogenation and methyl and adsorbed carbon direct coupling followed by two hydrogenations. All of these mechanisms have been merged into a single one showing the complexity of unraveling a (simplified) reaction mechanism for even the smallest molecules in the FT reaction. Our results seem to indicate that traditional mechanisms of the FT reaction as described in books and papers are definitely oversimplified.

The fact that all pathways towards methane, ethylene and ethane are endothermic illustrates how the iron (100) surface is too reactive towards the adsorbed hydrocarbon species, such that decomposition of the molecules is favored. Of course, in practice, iron catalysts convert into iron carbides, with a surface of reduced activity compared to that of the metal. In this respect, the present study cannot claim to represent a model for the Fischer-Tropsch active surface, and only presents a reference case for the initial stages of the synthesis with a freshly reduced catalyst.

ACKNOWLEDGMENTS

This work has been sponsored by the National Computing Facilities Foundation NCF (Grant SH-034-11) by granting access to the HUYGENS supercomputer at SARA with the financial support from the Netherlands Organisation for Scientific Research (Grant ECHO 700.59.041) and Sasol Technology, R&D Division.

REFERENCES

1. Chorkendorff, I.; Niemantsverdriet, J.W. Introduction to Catalysis. In *Concepts of Modern Kinetics and Catalysis*; Wiley-VCH: Weinheim, Germany, 2003; Volume 1, pp. 1–21.

2. Claeys, M.; van Steen, E. Basic Studies. In *Vol. 152: Fischer-Tropsch Technology*; Steynberg, A.P., Dry, M.E., Eds.; Elsevier B.V: Amsterdam, The Netherlands, 2004; Volume 1, pp. 601–680.

3. Van der Laan, G.P.; Beenackers, A.A.C.M. Kinetics and selectivity of the Fischer-Tropsch synthesis: A literature review.*Catal. Rev.-Sci. Eng.* 1999, *41*, 255–318.

4. Schulz, H. Short history and present trends of Fischer-Tropsch synthesis. *Appl. Catal. A-Gen.* 1999, *186*, 3–12.

5. Schulz, H.; Claeys, M. Reactions of α-olefins of different chain length added during Fischer-Tropsch synthesis on a cobalt catalyst in a slurry reactor. *Appl. Catal. A-Gen.* 1999, *186*, 91–107.

6. Nijs, H.H.; Jacobs, P.A. New evidence for the mechanism of the Fischer-Tropsch synthesis of hydrocarbons. *J. Catal.* 1980, *66*, 401–411.

7. Henrici-Olive, G.; Olive, S. Mechanism of the Fischer-Tropsch synthesis: Origin of oxygenates. *J. Mol. Catal.* 1984, *24*, 7–13.

8. Joyner, R.W. The role of surface science studies in elucidating the mechanism of the Fischer-Tropsch hydrocarbon synthesis. *Vacuum* 1988, *38*, 309–315.

9. Adesina, A.A.; Hudgins, R.R.; Silveston, P.L. Effect of ethene addition during the Fischer-Tropsch reaction. *Appl. Catal.* 1990, *62*, 295–308.

10. Carter, M.K. A molecular mechanism for Fischer–Tropsch catalysis. *J. Mol. Catal. A-Chem.* 2001, *172*, 193–206.

11. Ndlovu, S.B.; Phala, N.S.; Hearshaw-Timme, M.; Beagly, P.; Moss, J.R.; Claeys, M.; van Steen, E. Some evidence refuting the alkenyl mechanism for chain growth in iron-based Fischer–Tropsch synthesis. *Catal. Tod.* 2002, *71*, 343–349.

12. Lin, Y.-C.; Fan, L.T.; Shafie, S.; Bertok, B.; Friedler, F. Generation of light hydrocarbons through Fischer–Tropsch synthesis: Identification of potentially dominant catalytic pathways via the graph–theoretic method and energetic analysis. *Comput. Chem. Eng.* 2009, *33*, 1182–1186.

13. Biloen, P.; Sachtler, W.M.H. Mechanism of hydrocarbon synthesis over Fischer-Tropsch catalysts. *Adv. Catal.* 1981, *30*, 165–216.

14. Gaube, J.; Klein, H.F. Studies on the reaction mechanism of the Fischer–Tropsch synthesis on iron and cobalt. *J. Mol. Catal. A-Chem.* 2008, *283*, 60–68.

15. Gaube, J.; Klein, H.F. Further support for the two-mechanisms hypothesis of Fischer–Tropsch synthesis. *Appl. Catal. A-Gen.* 2010, *374*, 120–125.

16. Schulz, H.; Riedel, T.; Schaub, G. Fischer–Tropsch principles of co-hydrogenation on iron catalysts. *Top. Catal.* 2005, *32*, 117–124.

17. Maitlis, P.M.; Zanotti, V. The role of electrophilic species in the Fischer–Tropsch reaction. *Chem. Commun.* 2009, 1619–1634.

18. Maitlis, P.M.; Quyoum, R.; Long, H.C.; Turner, M.L. Towards a chemical understanding of the Fischer–Tropsch reaction: alkene formation. *Appl. Catal. A-Gen.* 1999, *186*, 363–374.

19. Dry, M.E. Practical and theoretical aspects of the catalytic Fischer-Tropsch process. *Appl. Catal. A-Gen.* 1996, *138*, 319–344.

20. Dry, M.E. Catalytic aspects of industrial Fischer-Tropsch synthesis. *J. Mol. Catal.* 1982, *17*, 133–144.

21. Davis, B.H. Fischer–Tropsch synthesis: Current mechanism and futuristic needs. *Fuel Process. Technol.* 2001, *71*, 157–166.

22. Davis, B.H. Fischer–Tropsch Synthesis: Reaction mechanisms for iron catalysts. *Catal. Tod.* 2009, *141*, 25–33.

23. Van der Laan, G.P. Beenackers, A.A.C.M. Intrinsic kinetics of the gas–solid Fischer–Tropsch and water gas shift reactions over a precipitated iron catalyst. *Appl. Catal. A-Gen.* 2000, *193*, 39.

24. Wang, Y.N.; Ma, W.P.; Lu, Y.J.; Yang, J.; Xu, Y.Y.; Xiang, H.W.; Li, Y.W.; Zhao, Y.L.; Zhang, B.J. Kinetics modelling of Fischer–Tropsch synthesis over an industrial Fe–Cu–K catalyst. *Fuel* 2003, *82*, 195–213.

25. Ciobîca, I.M.; Kramer, G.J.; Ge, Q.; Neurock, M.; van Santen, R.A. Mechanisms for chain growth in Fischer–Tropsch synthesis over Ru(0001). *J. Catal.* 2002, *212*, 136–144.

26. Ciobîca, I.M. The Molecular Basis of the Fischer Tropsch Reaction. Ph.D. thesis, Eindhoven University of Technology, Eindhoven, The Netherlands, 2002.

27. Van Santen, R.A.; Ciobîca, I.M.; van Steen, E.; Ghouri, M.M. Mechanistic issues in Fischer-Tropsch catalysis. *Adv. Catal.* 2011, *54*, 127–187.

28. Cheng, J.; Gong, X.Q.; Hu, P.; Lok, C.M.; Ellis, P.; French, S. A quantitative determination of reaction mechanisms from density functional theory calculations: Fischer–Tropsch synthesis on flat and stepped cobalt surfaces. *J. Catal.* 2008, *254*, 285–295.

29. Cheng, J.; Hu, P.; Ellis, P.; French, S.; Kelly, G.; Lok, C.M. Chain growth mechanism in Fischer–Tropsch synthesis: A DFT study of C–C coupling over Ru, Fe, Rh, and Re surfaces. *J. Phys. Chem. C* 2008, *112*, 6082–6086.

30. Cheng, J.; Hu, P.; Ellis, P.; French, S.; Kelly, G.; Lok, C.M. A DFT study of the chain growth probability in Fischer–Tropsch synthesis. *J. Catal.* 2008, *257*, 221–228.

31. Liu, Z.P.; Hu, P. A new insight into Fischer-Tropsch synthesis. *J. Am. Chem. Soc.* 2002, *124*, 11568–11569.

32. Zhuo, M.; Tan, K.F.; Borgna, A.; Saeys, M. Density functional theory study of the CO insertion mechanism for Fischer–Tropsch synthesis over Co catalysts. *J. Phys. Chem. C* 2009, *113*, 8357–8365.

33. Zhao, Y.H.; Sun, K.J.; Ma, X.F.; Liu, J.X.; Sun, D.P.; Su, H.Y.; Li, W.X. Carbon chain growth by formyl insertion on rhodium and cobalt catalysts in syngas conversion. *Angew. Chem. Int. Ed.* 2011, *50*, 5335–5338.

34. Shetty, S.G.; Ciobîca, I.M.; Hensen, E.J. M.; van Santen, R.A. Site regeneration in the Fischer–Tropsch synthesis reaction: a synchronized CO dissociation and C–C coupling pathway. *Chem. Commun.* 2011, *47*,

9822–9824.

35. Cao, D.-B.; Li, Y.-W.; Wang, J.; Jiao, H. Chain growth mechanism of Fischer–Tropsch synthesis on $Fe_5C_2(001)$. *J. Mol. Catal. A-Chem.* 2011, *346*, 55–69.

36. Cheng, J.; Hu, P.; Ellis, P.; French, S.; Kelly, G.; Lok, C.M. Density functional theory study of Iron and Cobalt carbides for Fischer-Tropsch synthesis. *J. Phys. Chem. C* 2010, *114*, 1085–1093.

37. Lo, J.M.H.; Ziegler, T. Theoretical studies of the formation and reactivity of C2 hydrocarbon species on the Fe (100) surface. *J. Phys. Chem. C* 2007, *111*, 13149–13162.

38. Lee, G.D.; Han, S.W.; Yu, J.J.; Ihm, J. Catalytic decomposition of acetylene on Fe (001): A first-principles study. *Phys. Rev. B* 2002, *66*, 081403.

39. Anderson, A.B.; Mehandru, S.P. Acetylene adsorption to Fe (100), (110), and (111) surfaces; structures and reactions.*Surf. Sci.* 1984, *136*, 398–418.

40. Mehandru, S.P.; Anderson, A.B. Dependence of carbon-carbon and carbon-hydrogen bond activation on d band position: acetylene on platinum(111) and iron(100). An electrochemical model. *J. Am. Chem. Soc.* 1985, *107*, 844–849.

41. Cheng, J.; Hu, P.; Ellis, P.; French, S.; Kelly, G.; Lok, C.M. A DFT study of the transition metal promotion effect on ethylene chemisorption on Co (0001). *Surf. Sci.* 2009, *603*, 2752–2758.

42. Xu, L.S.; Ma, Y.S.; Wu, Z.F.; Chen, B.H.; Yuan, Q.; Huang, W.X. A photoemission study of ethylene decomposition on a Co(0001) surface: Formation of different types of carbon species. *J. Phys. Chem. C* 2012, *116*, 4167–4174.

43. Swart, J.C.W.; Ciobîca, I.M.; van Santen, R.A.; van Steen, E. Intermediates in the formation of graphitic carbon on a flat FCC-Co (111) surface. *J. Phys. Chem. C* 2008, *112*, 12899–12904.

44. Swart, J.C.W.; van Steen, E.; Ciobîca, I.M.; van Santen, R.A. Interaction of graphene with FCC-Co(111). *Phys. Chem. Chem. Phys.* 2009, *11*, 803–807.

45. Ciobîca, I.M.; van Santen, R.A.; van Berge, P.J.; de Loosdrecht, J.V. Adsorbate induced reconstruction of cobalt surfaces. *Surf. Sci.* 2008, *602*, 17–27.

46. van Helden, P.; Ciobîca, I.M. A DFT study of carbon in the subsurface layer of cobalt surfaces. *ChemPhysChem.* 2011,*12*, 2925–2928.

47. Bernardo, C.G.P.M.; Gomes, J.A.N.F. The adsorption of ethylene on the (100) surfaces of platinum, Palladium and nickel: A DFT study. *J. Mol. Struct.(Theochem)* 2001, *542*, 263–271.

48. Whitten, J.L.; Yang, H. Theoretical studies of surface reactions on metals: I. Ethyl to ethylene conversion on Ni(100). II. Photodissociation of methane on platinum. *Catal. Tod* 1999, *50*, 603–612.

49. Ostrom, H.; Foehlisch, A.; Nyberg, M.; Weinelt, M.; Heske, C.; Pettersson, L.G.M.; Nilsson, A. Ethylene on Cu (110) and Ni (110): electronic structure and bonding derived from X-ray spectroscopy and theory. *Surf. Sci.* 2004, *559*, 85–99.

50. Li, B.; Bao, S.N.; Zhuang, Y.Y.; Cao, P.L. The adsorption geometry of ethylene on the Ni (110) surface. *Acta Phys. Sin.*2003, *52*, 202–206.

51. Li, B.; Bao, S.N.; Cao, P.L. Adsorption geometry of C_2H_4 and C_2H on Ni (110) surface. *Acta Phys. Sin.* 2005, *54*, 5784–5790.

52. Gutdeutsch, U.; Birkenheuer, U.; Bertel, E.; Cramer, J.; Boettger, J.C.; Roesch, N. On the adsorption site of ethylene at the Ni (110) surface: A combined experimental and theoretical study involving the unoccupied band structure. *Surf. Sci.* 1996, *345*, 331–346.

53. Vang, R.T.; Honkala, K.; Dahl, S.; Vestergaard, E.K.; Schnadt, J.; Laegsgaard, E.; Clausen, B.S.; Nørskov, J.K.; Besenbacher, F. Ethylene dissociation on flat and stepped Ni (111): A combined STM and DFT study. *Surf. Sci.* 2006,*600*, 66–77.

54. Bromfield, T.C.; Curulla-Ferré, D.; Niemantsverdriet, J.W. A DFT study of the adsorption and dissociation of CO on Fe (100): Influence of surface coverage on the nature of accessible adsorption states. *ChemPhysChem* 2005, *6*, 254–260.

55. Scheijen, F.J.E.; Curulla-Ferré, D.; Niemantsverdriet, J.W. Adsorption and dissociation of CO on body-centered cubic transition metals and alloys: Effect of coverage and scaling relations. *J. Phys. Chem. C* 2009, *113*, 11041–11049.

56. Elahifard, M.R.; Perez Jigato, M.; Niemantsverdriet, J.W. Direct *versus* hydrogen-assisted CO dissociation on the Fe (100) Surface: a DFT study. *ChemPhysChem* 2012, *13*, 89–91.

57. Govender, A.; Curulla-Ferré, D.; Niemantsverdriet, J.W. The surface chemistry of water on Fe (100): A density functional theory study. *ChemPhysChem* 2012, *13*, 1583–1590.

58. Govender, A.; Curulla-Ferré, D.; Niemantsverdriet, J.W. A density functional theory study on the effect of zero-point energy corrections on

the methanation profile on Fe (100). *ChemPhysChem* 2012, *13*, 1591–1596.

59.   Niemantsverdriet, J.W.; van der Kraan, A.M.; van Dijk, W.L.; van der Baan, H.S. Behavior of metallic iron catalysts during Fischer-Tropsch synthesis studied with Mössbauer spectroscopy, x-ray diffraction, carbon content determination, and reaction kinetic measuremen. *J. Phys. Chem.* 1980, *84*, 3363–3370.

60.   de Smit, E.; Cinquini, F.; Beale, A.M.; Safonova, O.V.; van Beek, W.; Sautet, P.; Weckhuysen, B.M. Stability and Reactivity of ε-χ- θ Iron Carbide Catalyst Phases in Fischer-Tropsch Synthesis: Controlling μC. *J. Am. Chem. Soc.* 2010,*132*, 14928–14941.

61.   Steynberg, P.J.; van den Berg, J.A.; van Rensburg, W.J. Bulk and surface analysis of Hägg Fe carbide (Fe5C2): A density functional theory study. *J. Phys. Condens. Matter* 2008, *20*, 064238.

62.   Gracia, J.M.; Prinsloo, F.F.; Niemantsverdriet, J.W. Mars-van Krevelen-like mechanism of CO hydrogenation on an iron carbide surface. *Catal. Lett.* 2009, *133*, 257–261.

63.   Blochl, P.E. Projector augmented-wave method. *Phys. Rev. B* 1994, *50*, 17953–17979.

64.   Kresse, G.; Furthmuller, J. Efficiency of ab-initio total energy calculations for metals and semiconductors using a plane-wave basis set. *Comput. Mater. Sci.* 1996, *6*, 15–50.

65.   Kresse, G.; Joubert, D. From ultrasoft pseudopotentials to the projector augmented-wave method. *Phys. Rev. B* 1999,*59*, 1758–1775.

66.   Henkelman, G.; Uberuaga, B.P.; Jonsson, H. A climbing image nudged elastic band method for finding saddle points and minimum energy paths. *J. Chem. Phys.* 2000, *113*, 9901–9904.

67.   Perdew, J.P.; Chevary, J.A.; Vosko, S.H.; Jackson, K.A.; Pederson, M.R.; Singh, D.J.; Fiolhais, C. Atoms, molecules, solids, and surfaces: Applications of the generalized gradient approximation for exchange and correlation. *Phys. Rev. B* 1992, *46*, 6671–6687.

68.   Methfessel, M.; Paxton, A.T. High-precision sampling for Brillouin-zone integration in metals. *Phys. Rev. B* 1989, *40*, 3616–3621.

69.   Monkhorst, H.J.; Pack, J.D. Special points for Brillouin-zone integrations. *Phys. Rev. B* 1976, *13*, 5188–5192.

70.   Govender, A. Towards a mechanism for the Fischer-Tropsch synthesis on Fe(100) using density functional theory. Ph.D. thesis, Eindhoven University of Technology, Eindhoven, The Netherlands, 2010.

71. Bernasek, S.L. Reaction of small molecules at well-characterized iron surfaces. *Ann. Rev. Phys. Chem.* 1993, *44*, 265–298.

72. Hung, W.H.; Bernasek, S.L. Adsorption and decomposition of ethylene and acetylene on Fe (100). *Surf. Sci.* 1995, *339*, 272–290.

Chapter 14

THE STRUCTURES AND PROPERTIES OF Y-SUBSTITUTED MG$_2$NI ALLOYS AND THEIR HYDRIDES: A FIRST-PRINCIPLES STUDY

Yuanyuan Li, Gaili Sun, Yiming Mi

School of Chemistry and Chemical Engineering, Shanghai University of Engineering Science, Shanghai, China

ABSTRACT

The structures and properties of Y-substituted Mg2Ni alloys and the corresponding hydrides are investigated by a first-principles plane-wave pseudopotential method within density functional theory. Results show that Mg2Ni has the best structural stability when Y atom occupies the Mg(6f) lattice sites. The calculated enthalpies of formation for Mg2Ni, Mg2NiH4 and Mg15YNi8H32 are −51.612, −64.667 and −62.554 kJ/mol, respectively. It is implied that the substitution of Y alloying destabilizes the stability of the hydrides. Moreover, the dissociated energies of H atoms are decreased significantly, indicating that Y alloying benefits the improvement of the dehydrogenating properties of Mg2Ni hydrides. The calculation and analysis of the electronic structures suggest that there is a stronger interaction between H and Ni atoms than the interaction between H and Mg atoms in Mg2NiH4. However, the Ni-H bond is weakened by the substitution of Y. Therefore, the substitution is an effective technique to decrease the structural stability of the hydrides and benefit for hydrogen storage.

INTRODUCTION

Due to rich reserves in the earth's crust, high hydrogen capacity (3.6 wt%), light weight and low cost, Mg$_2$Ni- type alloy hydrides remain as attractive hydrogen storage materials [1] [2] . However, the practical application of the alloy materials has not been achieved because of unfavorable thermodynamics, poor hydrogenation/dehy- drogenation kinetics and releasing undesirable by-products [3] .

Many researches have been devoted to overcoming these drawbacks and improving the properties of hydrogen storage via modifying microstructure by mechanical alloying [4] , alloying with other elements [5] [6] , adding catalysts [7] and composite structures [8] . The effects of transition metals including Cu, Co, Mn, Y, Ti, N band Crelements [9] -[12] on the hydrogen storage properties of Mg-based metal hydrides are investigated and discovered that the properties of hydrogen storage are improved by alloying with a small amount of transition metals in different degrees.

It is believed that alloying of Mg_2Ni with transition metals is beneficial to improve the hydrogenating and dehydrogenating kinetics. The electronic structure of element Y is $4d^15s^2$ and it can be incorporated into the metal boride. In addition, its chemical properties and physical performance are similar to La which can be used as an alloy element for hydrogen storage. The density and cohesive energy of Y atom are also relatively small. Therefore, Y has great potential to improve the performance of Mg_2Ni alloy and its hydride. Kalinichenka et al. [13] studied that Y can be solved in Mg_2Ni and the Mg-Ni-Y alloy exhibits higher dehydrogenation rates comparing with that of the Mg-Ni alloy. Song et al. [14] reported the microstructure and the hydrogenation properties of melt-spun $Mg_{67}Ni_{33-x}Y_x$ alloys and found that the hydrogen storage capacity and kinetics of Mg_2Ni are improved with Y doping. Zhang et al. [15] investigated that the substitution of Y for Mg had an insignificant effect on the activation ability of the Mg_2Ni-type alloys, but it dramatically improved the cycle stability of the as- milled alloys. These experiments proved that Y plays an important role in improving the properties of Mg_2Ni alloy for hydrogen storage. Thus, my understanding is that, alloying of Mg_2Ni with Y can be expected to improve some performances of hydrogen absorption/desorption capacity and kinetics significantly.

In recent years, a number of theoretical investigations about the doped/substituted complex hydrides using first-principles calculations have been reported [16] -[19] . A first-principles study on the structures and properties of hydrogen storage alloy Mg_2Ni, of aluminum and silver substituted alloys $Mg_{2-x}M_xNi$ (M = Al and Ag), and of their hydrides Mg_2NiH_4, $Mg_{2-x}M_xNiH_4$ was performed by Zeng et al. [20] . Their results show that the hydrogen storage capacity is decreased by the substitution and the substitution destabilizes the hydrides. However, there are no available theoretical reports about the structures and properties of Y substituted Mg_2Ni alloys and their respective hydrides to the authors' knowledge. The models are new for the materials to store hydrogen.

We focus primarily on the stable configuration of Mg$_2$Ni alloys with Y substitution and determine the optimum position of Y. Furthermore, the energies, enthalpies of formation and electronic structures of Y alloying Mg$_2$Ni and its hydrides are also calculated and analyzed using a first-principles plane-wave pseudo potential simulations based on the density functional theory in this paper. These simulations are beneficial to improve our understanding of the effects of substitution on the properties of Mg$_2$Ni, and of the design about advanced magnesium-based hydrogen storage materials.

COMPUTATIONAL DETAILS

Computational Model

The crystal structure of Mg$_2$Ni is hexagonal and its space group is P6$_2$22 (No.180) [21] , as shown inFigure 1(a). The lattice constants of Mg$_2$Ni are a = b = 5.205 Å, c = 13.236 Å, $\alpha = \beta = 90°$, $\gamma = 120°$. There are 12 Mg and 6 Ni atoms existing in the unit cell of Mg$_2$Ni. The spatial positions Mg and Ni atoms are respectively 6f(0.5, 0, 0.1187), 6i(0.16, 0.324, 0) and 3b(0, 0, 0.5), 3d(0.5, 0, 0.5). Single Y atom substituting for Mg and Ni atoms are investigated respectively. Moreover, it has been shown that Mg$_2$NiH$_4$ forms readily by hydrogenating the alloy Mg$_2$Ni [22] . The space group of Mg$_2$NiH$_4$ is monoclinic C2/c (No.15) and the lattice constants are a = 14.343 Å, b = 6.404 Å, c = 6.483 Å, $\beta = 113.52°$, as shown in Figure 1(b). 16 Mg, 8 Ni and 32 H atoms are in the unit cell of Mg$_2$NiH$_4$ where Mg occupying the 8f, 4e, 4e sites and Ni the 8f site and H the 8f, 8f, 8f, 8f sites [23] [24] . The new systems of Y alloying Mg$_2$NiH$_4$ are studied.

Computational Method

All the density-functional theory (DFT) calculations are performed using a plane-wave basis set with the projector augmented plane wave (PAW) method as implemented in the Vienna ab initio simulation package (VASP)

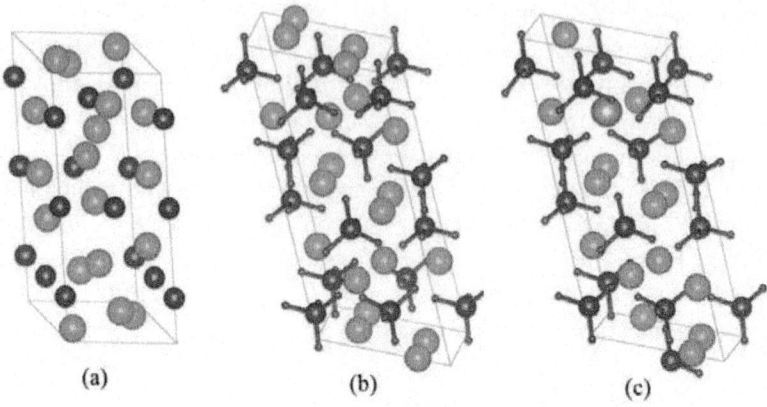

(a)   (b)   (c)

Figure 1: Structures of (a) Mg_2Ni, (b) Mg_2NiH_4, (c) $Mg_{15}YNi_8H_{32}$ (where green, purple, red and orange balls denote Mg, Ni, H and Y atoms, respectively).

[25] - [27] . Projector Augmented Wave (PAW) potentials are used to treat the core-valence interaction [28] . The PW91 [29] [30] generalized gradient approximation (GGA) is employed for the exchange-correlation functional. The electronic wave functions are expanded by plane waves with a kinetic energy cutoff of 350 eV to attain the required convergence. All of the self-consistent loops are iterated until the total energy difference of the systems between the adjacent iterating steps is less than 10^{-7} eV. The Brillouin zone is sampled by $6 \times 6 \times 2$ mesh points in k-space based on Monkhorst-Pack scheme [31] for all systems. The valence electrons of 1s for H, 2p and 3s for Mg, 3p, 3d and 4s for Ni, and 4d and 5s for Y are considered in the calculations.

RESULTS AND DISCUSSIONS

The Structure of Substituted Mg_2Ni by Y

In order to check the accuracy of the calculations, we first optimize the structure of Mg_2Ni alloy and its hydride and compare the calculated lattice parameters with those determined experimentally. Then we consider the substitution of Mg and Ni by Y in independent spatial positions respectively. To single out a scenario that is most likely responsible for the stabilization of the crystal structure, the lattice parameters and enthalpies of formation ΔH for each case are calculated. The ΔH is calculated by taking the difference in total electronic energy of the products and the reactants [32] :

$$\Delta H = \sum_{products} E_p - \sum_{reactants} E_r$$

(1)

In the case of the crystal structure $^{Mg_x Y_y Ni_z}$ which including xMg, yY, zNi, the enthalpies of formation are calculated by the following equation:

$$\Delta H\left(Mg_x Y_y Ni_z\right) = E\left(Mg_x Y_y Ni_z\right) - xE\left(Mg\right) - yE\left(Y\right) - zE\left(Ni\right) \tag{2}$$

where $E\left(Mg_x Y_y Ni_z\right)$ refers to the total energy of substituted Mg₂Ni by Y.

$E(Mg)$, $E(Y)$ and $E(Ni)$ are the energy of every atom in HCP Mg, HCP Y and FCC Ni crystals, respectively. x, y, z are the numbers of Mg, Y and Ni atoms, respectively. Through the calculation, the values of $E(Mg)$, $E(Y)$ and $E(Ni)$ are −1.595, −6.379 and −5.415 eV, respectively.

Table 1 displays the volume, lattice constant, total energy and enthalpies of formation of all the structures including Mg₂Ni, substituted Mg₂Ni by Y and their hydrides. The lattice constants of Mg₂Ni after geometry optimization are a = b = 5.180 Å, c = 13.232 Å, which agree well with the experimental data a = b = 5.205 Å, c = 13.236 Å [21] . The enthalpy of formation of Mg₂Ni is −3.211 eV, which means that the unit cell of Mg₂Ni is −51.612 kJ/mol. It is very close to the experimental values −51.9 kJ/mol [33] . When Y atom is added into Mg₂Ni, all the volumes of crystal structures will increase compared with the original structures. Moreover, it can be clearly observed that when the position of Mg (6f) is occupied by Y atom in Mg₂Ni, the total energy and the enthalpy of formation are the minimum. It indicates that the structure of Mg₁₁Y(6f) Ni₆ has the optimal stabilization among all the substituted structures.

Table 1: Volume, lattice constant, total energy, enthalpy of formation of Mg₂Ni, Y-substituted Mg₂Ni and their hydrides

| Alloy model | Volume (Å³) | Lattice constant (Å) | | | Total energy (eV) | Enthalpy of formation (eV) |
		a	b	c		
Mg₂Ni(exp.) [21]	310.55	5.205	5.205	13.236	—	—
Mg₂Ni(cal.)	307.43	5.180	5.180	13.232	−54.836	−3.211
Mg₂NiH₄(exp.) [23]	545.91	14.343	6.404	6.483	—	—
Mg₂NiH₄(cal.)	534.63	14.234	6.352	6.434	−192.248	−0.670
Mg₁₁Y(6f)Ni₆	316.88	5.222	5.193	13.437	−59.975	−3.565
Mg₁₁Y(6i)Ni₆	317.74	5.263	5.239	13.328	−59.954	−3.544
Mg₁₂Y(3b)Ni₅	341.89	5.135	5.266	14.481	−53.511	−0.921
Mg₁₂Y(3d)Ni₅	341.23	5.130	5.253	14.510	−53.730	−1.140
Mg₁₅YNi₈H₃₂	543.75	14.303	6.396	6.478	−197.148	−0.649

The Properties of Substituted Mg_2NiH_4 by Y

Based on the stable structure of $Mg_{11}Y(6f)Ni_6$, we study the properties of substituted Mg_2NiH_4 by Y. Firstly, We have proved that the theoretical lattice constants and internal atomic positions of Mg_2NiH_4 are in good agreement with experimental results [23] . The states are displayed in Table 1. Various substitutive positions of Mg are considered. We find that the total energy of each new structure is very close. Thereby, a reasonable structure $Mg_{15}YNi_8H_{32}$ is selected to be investigated in detail, as shown in Figure 1(c).

In order to research the effects of Y on the properties of Mg_2NiH_4, We calculate the enthalpies of formation of Mg_2NiH_4 and $Mg_{15}YNi_8H_{32}$ respectively. In general, the formation of Mg_2NiH_4 can be expressed by the following reaction:

$$Mg_2Ni + 2H_2 = Mg_2NiH_4 \tag{3}$$

The enthalpy of formation of Mg_2NiH_4 can be expressed in Equation (4):

$$\Delta H\left(Mg_{16}Ni_8H_{32}\right) = \frac{1}{16}E\left(Mg_{16}Ni_8H_{32}\right) - \frac{1}{12}E\left(Mg_{12}Ni_6\right) - E\left(H_2\right) \tag{4}$$

In the same way, the reaction of formation and the enthalpy of formation of $Mg_{15}YNi_8H_{32}$ can be respectively written as Equations (5) and (6):

$$\frac{1}{12}Mg_{11}YNi_6 + \frac{1}{12}Mg + \frac{11}{12}Y + H_2 = \frac{1}{16}Mg_{15}YNi_8H_{32} + \frac{1}{16}Mg + \frac{15}{16}Y \tag{5}$$

$$\Delta H\left(Mg_{15}YNi_8H_{32}\right) = \frac{1}{16}E\left(Mg_{15}YNi_8H_{32}\right) + \frac{1}{48}E(Y) - \frac{1}{12}E\left(Mg_{11}YNi_6\right) - \frac{1}{48}E(Mg) - E\left(H_2\right) \tag{6}$$

where $E\left(Mg_{16}Ni_8H_{32}\right)$, $E\left(Mg_{12}Ni_6\right)$, $E\left(Mg_{15}YNi_8H_{32}\right)$ and $E(Mg_{11}YNi_6)$ are the total energy of Mg_2NiH_4, Mg_2Ni, $Mg_{15}YNi_8H_{32}$ and $Mg_{11}Y(6f)Ni_6$, respectively.

$E(H_2)$ is the energy of free H_2 molecule. The calculated results are also shown in Table 1. For pure Mg_2NiH_4, the enthalpy of formation is −64.667 kJ/mol which coincides closely with the experimental result −64.4 ± 4.2 kJ/mol reported by Reilly et al. [22] . Furthermore, the enthalpy of formation of $Mg_{15}YNi_8H_{32}$ is −62.554 kJ/mol which is higher than that of pure Mg_2NiH_4. It can be clearly seen that the introduction of Y atom has effects on the destabilization of Mg_2NiH_4 in terms of energy. This is energetically favorable to perform the dehydrogenation reaction of substituted Mg_2NiH_4 by Y.

To make further investigation about the performance of dehydrogenation, we calculate the energies of Mg_2NiH_4 and $Mg_{15}YNi_8H_{32}$ which dissociate the nearest 2 H atoms around Ni atoms. The dehydrogenation energy is calculated by Equation (7):

$$\Delta E = E\left(Mg_{15}\,M\,Ni_8H_{30}\right) - E\left(Mg_{15}\,M\,Ni_8H_{32}\right) + E\left(H_2\right), \left(M = Mg, Y\right) \tag{7}$$

The results are shown in Table 2. From Table 2 we can see that the addition of Y clearly decreases the dehydrogenation energy of Mg$_2$NiH$_4$ by about 47% to 0.983 eV. It suggests that although Y atom has poor effects on the destabilization of Mg$_2$Ni, it breaks down the stability of Mg$_2$NiH$_4$ positively and improve the dehydrogenation kinetics of Mg$_2$NiH$_4$ which as one of the hydrogen storage materials.

Electronic Structure

In order to further understand the effects of Y atom on the dehydrogenation properties of Mg$_2$NiH$_4$ alloy, the electronic properties of Mg$_2$Ni and Mg$_{15}$YNi$_8$H$_{32}$ are studied by calculating total density of states (DOS) and partial density of states (PDOS). Figure 2 displays the DOS and PDOS of Mg$_2$Ni and Mg$_{15}$YNi$_8$H$_{32}$ alloys.

Form Figure 2(a) we can see that there are two main peaks in total density of states below Fermi level. The bonding electron of the energy region between −9.2 and −3.7 eV is mainly dominated by Hs, Nis and Nid orbits, partial Mgs orbit. It is implied that H atoms tend to bond with Ni rather than Mg atoms in the structure of Mg$_2$NiH$_4$. The result is in correspondence with the conclusion that the interaction Ni-H is stronger than that of Mg-H which studied by Jasen [34] . There is a major contribution with Ni p, Ni d and Mg s orbits in the region from −2.4 eV to Fermi level. This indicates that Mg and Ni atoms have hybridization which keeps the structure of Mg$_2$NiH$_4$ stable. In addition, Ni d orbit plays the dominating role in the bonding electron.

Table 2: Calculated dehydrogenation energies of Mg$_{15}$MNi$_8$H$_{32}$ (M = Mg, Y) (eV).

M	$E(Mg_{15}MNi_8H_{32})$	$E(Mg_{15}MNi_8H_{30})$	ΔE
Mg	−192.248	−183.617	1.856
Y	−197.148	−189.390	0.983

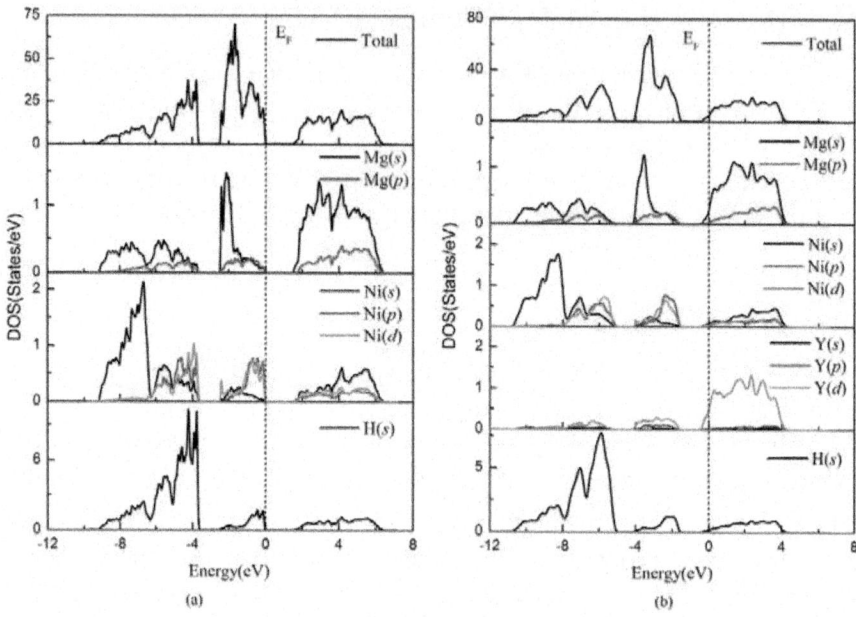

Figure 2: Total density of states and partial density of states of (a) Mg_2NiH_4, (b) $Mg_{15}YNi_8H_{32}$.

Compared to pure Mg_2NiH_4, the enthalpy of formation and dehydrogenation energy change markedly due to the substituted Mg_2NiH_4 by Y. Figure 2(b) displays that below Fermi level $Mg_{15}YNi_8H_{32}$ has two main bonding peaks from -10.7 to -5.2 eV and -4.1 to -1.6 eV. It is not difficult to find that all the bonding peaks in total density of states move to the energy of deep potential well and the number of bonding electron reduces comparing to Mg_2NiH_4. It demonstrates that the substitution of Y alloying weakens the interaction of the atoms and destabilizes the structure of the hydride. The effects of Yd orbit on the bonding electron are significant especially for the energy region from -4.1 to -1.6 eV. What is more, Yp and d orbits contribute to the bonding electron and have mutual interaction with Nip and d orbits. It is also worth noting that the overlapping region between Nid and Hs orbits decreases obviously. It means that the interaction between Ni and H atoms become weak.

CONCLUSION

We have investigated the structure and properties of substituted Mg_2Ni alloys by Y and the corresponding hydrides. The structure parameter, enthalpy of formation, dehydrogenation energy and electronic structure are calculated by the first-principles method based on density functional theory in this paper.

Through analyzing the simulation results, we can draw the conclusions that when Y atom occupies the Mg(6f) lattice site, the structure of Mg$_2$Ni is the optimal stable. The substitution of Y destabilizes the stability of Mg$_2$NiH$_4$ and decreases the dissociated energies of H atoms due to the Ni-H bond weakened by Y. Therefore, the method of substitution is in favor of the dehydrogenation reaction for Mg-based hydrides as hydrogen storage materials. Moreover, we will continue to perfect this respect, for instance, whether the effect of Y elements in the case of different numbers of Y metals and different substituents will change.

ACKNOWLEDGEMENTS

This work was supported by Innovation Program of Shanghai Municipal Education Commission, China (10YZ172) and Subjects Construction Program of Shanghai University of Engineering Science, China (2012gp43) and Graduated Innovative Research Project of Shanghai University of Engineering Science (E1-0903-14-01107- 14KY0411).

REFERENCES

1. Schlapbach, L. and Züttel, A. (2001) Hydrogen-Storage Materials for Mobile Applications. Nature, 414, 353-358. http://dx.doi.org/10.1038/35104634

2. Jain, I.P., Lal, C. and Jain, A. (2010) Hydrogen Storage in Mg: A Most Promising Material. International Journal of Hydrogen Energy, 35, 5133-5144. http://dx.doi.org/10.1016/j.ijhydene.2009.08.088

3. Liu, T., Wang, C.X. and Wu, Y. (2014) Mg-Based Nano-composites with Improved Hydrogen Storage Performances. International Journal of Hydrogen Energy,39, 14262-14274. http://dx.doi.org/10.1016/j.ijhydene.2014.03.125

4. Shang, C.X. and Guo, Z.X. (2007) Structural and Desorption Characterisations of Milled (MgH2+ Y, Ce) Powder Mixtures for Hydrogen Storage. International Journal of Hydrogen Energy, 32, 2920-2925. http://dx.doi.org/10.1016/j.ijhydene.2006.11.035

5. Shao, H., Asano, K., Enoki, H. and Akiba, E. (2009) Preparation and Hydrogen Storage Properties of Nanostructured Mg-Ni BCC Alloys. Journal of Alloys and Compounds, 477, 301-306. http://dx.doi.org/10.1016/j.jallcom.2008.11.004

6. Kim, H., Nakamura, J., Shao, H., Nakamura, Y., Akiba, E., Chapman, K.W., Chupas, P.J. and Proffen, T. (2011) Insight into the Hydrogenation Properties of Mechanically Alloyed Mg50Co50 from the Local Structure.

Journal of Physical Chemistry C, 115, 20335-20341. http://dx.doi.org/10.1021/jp207197k

7. Barkhordarian, G., Klassen, T. and Bormann, R. (2003) Fast Hydrogen Sorption Kinetics of Nanocrystalline Mg Using Nb2O5 as Catalyst. Scriptamaterialia,49,213-217.http://dx.doi.org/10.1016/s1359-6462(03)00259-8

8. Oelerich, W., Klassen, T. and Bormann, R. (2001) Metal Oxides as Catalysts for Improved Hydrogen Sorption in Nanocrystalline Mg-Based Materials. Journal of Alloys and Compounds, 315, 237-242. http://dx.doi.org/10.1016/S0925-8388(00)01284-6

9. Zhang, Y., Zhao, C., Yang, T., Shang, H.W., Xu, C. and Zhao, D.L. (2013) Comparative Study of Electrochemical Performances of the As-Melt Mg20Ni10-xMx (M = None, Cu, Co, Mn; x = 0, 4) Alloys Applied to Ni/Metal Hydride (MH) Battery. Journal of Alloys and Compounds, 555, 131-137. http://dx.doi.org/10.1016/j.jallcom.2012.12.016

10. Zhang, Q.A., Zhang, L.X. and Wang, Q.Q. (2013) Crystallization Behavior and Hydrogen Storage Kinetics of Amorphous Mg11Y2Ni2 Alloy. Journal of Alloys and Compounds, 551, 376-381. http://dx.doi.org/10.1016/j.jallcom.2012.11.046

11. Cui, J., Liu, J., Wang, H., Ouyang, L.Z., Sun, D.L., Zhu, M. and Yao, X.D. (2014) Mg-TM (TM: Ti, Nb, V, Co, Mo or Ni) Core-Shell Like Nanostructures: Synthesis, Hydrogen Storage Performance and Catalytic Mechanism. Journal of Materials Chemistry A, 2, 9645-9655. http://dx.doi.org/10.1039/c4ta00221k

12. Vyas, D., Jain, P., Agarwal, G., Ankur, J. and Jain, I.P. (2012) Hydrogen Storage Properties of Mg2Ni Affected by Cr Catalyst. International Journal of Hydrogen Energy, 37, 16013-16017. http://dx.doi.org/10.1016/j.ijhydene.2012.08.039

13. Kalinichenka, S., Rontzsch, L., Baehtz, C. and Kieback, B. (2010) Hydrogen Desorption Kinetics of Melt-Spun and Hydrogenated Mg90Ni10 and Mg80Ni10Y10 Using in Situ Synchrotron, X-Ray Diffraction and Thermogravimetry. Journal of Alloys and Compounds, 496, 608-613.http://dx.doi.org/10.1016/j.ijhydene.2012.08.039

14. Song, W., Li, J., Zhang, T., Kou, H. and Xue, X. (2014) Microstructure and Tailoring Hydrogenation Performance of Y-Doped Mg2Ni Alloys. Journal of Power Sources, 245, 808-815. http://dx.doi.org/10.1016/j.jpowsour.2013.07.049

15. Zhang, Y., Zhang, P., Yuan, Z., Yang, T., Qi, Y. and Zhao, D.L. (2015) An Investigation on Electrochemical Hydrogen Storage Performances

of Mg-Y-Ni Alloys Prepared by Mechanical Milling. Journal of Rare Earths, 33, 874-883.http://dx.doi.org/10.1016/S1002-0721(14)60499-3

16. Haussermann, U., Blomqvist, H. and Noréus, D. (2002) Bonding and Stability of the Hydrogen Storage Material Mg2NiH4. Inorganic Chemistry, 41, 3684-3692. http://dx.doi.org/10.1021/ic0201046

17. Van Setten, M.J., De Wijs, G.A. and Brocks, G. (2007) Ab Initio Study of the Effects of Transition Metal Doping of Mg2NiH4. Physical Review B, 76, Article ID: 075125. http://dx.doi.org/10.1103/PhysRevB.76.075125

18. Jezierski, A., Jurczyk, M. and Szajek, A. (2009) Electronic Structure of Mg2Ni1-xCux. Actaphysicapolonica A, 115, 223-225.

19. Huang, L.W., Elkedim, O., Nowak, M., Chassagnond, R. and Jurczyk, M. (2012) Mg2-xTixNi (x = 0, 0.5) Alloys Prepared by Mechanical Alloying for Electrochemical Hydrogen Storage: Experiments and First-Principles Calculations. International Journal of Hydrogen Energy, 37, 14248-14256.http://dx.doi.org/10.1016/j.ijhydene.2012.07.036

20. Zeng, Y., Fan, K., Li, X., Xu, B., Gao, X. and Meng, L. (2010) First-Principles Studies of the Structures and Properties of Al- and Ag-Substituted Mg2Ni Alloys and Their Hydrides. International Journal of Hydrogen Energy, 35, 10349-10358. http://dx.doi.org/10.1016/j.ijhydene.2010.07.131

21. Darriet, B., Soubeyroux, J.L., Pezat, M. and Fruchart, D. (1984) Structural and Hydrogen Diffusion Study in the Mg2Ni-H2 System. Journal of the Less Common Metals, 103, 153-162. 22. Reilly Jr., J.J. and Wiswall Jr., R.H. (1968) Reaction of Hydrogen with Alloys of Magnesium and Nickel and the Formation of Mg2NiH4. Inorganic Chemistry, 7, 2254-2256. http://dx.doi.org/10.1021/ic50069a016

22. Zolliker, P., Yvon, K., Jorgensen, J.D. and Rotella, F.J. (1986) Structural Studies of the Hydrogen Storage Material Magnesium Nickel Hydride (Mg2NiH4). 2. Monoclinic Low-Temperature Structure. Inorganic Chemistry, 25, 3590-3593. http://dx.doi.org/10.1021/ic00240a012

23. Zhang, J., Huang, Y.N., Peng, P., Mao, C., Shao, Y.M. and Zhou, D.W. (2011) First-Principles Study on the Dehydrogenation Properties and Mechanism of Al-Doped Mg2NiH4. International Journal of Hydrogen Energy, 36, 5375-5382. http://dx.doi.org/10.1021/ic00240a012

24. Kresse, G. and Hafner, J. (1993) Ab initio Molecular Dynamics for Open-Shell Transition Metals. Physical Review B, 48, 13115-13118. http://dx.doi.org/10.1103/PhysRevB.48.13115

25. Kresse, G. and Furthmüller, J. (1996) Efficiency of Ab-Initio Total Energy Calculations for Metals and Semiconductors Using a Plane-Wave

Basis Set. Computational Materials Science, 6, 15-50. http://dx.doi.org/10.1016/0927-0256(96)00008-0

26. Kresse, G. and Furthmüller, J. (1996) Efficient Iterative Schemes for Ab initio Total-Energy Calculations Using a Plane-Wave Basis Set. Physical Review B, 54, 11169-11186.http://dx.doi.org/10.1103/PhysRevB.54.11169

27. Perdew, J.P., Burke, K. and Wang, Y. (1996) Generalized Gradient Approximation for the Exchange-Correlation Hole of a Many-Electron System. Physical Review B, 54, 16533-16539.http://dx.doi.org/10.1103/PhysRevB.54.16533

28. Perdew, J.P., Chevary, J.A., Vosko, S.H., Jackson, K.A., Pederson, M.R., Singh, D.J. and Fiolhais, C. (1992) Atoms, Molecules, Solids, and Surfaces: Applications of the Generalized Gradient Approximation for Exchange and Correlation. Physical Review B, 46, 6671-6687. http://dx.doi.org/10.1103/PhysRevB.46.6671

29. Perdew, J.P., Burke, K. and Ernzerhof, M. (1996) Generalized Gradient Approximation Made Simple. Physical Review Letters, 77, 3865-3868. http://dx.doi.org/10.1103/PhysRevLett.77.3865

30. Monkhorst, H.J. and Pack, J.D. (1976) Special Points for Brillouin-Zone Integrations. Physical Review B, 13, 5188-5192. http://dx.doi.org/10.1103/PhysRevB.13.5188

31. Broedersz, C.P., Gremaud, R., Dam, B., Griessen, R. and Lovvik, O.M. (2008) Highlydestabilized Mg-Ti-Ni-H System Investigated by Density Functional Theory and Hydrogenography. Physical Review B, 77, Article ID: 024204. http://dx.doi.org/10.1103/PhysRevB.77.024204

32. Kubaschewski, O. and Alcock, C.B. (1979) International Series on Materials Science and Technology. Pergamon Press, Oxford, 294.

33. Jasen, P.V., Gonzalez, E.A., Brizuela, G., Nagel, O.A., González, G.A. and Juan, A. (2007) A Theoretical Study of the Electronic Structure and Bonding of the Monoclinic Phase of Mg2NiH4. International Journal of Hydrogen Energy, 32, 4943-4948.http://dx.doi.org/10.1016/j.ijhydene.2007.08.011

CITATION

CHAPTER 1

Miguel Valcárcel (2012). Analytical Chemistry Today and Tomorrow, Analytical Chemistry, Dr. Ira S. Krull (Ed.), ISBN: 978-953-51-0837-5, InTech, DOI: 10.5772/50497.

CHAPTER 2

Aline Thaís Bruni and Vitor Barbanti Pereira Leite (2012). Quantum Chemistry and Chemometrics Applied to Conformational Analysis, Quantum Chemistry - Molecules for Innovations, Dr. Tomofumi Tada (Ed.), ISBN: 978-953-51-0372-1, InTech, DOI: 10.5772/34994.

CHAPTER 3

Durrant JD, McCammon JA (2012) AutoClickChem: Click Chemistry in Silico. PLoS Comput Biol 8(3): e1002397. doi:10.1371/journal.pcbi.1002397

CHAPTER 4

Kolluru B, Hawizy L, Murray-Rust P, Tsujii J, Ananiadou S (2011) Using Workflows to Explore and Optimise Named Entity Recognition for Chemistry. PLoS ONE 6(5): e20181. doi:10.1371/journal.pone.0020181

CHAPTER 5

Misra SK, Ye M, Kim S, Pan D (2015) Defined Nanoscale Chemistry Influences Delivery of Peptido-Toxins for Cancer Therapy. PLoS ONE 10(6): e0125908. doi:10.1371/journal.pone.0125908

CHAPTER 6

A. Sarwar, M. Khan and K. Azhar, "Coal Chemistry and Morphology of Thar Reserves,Pakistan," *Journal of Minerals and Materials Characterization and Engineering*, Vol. 11 No. 8, 2012, pp. 817-824. doi:10.4236/jmmce.2012.118072.

CHAPTER 7

K. Valsaraj, "A Review of the Aqueous Aerosol Surface Chemistry in the Atmospheric Context," Open Journal of Physical Chemistry, Vol. 2 No. 1, 2012, pp. 58-66. doi: 10.4236/ojpc.2012.21008.

CHAPTER 8

Francesco G. Mutti, Roberta Pievo, Maila Sgobba, Michele Gullotti, and Laura Santagostini, "Biomimetic Modeling of Copper Complexes: A Study of Enantioselective Catalytic Oxidation on D-(+)-Catechin and L-(-)-Epicatechin with Copper Complexes," Bioinorganic Chemistry and Applications, vol. 2008, Article ID 762029, 9 pages, 2008. doi:10.1155/2008/76202

CHAPTER 9

McEnroe NA, Williams CJ, Xenopoulos MA, Porcal P, Frost PC (2013) Distinct Optical Chemistry of Dissolved Organic Matter in Urban Pond Ecosystems PLoS ONE 8(11): e80334. doi:10.1371/journal.pone.0080334

CHAPTER 10

Weisinger RM, Marinelli RJ, Wrenn SJ, Harbury PB (2012) Mesofluidic Devices for DNA-Programmed Combinatorial Chemistry. PLoS ONE 7(3): e32299. doi:10.1371/journal.pone.0032299

CHAPTER 11

Ficker M, Petersen JF, Hansen JS, Christensen JB (2015) Guest-Host Chemistry with Dendrimers—Binding of Carboxylates in Aqueous Solution. PLoS ONE 10(10): e0138706. doi:10.1371/journal.pone.0138706 Ficker M, Petersen JF,

Hansen JS, Christensen JB (2015) Guest-Host Chemistry with Dendrimers—Binding of Carboxylates in Aqueous Solution. PLoS ONE 10(10): e0138706. doi:10.1371/journal.pone.0138706.

CHAPTER 12

Minmin Chu, Xin Liu, Yanhui Sui, Jie Luo and, Changgong Meng, Unique Reactivity of Transition Metal Atoms Embedded in Graphene to CO, NO, O_2 and O Adsorption: A First-Principles Investigation, doi:10.3390/molecules201019540

CHAPTER 13

Ashriti Govender, Daniel Curulla-Ferré , Manuel Pérez-Jigato and Hans Niemantsverdriet, First-Principles Elucidation of the Surface Chemistry of the C2Hx (x = 0–6) Adsorbate Series on Fe(100), doi:10.3390/molecules18043806

CHAPTER 14

YuanyuanLi,GailiSun,YimingMi, (2016) The Structures and Properties of Y-Substituted Mg_2Ni Alloys and Their Hydrides: A First-Principles Study. *American Journal of Analytical Chemistry*,**07**,67-74. doi: 10.4236/ajac.2016.71007

INDEX